Springer Tracts in Natural Philosophy

Volume 27

Daniel D. Joseph

Stability of Fluid Motions I

With 57 Figures

Springer-Verlag
Berlin Heidelberg New York 1976

DANIEL D. JOSEPH

University of Minnesota, Department of Aerospace Engineering and Mechanics, Minneapolis, Minnesota 55455/U.S.A.

AMS Subject Classification (1970): 34 Dxx, 35 A 20, 35 B 10, 35 B 15, 35 B 30, 35 B 35, 35 Cxx, 35 P 15, 35 P 20, 35 Q 10, 49 F 25, 49 G 05, 49 Gxx, 49 H 05, 73 Bxx, 76 Axx, 76 Dxx, 76 S 05, 85 A 30

ISBN 3-540-07514-3 Springer-Verlag Berlin Heidelberg New York
ISBN 0-387-07514-3 Springer-Verlag New York Heidelberg Berlin

Library of Congress Cataloging in Publication Data. Joseph, Daniel D. Stability of fluid motions. (Springer tracts in natural philosophy; v. 27–28) Includes bibliographical references and indexes. 1. Fluid dynamics. 2. Stability. I. Title. QA911.J67. 532'.053. 76-4887.

Typesetting and printing: Zechnersche Buchdruckerei, Speyer, Bookbinding: Konrad Triltsch, Würzburg.

Preface

The study of stability aims at understanding the abrupt changes which are observed in fluid motions as the external parameters are varied. It is a demanding study, far from full grown, whose most interesting conclusions are recent. I have written a detailed account of those parts of the recent theory which I regard as established.

Acknowledgements

I started writing this book in 1967 at the invitation of Clifford Truesdell. It was to be a short work on the energy theory of stability and if I had stuck to that I would have finished the writing many years ago. The theory of stability has developed so rapidly since 1967 that the book I might then have written would now have a much too limited scope. I am grateful to Truesdell, not so much for the invitation to spend endless hours of writing and erasing, but for the generous way he has supported my efforts and encouraged me to higher standards of good work. I have tried to follow Truesdell's advice to write this work in a clear and uncomplicated style. This is not easy advice for a former sociologist to follow; if I have failed it is not due to a lack of urging by him or trying by me.

My research during the years 1969–1970 was supported in part by a grant from the Guggenheim foundation to study in London. I enjoyed the year in London and I am grateful to Trevor Stuart, Derek Moore and my other friends at Imperial College for their warm hospitality. I welcome the opportunity to acknowledge the unselfish assistance of the world's best Maths librarian, Miss J. Pindelska of Imperial College. In the summer and fall of 1973 a grant from the British Science Research Council made it possible for me to work in England again, this time with L. A. Peletier, D. Edmunds and other mathematicians in Sussex. In the summer of 1974 I taught a short course on stability at L'École d'Eté in Bréau Sans Nappe, France. The French students were able and well-trained engineers with graduate degrees and a good background in mathematics. It was a stimulating group and some of the better results which are set down in § 15, Chapters XI and XIV and in the Addendum to Chapter X stem directly from questions raised at the summer school.

My research in stability theory has been funded from the beginning by the fluid mechanics branch of the National Science Foundation. The contribution which their funding has made to this book can scarcely be overestimated. My work was also made easier by the good conditions which prevail in the mechanics department of the University of Minnesota and by the presence there of good friends and fine colleagues, by an understanding chairman, P. Sethna and by the two cheerful and efficient secretaries, Jean Jindra and Susan Peterson, who typed the various drafts of this manuscript.

Many persons have contributed to this book in different ways. Fritz Busse, Ta Shen Chen, Stephen Davis, Daniel Jankowski, Klaus Kirchgässner, Simon Rosenblat, Robert Sani, William Warner and Hans Weinberger read and criticized various parts of the text. The contributions of W. Hung to Chapters V and VI, of Bruce Munson to Chapter VII, of Ved Gupta to Chapter XII and Appendix C, and of E. Dussan V to Chapter XIV deserve special mention. Chapter XI is based on joint work with D. A. Nield. Nield also worked through the whole book in the final stages of preparation for printing. I do not have words sufficient to convey the depth of gratitude I feel for Nield's careful and efficient assistance with the onerous task of reading proofs.

As a graduate student I was strongly influenced by James Serrin's beautiful article on fluid mechanics in the Handbuch der Physik. I later had the good fortune to find a position at the University of Minnesota where, through contact with Serrin, I became interested in the energy theory of stability. I am really indebted to Serrin, for his support in the early days, and for the continued inspiration which I derive from seeing good mathematics applied to problems at the foundations of mechanics.

Finally I want to acknowledge all that I have learned from my colleagues, F. H. Busse, T. S. Chen, E. Dussan V, V. Gupta, W. Hung, B. Munson, D. A. Nield and D. Sattinger, with whom I have collaborated in stability studies. My view of stability theory as a branch of mathematics has been particularly influenced by David Sattinger, and as a branch of physics, by Fritz Busse.

"Like most philosophers, I am much indebted to conversations I have had with others over the years... I will not indulge in the conventional fatuity of remarking that they are not responsible for the errors this book may contain. Obviously, only I can be *held* responsible for these: but, if I could recognize the errors, I should have removed them, and, since I cannot, I am not in a position to know whether any of them can be traced back to the opinions of those who have influenced me."[1]

My wife, Ellen, read the entire manuscript with me and together in our effort to achieve good writing we studied Truesdell's letters and Gowers' splendid little book[2] "The Complete Plain Words". Ellen has agreed to take responsibility for lack of clarity and precision in the writing and for all mathematical and conceptual mistakes in the presentation. It is a pleasure for me to dedicate my first book to this lovely lady.

---

[1] Michael Dummett "Frege: Philosophy of Language" Duckworth: London, 1973.

[2] Sir Ernest Gowers "The Complete Plain Words" Penguin Books Ltd.: Middlesex, England, 1973.

Plan of the Work

The plan of this work is given in the table of contents. The book is divided into two parts. Part I gives the general theory of stability, instability, bifurcation and some discussion of the problem of repeated bifurcation and turbulence. The general theory is developed in Chapters I and II and is applied to flows between concentric cylinders (Chapters III—VI) and between concentric spheres (Chapter VII). Part I is self-contained. The six chapters (VIII—XIV) of Part II take up special topics of general interest. The topics are selected to develop extensions of the general theory, to introduce new techniques of analysis, to extend the scope of application of the theory and to demonstrate how stability theory is essential in understanding the mechanics of the motion of fluids. The purpose of each chapter in both parts is set out in the introduction to that chapter. Attributions are given in the text, where a new result first appears, or in bibliographical notes at the end of each chapter. Many results are given here for the first time. Other results, not known to most readers, are not new but are not well known or are not available in a form which can be understood by interested persons with training only in classical analysis. All results new and old, have been reworked to fit the plan of this work.

Remarks for Students

I expect readers of this book to know calculus and parts of the theory of differential equations. If you know more, so much the better. Students at the required level of preparation will greatly improve their knowledge of useful techniques of analysis which are required in the study of different aspects of the theory of hydrodynamic stability. Among these, bifurcation theory and variational methods for linear and nonlinear eigenvalue problems defined on solenoidal fields have received an especially thorough treatment. My explanation of these topics has been guided by my desire to make the theory accessible to a wide audience of potentially interested persons. The appendices are to help beginners who wish to learn the details of the mathematical procedures considered by me to be important for stability studies but inappropriate for the main line of development in the text. I have formulated about 230 exercises to help students learn and to elaborate and extend results which are developed in the text.

Table of Contents

Contents of Part II

Introduction

Stability theory is the body of mathematical physics which enables one to deduce from first principles the critical values which separate the different regimes of flow as well as the forms of the fluid motions in these different regimes. Stability theory leads to statements about *solutions* of equations. Given that the actual motion is described by these equations, statements about solutions are also statements about flows.

In the work which follows we shall study the stability of flows which satisfy equations of the Navier-Stokes type. When the flow domain, boundary conditions and external forces are fixed, the solutions of these equations vary with the viscosity and the initial conditions. It is useful to seek critical values of the viscosity which separate different regimes.

Flows satisfying the Navier-Stokes equations have an important property. When the viscosity of the fluid is greater than a critical value, all solutions of the initial value problem tend monotonically to a single flow—the basic flow—and the energy of any disturbance of this flow will decay from the initial instant. For less viscous fluids, disturbances of the basic flow can be found which grow at first; later they may die away or persist, depending on conditions. This critical value of the viscosity which separates the monotonically decaying disturbances from those whose energy increases initially is called the first critical viscosity of energy theory.

The stability of the basic flow depends on the behavior of solutions after a long time rather than on their initial behavior. In most situations there is a second critical value of the viscosity *not* larger than the first. When the fluid viscosity is larger than this value, the basic flow is stable because no disturbances, no matter how large, can persist after a long time. However, when the fluid viscosity is below this second critical value, the basic flow gives up its stability to a more complicated flow (or set of flows). As the viscosity is reduced still further, the flows which have replaced the basic flow may, in turn, lose their stability to other yet more complicated motions.

It is generally believed that the increasing complexity of fluid motions which is observed as the viscosity is decreased, and which is frequently called "turbulence", is a manifestation of the successive loss of stability of flows of less complicated structure to those with a more complicated structure. The process which is frequently identified with the conjecture of L. Landau (1944) and E. Hopf (1948) is sometimes identified as the "transition to turbulence through repeated branching". Landau and Hopf regard repeated branching as a process involving continuous bifurcation of manifolds with n frequencies into manifolds with $n+1$ frequencies. Here the attractive property of the stable solution is replaced with the attractive property of the manifold. For example, when the data is steady and the viscosity is high, all solutions are attracted to the steady basic flow. For lower viscosities, the steady flow is unstable and stability is supposed now to be claimed by an attracting manifold of time-periodic motions differing from one another in phase alone. Arbitrary solutions of the initial value problem will be attracted to one or another of the members of the attracting set according to their initial values. At still lower values of the viscosity, the manifold of periodic solutions loses its stability to a larger manifold of quasi-periodic solutions with two frequencies of independently arbitrary phase. Now arbitrary solutions of the initial value problem are attracted to the manifold with two frequencies. In the same way, as the viscosity is decreased, the two-frequency manifold loses its stability to a three-frequency manifold, and so on.

The idea of Landau and Hopf that the branching of manifolds is always continuous is certainly incorrect (see § 16) and their conjectures about the quasi-periodic characterization of the branching manifolds probably needs revision. But some of the basic notions about branching sets of solutions have a foundation in recent results of bifurcation theory (see notes to § 16).

It follows from the description of repeated branching that one general aim of stability theory should be to give an accurate description of the process by which solutions lose their stability to other solutions as the viscosity is decreased. At present only parts of this process are understood and the rest lies in the realm of conjecture. Of course, a complete mathematical theory is unthinkable in a situation in which the qualitative processes are only partially understood.

In view of this difficulty, it is perhaps understandable that the attention of research in hydrodynamic stability theory should focus on a quantitative treatment of special problems which can be solved and compared with experiments. It is undoubtedly for this reason that most of the existing monographs on hydrodynamic stability theory, and most of the research, are restricted to a discussion of the stability of steady laminar flows to very small disturbances. In this monograph, I have set the problem of stability more generally in order to relate various kinds of stability rather than to proceed with a restricted analysis of disconnected special problems. The problem of stability is intrinsically nonlinear and even linearized analysis of it can be understood only as an approximation of the true problem. When the nonlinear analysis of the true problem is hooked on the linearized theories for small disturbances we say that the nonlinear analysis is local. In global analysis the size of disturbances is unrestricted or, in a looser use of "global", less restricted. Both kinds of analysis play a role in the study of stability of solutions of the Navier-Stokes equations in general circumstances.

(a) Energy Methods

In the global theory, energy methods have an important place. These methods lead to a variational problem for the first critical viscosity of energy theory and to a definite criterion which is sufficient for the global stability of the basic flow.

It is sometimes possible to find positive definite functionals of the disturbance of a basic flow, other than the energy, which decrease on solutions when the viscosity is larger than a critical value. Such functionals, which may be called generalized energy functionals of the Liapounov type, are of interest because they can lead to a larger interval of viscosities on which the global stability of the basic flow can be guaranteed. Examples of the construction and use of generalized energy functionals are given in §§ 40, 51, 57, 69, in the addendum to Chapter IX and in Chapter XIV.

The essential physical fact which makes the analysis of generalized energy functionals possible is that the ratio of the total production of the generalized energy to the total dissipation is bounded above for all disturbance fields which are kinematically admissible as initial values for the disturbance. The essential mathematical tool is the calculus of variations for functionals defined on solenoidal vector fields.

The variational analysis of bounded functionals also forms the basis for my analysis of turbulent flow. This type of analysis leads, for example, to bounds on the pressure gradient needed to drive a given mass flow under turbulent conditions (Chapter IV) or to bounds on the heat which is transported by a fluid between horizontal walls heated from below when the temperature contrast is given (Chapter XII). The nonlinear variational equations for the turbulence problem are simpler than the Navier-Stokes equations but model these equations in many important respects. In particular, the variational equations have instability and bifurcation properties which model the process of transition to turbulence through repeated branching. A very thorough exposition of this theory is given for the convection problems studied in Chapter XII.

All of the methods which lead to variational problems for bounded functionals can be considered "energy" methods in a generalized sense. The starting place for generalized energy analysis is with functionals derived from the Navier-Stokes equations, defined first over solutions of these equations, and then extended over a larger class of fields which are kinematically admissible as initial velocity fields. These functionals are bounded in the wider class of fields, and the resulting mathematical problem is to find the best bound.

Energy analysis leads to global statements about stability which take form as criteria sufficient for stability[1]. Two typical examples are (1) steady flow is stable to arbitrary disturbances when the viscosity is larger than a critical value, and (2) statistically stationary turbulence of a given intensity cannot exist when the viscosity is larger than another critical value (which depends on the intensity of the turbulence).

Energy analysis also leads to "symmetrized" mathematical problems which are simpler than the Navier-Stokes problems but which one hopes will abstract

[1] Typically, the energy theory is silent about instability; it gives a critical viscosity and a critical initial disturbance. When the viscosity of the fluid is greater than critical the energy of any disturbance decays at each and every instant; when smaller than critical the energy of the critical disturbance will increase for a time. But this energy may ultimately decay so the criterion based on the critical viscosity of energy theory does not guarantee instability. There are problems, like the interfacial stability problem, discussed in Chapter XIV where energy theory leads to global statements of *instability* which cannot be obtained from local (linear or nearly linear) theories.

from the many details of the problem some of the essential processes in purer form.

The price paid for the substantial advantages which accrue to energy analysis is in the currency of losses incurred in going from the exact Navier-Stokes problem to a derived problem. Inevitably, in the process of "symmetrization" the mathematical consequences of some important dynamic processes (those, for example, which lead to "asymmetry" in the form of nonselfadjoint operators) are lost. In particular, such a theory inevitably loses its capacity to make precise the exact nature of the mechanics of the loss of stability.

It goes almost without saying that the results of energy analysis describe true solutions more adequately when the functionals and classes of admissible functions allowed in the variational competition are made to share with *true solutions* an ever increasing number of properties. Whether this process of refinement can ever lead back to the exact problem is an interesting question. From a practical point of view, this refinement has a limited application since the incorporation of new constraints rapidly leads to "symmetrized" problems which are as intractable as the exact problem.

(b) Linearized Theories of Instability

To draw a fuller picture of the stability properties of fluid motions, we should have procedures adequate to deduce exact criteria for instability and to describe with mathematics the transition from one stable form of motion to another. For this, local analysis can sometimes suffice. Indeed, the principle of linearized stability (sometimes called the "conditional stability theorem" or Liapounov's theorem) states that given mild conditions, critical values of the viscosity may be established by analysis of the linearized equations for the disturbance of a basic steady or time-periodic solution of the Navier-Stokes equations. It ought then suffice for instability to consider disturbances of indefinitely small size in which the linearized equations apply.

Of course, the criteria of linearized theory can only give sufficient conditions for instability since a flow which is judged stable by linear theory may actually be unstable to disturbances of finite size. In this sense, linear and energy theory complement each other with the former leading to sufficient conditions for instability and the latter to sufficient conditions for stability.

It is to be expected that the predictions of stability following from linearized theory will frequently be in disagreement with experiment and it can hardly be a surprise that this is in fact the case. In many cases, however, the results of analysis of the linearized equations are in good agreement with experiments.

(c) Bifurcation Theory

This superficially anomalous relation of linear theory to experiment is intimately related to a property of the solutions which bifurcate from the basic flow. This bifurcation is by definition a process which is continuous in a parameter measuring the amplitude ε of the bifurcating solution. Bifurcating solutions can exist only when the viscosity ν and the amplitude ε lie on the bifurcation curve $\nu(\varepsilon)$ where $\nu(0)$ is the critical viscosity given by linearized analysis of the stability of the basic flow. If $\nu(\varepsilon)>\nu(0)$, then a bifurcating solution will exist for values of ν

for which infinitesimal disturbances of the basic flow decay. This situation which "contradicts" linear theory is called subcritical bifurcation. The other possibility is described as supercritical bifurcation.

The important discriminating property to which I alluded at the beginning of the previous paragraph is the property of instability of the subcritical bifurcating solutions. Since subcritical bifurcating solutions are unstable when their amplitude is small, it would not be possible to observe a continuous subcritical bifurcation in nature. Instead we might *expect* a discontinuous process in which a disturbance which escapes the domain of attraction of the basic flow will snap through the unstable bifurcating solution to a stable solution with a much larger norm. This solution need not have a strong resemblance to the critical eigenfunctions of the linearized theory.

On the other hand, the case of supercritical bifurcations is different. Since supercritical bifurcating solutions can be stable, a continuous bifurcation is possible, and for small amplitudes, the stable bifurcating solutions are essentially those given by the linearized stability theory.

For flow through pipes and channels, the bifurcation is subcritical and the linear theory of stability does not describe the physically realized flows. On the other hand, in the Couette flow problems described in Chapters V and VI and in some of the convection problems described in Chapters VIII, IX, X and XI, the bifurcation of laminar flow is supercritical and the linearized theory of stability is in good agreement with experiments.

Among the different (mostly equivalent) theories of bifurcation which are discussed in the notes for Chapter II, I have singled out one for extended discussion. This bifurcation theory, whose antecedents may be traced back to Lindstedt (1882) and Poincaré (1892), is derived in detail in the body of Chapter II and applied repeatedly to applications which are studied in subsequent chapters. The results about the stability of supercritical bifurcating solutions and the instability of subcritical bifurcating solutions, which were originally derived by Hopf (1942) for solutions of ordinary differential equations which bifurcate at a simple eigenvalue, are the most important general results of stability theory. They enable one to describe and categorize the loss of stability for whole classes of motions and not only in the very special cases where one can carry out explicit computations. Hopf extends methods which were used by Poincaré to study conservative equations to the study of dissipative equations; he introduces an amplitude-dependent frequency and expands it, the solutions and an additional (Hopf) bifurcation parameter into power series in the amplitude. The method and the results of Hopf's analysis are local, restricted to small amplitudes. In §§ 14 and 15 a local analysis is used in the neighborhood of points of finite, and not necessarily small amplitude to extend Hopf's local stability results. The same type of local analysis leads, under conditions, to the form which Hopf's results must take when the restriction on the size of the amplitude of the bifurcating solution is removed. The local statement is: "Subcritical solutions branching at a simple eigenvalue are unstable; supercritical solutions are stable." The global statement is: "Solutions for which the response decreases with increasing amplitude are unstable; solutions for which the response increases with amplitude are stable." Expressed in physical terms, the global statement asserts that pipe flows for which the mass flux increases as the pressure gradient decreases are unstable or, for another example, convection for which the heat transported decreases as the temperature is increased is unstable.

(d) Arrangements of Topics in the Text

The arrangement of topics in the text is as follows: the classical energy theory of stability is given in Chapter I. Chapter II treats the problem of instability and time-periodic or steady (symmetry-breaking) bifurcation of steady solutions of the Navier-Stokes equations at a simple eigenvalue[2]. Chapters I and II give a short course on stability and bifurcation. In Chapter III, the methods and physical significance of energy theory are illustrated by studying the disturbance of laminar pipe flow whose energy will increase initially at the smallest Reynolds number. The same flow is studied in Chapter IV using linear theory, bifurcation theory and the variational theory of turbulence. Chapter V treats the well-known Taylor problem of stability of Couette flow between rotating cylinders. In Chapter VI we consider the stability of Couette flow between concentric sliding and rotating cylinders. In Chapter VII we consider the stability of flow between concentric rotating spheres. The flows studied in V and VI allow a generalization of classical energy analysis, which we study in order to learn how to treat at least one example of every possible type of energy analysis. Four of the five appendices in Part I are about mathematics used in stability analysis. Appendix E gives a brief critical analysis and discussion of aspects of the popular theory of stability of "nearly-parallel" flow.

The first four of the six chapters of Part II are about stability problems in convecting fluids in which motion is induced by density differences associated with gradients of temperature and chemical composition. The equations governing such motion, motionless solutions of these equations, and criteria for their stability, are given in Chapter VIII. In IX, we consider the stability of some motionless solutions in a heterogeneous fluid like salt water. The large difference in the values of the diffusion constants for heat and salt allow for a new mechanism of instability and the equations may be studied by an interesting type of generalized energy analysis. A similar generalization applies to magnetohydrodynamic flows. Chapters X, XI and XII treat the problem of convection in porous materials. In X, the problem is posed in an impermeable container with insulated side walls and admits an elementary separation of variable which leads to an equally elementary, analysis of bifurcation and stability at eigenvalues of higher multiplicity. The results in X are all new. Many of the results in Chapter XI, which treats the problem of wave number selection through stability, are new results based on joint work with D.A. Nield. Chapter XII gives a fairly complete discussion of the variational theory of turbulence applied to convection in porous materials. Chapter XIII gives new methods of analysis for studying the flow and stability of flow of viscoelastic fluids. In Chapter XIV we consider the new nonlinear dynamic theory of interfacial stability due to E. Dussan V.

[2] I am not going to treat the important problem of bifurcation from unsteady motion—this subject is forced on us by the problem of repeated bifurcations even when the external data is steady. Apart from the interesting results about bifurcating tori and quasi-periodic bifurcation which are discussed in the notes for Chapters II and IV very little is presently known.

Chapter I

Global Stability and Uniqueness

§ 1. The Initial Value Problem and Stability

Suppose that a viscous fluid is set into motion by external forces or by the motion of the boundary $S(t)$ of a closed container $\mathscr{V}(t)$. The velocity field $\mathbf{U}(\mathbf{x},t)$ is assumed to be governed by the initial boundary value problem (IBVP) for the Navier-Stokes equations. In the container, $\mathbf{U}$ satisfies the following set of equations:

$$\frac{\partial \mathbf{U}}{\partial t}+\mathbf{U}\cdot\nabla\mathbf{U}-\nu\nabla^2\mathbf{U}+\nabla\pi-\mathbf{F}(\mathbf{x},t)=0\,,\qquad \operatorname{div}\mathbf{U}=0 \tag{1.1 a, b}$$

where $\rho_0\pi$ is the pressure, $\mathbf{F}(\mathbf{x},t)$ is an external force field per unit mass, ρ_0 is the constant density and ν is the constant kinematic viscosity. Since the fluid is assumed to be incompressible the motion of the boundary must be such as to leave the total volume $\mathscr{V}(t)$ unchanged even though its shape *can* change. The external force field is to be regarded as "given" but is otherwise arbitrary. The boundary values

$$\mathbf{U}(\mathbf{x},t)=\mathbf{U}_S(\mathbf{x},t)\,,\qquad \mathbf{x}\in S(t)\,,\qquad t\geqslant 0 \tag{1.1 c}$$

are also prescribed as are the initial values

$$\mathbf{U}(\mathbf{x},0)=\mathbf{U}_0(\mathbf{x})\,,\qquad \mathbf{x}\in\mathscr{V}(0) \tag{1.1 d}$$

where $\operatorname{div}\mathbf{U}_0(\mathbf{x})=0$ for $\mathbf{x}\in\mathscr{V}(0)$.

The solutions of (1.1) corresponding to given data

$$\{\mathbf{U}_S(\mathbf{x},t),\ \mathbf{F}(\mathbf{x},t),\ \mathscr{V}(t)\} \tag{1.2}$$

are to be designated as $\mathbf{U}(\mathbf{x},t;\nu,\mathbf{U}_0)$. We wish to study the stability of these solutions as $\mathbf{U}_0(\mathbf{x})$ and ν are varied. The notation $\mathbf{U}(\mathbf{x},t;\mathbf{U}_0)$ is used to designate the family of solutions which are obtained from the set $\mathbf{U}(\mathbf{x},t;\nu,\mathbf{U}_0)$ when ν is fixed and the initial condition $\mathbf{U}_0(\mathbf{x})$ is varied.

Consider another motion

$$(\mathbf{U}^a,\pi^a)=(\mathbf{U}(\mathbf{x},t;\mathbf{U}_0+\mathbf{u}_0),\ \pi(\mathbf{x},t;\mathbf{U}_0+\mathbf{u}_0)) \tag{1.3}$$

which also satisfies (1.1 a, b, c) when ν and (1.2) are fixed but which differs from $(\mathbf{U}(\mathbf{x}, t; \mathbf{U}_0), \pi(\mathbf{x}, t; \mathbf{U}_0))$ initially:

$$\mathbf{U}^a(\mathbf{x},0)=\mathbf{U}(\mathbf{x}, 0; \mathbf{U}_0+\mathbf{u}_0)=\mathbf{U}_0(\mathbf{x})+\mathbf{u}_0(\mathbf{x}).$$

One wishes to know if and under what conditions the two solutions "come together" (stability) or "stay apart" (instability). To answer this question we form the IBVP governing the evolution of a *disturbance*

$$(\mathbf{u},p)=(\mathbf{U}^a-\mathbf{U}, \pi^a-\pi)$$

of the flow $\mathbf{U}(\mathbf{x}, t; \mathbf{U}_0)$. Thus,

$$\frac{\partial \mathbf{u}}{\partial t}+\mathbf{U}\cdot\nabla\mathbf{u}+\mathbf{u}\cdot\nabla\mathbf{U}+\mathbf{u}\cdot\nabla\mathbf{u}-\nu\nabla^2\mathbf{u}+\nabla p=0, \tag{1.4a}$$

$$\operatorname{div}\mathbf{u}=0, \quad \mathbf{u}|_S=0 \quad \text{and} \quad \mathbf{u}|_{t=0}=\mathbf{u}_0(\mathbf{x}). \tag{1.4b, c, d}$$

We are interested in determining whether the forms which the fields $\mathbf{U}(\mathbf{x}, t; \mathbf{U}_0)$ take as $t\to\infty$ are stable to perturbations in the initial conditions. This question is conveniently formulated as a stability problem for the *null solution* $\mathbf{u}(\mathbf{x}, t)\equiv 0$ of (1.4 a, b, c). This solution is the only one which can evolve from the initial field $\mathbf{u}_0\equiv 0$ (this is a consequence of the uniqueness theorem for the initial value problem which is proved in § 5). The stability problem for the null solution arises from perturbing the zero initial values and is, therefore, governed by (1.4) with $\mathbf{u}_0\not\equiv 0$.

To assign a definite meaning to the word "stable" the *average energy of a disturbance*

$$\mathscr{E}(t)=\tfrac{1}{2}\langle|\mathbf{u}|^2\rangle \tag{1.5}$$

is introduced. Here the angle bracket designates the volume-averaged integral defined by

$$\langle\cdot\rangle=[\mathscr{M}(\mathscr{V})]^{-1}\int_{\mathscr{V}}(\cdot)$$

where $\mathscr{M}(\mathscr{V})$ is the constant measure of the volume $\mathscr{V}$; thus,

$$\langle c\rangle=c$$

for any constant c.[1]

[1] The region of integration is indicated by the symbol at lower right of the integral. The usual infinitesimal volume and surface elements are used only in situations where their use clarifies the formulas.

We shall call the null solution of (1.4) *stable* to perturbations of the initial conditions if

$$\lim_{t\to\infty} \mathscr{E}(t)/\mathscr{E}(0) \to 0\,. \tag{1.6}$$

This definition shows that our understanding of "stability" is what is conventionally designated as *asymptotic stability in the mean.*

If there exists a positive value δ such that the null solution is stable when $\mathscr{E}(0)<\delta$, then the null solution is said to be *conditionally stable.* The number δ defines a set of initial values which are attracted to the solution $\mathbf{u}\equiv 0$. We shall call δ an *attracting radius* for the conditionally stable disturbances of $\mathbf{u}=0$.

If $\delta\to\infty$, then the null solution is said to be *unconditionally* or *globally stable.* If (1.6) holds and $d\mathscr{E}/dt\leqslant 0$ for all $t>0$, then the null solution is called *monotonically stable.*

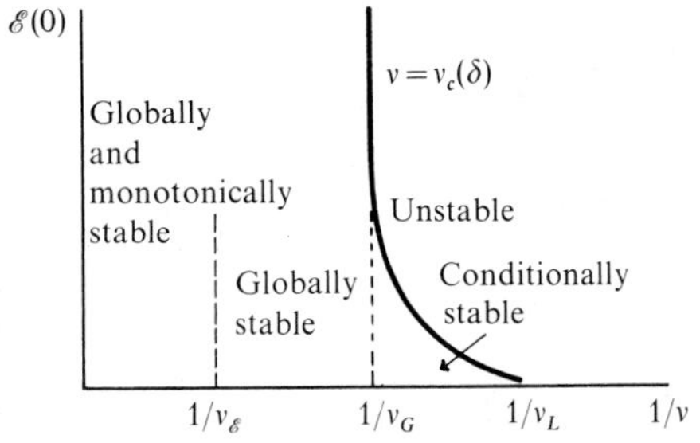

Fig. 2.1: Stability limits for the basic flow

There are two kinds of instability: (1) not conditionally stable, (2) conditionally stable but not unconditionally stable (see Fig. 2.1).

Though a more elaborate list of definitions is useful in general studies of stability, this simple list is sufficient for our needs.

§ 2. Stability Criteria—the Basic Flow

An important task of stability theory is to separate the stable solutions of (1.1) from the unstable solutions. The definition of conditional stability can be used for this purpose. Rephrasing the definition of conditional stability, we say that $\mathbf{U}(\mathbf{x}, t; \mathbf{U}_0)$ is stable if

$$\langle|\mathbf{U}(\mathbf{x}, t; \mathbf{U}_0)-\mathbf{U}(\mathbf{x}, t; \mathbf{U}_0+\mathbf{u}_0)|^2\rangle\to 0 \tag{2.1}$$

as $t\to\infty$ whenever $\langle|\mathbf{u}_0|^2\rangle<2\delta$.

To separate the stable solutions of (1.1) from the unstable solutions, we not only fix the data (1.2) but also fix the initial field $\mathbf{U}_0(\mathbf{x})$. We then organize the solutions $\mathbf{U}(\mathbf{x}, t; \nu, \mathbf{U}_0)$ of (1.1) into a one-parameter family of solutions depending on the viscosity. In the conditional stability problem we seek the values of ν and δ such that

$$\mathbf{u}(\mathbf{x}, t; \nu, \mathbf{u}_0) \text{ is a stable solution of (1.4)}$$

or, equivalently,

$$\mathbf{U}(\mathbf{x}, t; \nu, \mathbf{U}_0) \text{ is a stable solution of (1.1).}$$

A *stability limit* is the locus $F(\nu_c, \delta)=0$ of *critical* values (ν_c, δ) in the $(\nu, \mathscr{E}(0))$ plane which marks the boundary separating the stable from unstable flows (see Fig. 2.1). For example, we may generate a stability boundary $\delta(\nu)$ for conditional stability by fixing ν and testing for stability for $\mathscr{E}(0)\leq\delta$ for increasing δ. It is useful to obtain stability limits in the form $\nu=\nu_c(\delta)$.

A *stability criterion* is an inequality among the values $(\nu, \mathscr{E}(0))$ and the critical values (ν_c, δ) which guarantees stability. For example, $\mathbf{u}\equiv 0$ is stable to perturbations satisfying the criterion $\mathscr{E}(0)<\delta(\nu)$.

The most useful stability criteria for our purposes are in the form of inequalities between the viscosity and critical values of the viscosity. Indeed the existence of the following critical values of the viscosity implies the structure on which the stability theory is elaborated:

$\nu_{\mathscr{E}}(\nu)$: When $\nu>\nu_{\mathscr{E}}(\nu)$ the null solution is monotonically and globally stable. When $\nu<\nu_{\mathscr{E}}(\nu)$ a disturbance can be found for which $d\mathscr{E}(t)/dt>0$ at $t=0$.

$\bar{\nu}_{\mathscr{E}}$: This is the smallest critical viscosity of energy theory. We call $\bar{\nu}_{\mathscr{E}}=\nu=\nu_{\mathscr{E}}(\nu)=\nu_{\mathscr{E}}(\bar{\nu}_{\mathscr{E}})$ an *energy stability limit* (see § 4).

ν_G: When $\nu>\nu_G$ the null solution is globally stable and when $\nu\leq\nu_G$ the null solution is unstable even though it may be conditionally stable. We call ν_G the *global stability limit* (see § 6).

$\nu_L(\nu)$: When $\nu>\nu_L(\nu)$ there exists some attracting radius $\delta(\nu)$ for conditional stability.

$\bar{\nu}_L$: This is the smallest critical viscosity of linear theory. We call $\bar{\nu}_L=\nu=\nu_L(\nu)=\nu_L(\bar{\nu}_L)$ a *linear stability limit* (see §§ 7 and 8).

It is necessary to say that the definitions of $\nu_{\mathscr{E}}$ and $\bar{\nu}_{\mathscr{E}}$ suppress the fact that $\nu_{\mathscr{E}}(\nu, t)$ depends on time when $\mathbf{U}(\mathbf{x}, t, \nu, \mathbf{U}_0)$ does (see (4.6 and 4.7b)).

Other critical viscosities than the ones just defined do occur; they will be defined where they are first introduced.

A flow $\mathbf{U}(\mathbf{x}, t; \nu, \mathbf{U}_0)$ is called *absolutely stable* if it is stable when $\nu>0$. The critical viscosity of an absolutely stable flow is zero. Many flows are absolutely stable when the admissible initial perturbation fields are constrained by extra requirements of various kinds. For example, flow through pipes is absolutely and globally stable to disturbances which do not vary along the pipe axis (§ 24). Some unstable flows are absolutely stable to arbitrary disturbances of infinitesimal amplitude (§§ 35(a), 52). Rigid body motions are absolutely, globally and monotonically stable to all disturbances (Exercise 4.4).

Though it is desirable to formulate the stability problem in terms of arbitrary solutions of the initial value problem, this procedure does tend to hide the fact that there is a sense in which conditionally stable solutions of (1.1) are independent of the initial conditions. This follows from the assumption that the criterion (2.1) implies a common limit as $t\to\infty$ for all solutions,

$$\mathbf{U}(\mathbf{x},t;\nu,\mathbf{u}_0+\mathbf{U}_0)\to\mathbf{U}(\mathbf{x},t;\nu),$$

for which $\langle|\mathbf{u}_0|^2\rangle<2\delta$. We call $\mathbf{U}(\mathbf{x},t;\nu)$ a *limiting flow*.

A particularly important role in the global theory is taken on by limiting flows which are globally stable. We say that $\mathbf{U}(\mathbf{x},t;\nu)$ is a *basic flow* if when $\nu>\nu_G$,

$$\langle|\mathbf{U}(\mathbf{x},t;\nu)-\mathbf{U}(\mathbf{x},t;\nu,\mathbf{U}_0)|^2\rangle\to 0$$

as $t\to\infty$ for every initial field $\mathbf{U}_0(\mathbf{x})$.

The basic flow is completely determined by the data (1.2), or rather, by the asymptotic form ($t\to\infty$) of the data (1.2) after the effect of the initial conditions have died away. For example, in § 5 we shall show that $\mathbf{U}$ is steady if the data is steady, $\mathbf{U}$ is time periodic if the data is time periodic and $\mathbf{U}$ is almost periodic in time if the data is almost periodic in time[2].

The basic flow $\mathbf{U}(\mathbf{x},t;\nu)$ probably exists for all values of ν, but it is globally stable only when $\nu>\nu_G$. For example, the family of flows between cylinders which are considered in Chapters III—VII exist for all values of the viscosity, but they are stable only when $\nu>\nu_G$. In general (Leray, 1933), steady solutions of (1.1) exist for all values of $\nu>0$ but there is one and only one solution when $\nu>\nu_G$.

Stability analysis of the basic flow is, of course, greatly simplified by the fact that it is the only flow which is stable when $\nu>\nu_G$.[3]

It is natural to begin our discussion of the stability of the basic flow with the theory of the first critical viscosity. The existence of this critical number will guarantee the uniqueness and stability of a basic limiting flow. The energy theory does not start directly from the basic problem (1.4) but starts instead from a derived problem associated with the equation governing the evolution of the energy of a disturbance.

§ 3. The Evolution Equation for the Energy of a Disturbance

The evolution equation to be established has the form

$$\frac{d\mathscr{E}}{dt}=-\langle\mathbf{u}\cdot\nabla\mathbf{U}\cdot\mathbf{u}\rangle-\nu\langle|\nabla\mathbf{u}|^2\rangle. \tag{3.1}$$

---

[2] A brief description of the elementary properties of almost periodic functions is given in Appendix A.

[3] The study of steady incompressible motions of a viscous fluid moving slowly is equivalent, by dynamical similarity (see Notes to Chapter I), to the study of more rapid motions of very viscous fluids. The fact that "slow motions" are both stable and unique means that "slow motion fluid mechanics" is a comparatively simple subject in which correctly worked theoretical results and observed motions should be in good agreement. At higher speeds non-uniqueness is the rule; some solutions are stable and some are unstable. It is much more difficult to predict from analysis the motions which will be observed at higher speeds; sets of solutions must be described and the stable and observable subsets must be separated from the others.

Here

$$\langle|\nabla \mathbf{u}|^2\rangle = \langle \partial_i u_j \partial_i u_j \rangle$$

is the average dissipation and

$$\langle \mathbf{u}\cdot\nabla\mathbf{U}\cdot\mathbf{u}\rangle = \langle u_i \partial_i U_j u_j\rangle = \langle u_i D_{ij}[\mathbf{U}] u_j\rangle$$

is an energy production integral which couples the basic flow $\mathbf{U}$ (with stretching tensor[4] $D_{ij}[\mathbf{U}] = \frac{1}{2}(\partial_i U_j + \partial_j U_i)$), to the disturbance $\mathbf{u}$.

To derive the evolution equation (3.1), we shall need Reynolds' transport theorem. Let $\mathscr{V}(0)$ be the (bounded) flow region and $\mathbf{X}$ the position at the initial instant. The mappings

$$\mathbf{x} = \mathbf{x}(\mathbf{X}, t),$$

are particle paths for the motion $\mathbf{U} = d\mathbf{x}/dt$ and satisfy

$$\int_{\mathscr{V}(t)} F(\mathbf{x}, t) = \int_{\mathscr{V}(0)} J F(\mathbf{X}, t)$$

where J is the Jacobian from $\mathbf{x}$ to $\mathbf{X}$. Since $\mathbf{U}$ is solenoidal, $dJ/dt = J \operatorname{div} \mathbf{U} = 0$. It follows that

$$\frac{d}{dt}\langle F\rangle = \left\langle \frac{dF}{dt}\right\rangle = \left\langle \frac{\partial F}{\partial t}\right\rangle + \langle \mathbf{U}\cdot\mathbf{n}F\rangle_S$$

where

$$\langle\cdot\rangle_S \equiv \frac{1}{\mathscr{M}(\mathscr{V})}\int_S (\cdot).$$

In this way,

$$\frac{1}{2}\frac{d}{dt}\langle|\mathbf{u}|^2\rangle = \left\langle \mathbf{u}\cdot\frac{d\mathbf{u}}{dt}\right\rangle = \left\langle \mathbf{u}\cdot\left(\frac{\partial \mathbf{u}}{\partial t} + \mathbf{U}\cdot\nabla\mathbf{u}\right)\right\rangle. \tag{3.2}$$

The evolution equation now follows from the application of the divergence theorem to (3.2) using (1.4a, b, c).

Exercise 3.1: Show that

$$\langle|\nabla \mathbf{U}|^2\rangle \equiv \langle|\nabla\mathbf{U}:\nabla\mathbf{U}|\rangle = \langle|\operatorname{curl}\mathbf{U}|^2\rangle = 2\langle|\mathbf{D}:\mathbf{D}|\rangle \tag{3.3}$$

where $(\mathbf{D})_{ij} = \frac{1}{2}(\partial_i U_j + \partial_j U_i)$ for all vector fields $\mathbf{U}$ such that $\operatorname{div}\mathbf{U} = 0$ and $\mathbf{U}|_S = 0$.

---

[4] This tensor is sometimes called the strain rate or the deformation rate tensor.

Exercise 3.2: Derive Eq. (3.1). Show that the identity

$$\mathbf{u}\cdot\mathbf{D}[\mathbf{U}]\cdot\mathbf{u}=\operatorname{div}\{(\mathbf{u}\cdot\mathbf{U})\mathbf{u}\}-(\mathbf{u}\cdot\nabla)\mathbf{u}\cdot\mathbf{U}$$

$$=\operatorname{div}\{(\mathbf{u}\cdot\mathbf{U})\mathbf{u}\}-\mathbf{U}\cdot\nabla\frac{|\mathbf{u}|^2}{2}+\mathbf{U}\cdot(\mathbf{u}\wedge\operatorname{curl}\mathbf{u})$$

holds for solenoidal **u** and **U**. Show that

$$\langle\mathbf{u}\cdot D[\mathbf{U}]\cdot\mathbf{u}\rangle=\langle\mathbf{U}\cdot(\mathbf{u}\wedge\operatorname{curl}\mathbf{u})\rangle\,.$$

Give a physical interpretation of the energy production integral and show that an irrotational disturbance can produce no energy.

Exercise 3.3: Form the equations for a disturbance when the body force in the disturbed flow is not zero and is given by $\delta\mathbf{F}(\mathbf{x},t)=\mathbf{F}^a(\mathbf{x},t)-\mathbf{F}(\mathbf{x},t)$. Show that

$$\frac{d\mathscr{E}}{dt}=-\langle\mathbf{u}\cdot\nabla\mathbf{U}\cdot\mathbf{u}\rangle-\nu\langle|\nabla\mathbf{u}|^2\rangle+\langle\mathbf{u}\cdot\delta\mathbf{F}\rangle\,. \tag{3.4}$$

§ 4. Energy Stability Theorems

The existence of a first critical viscosity $\nu_{\mathscr{E}}$ separating the disturbances whose energy may increase initially from those whose energy must decay monotonically is established in this section. The first critical viscosity is bounded above if the ratio of the dissipation to the energy, $\langle|\nabla\mathbf{u}|^2\rangle/\langle|\mathbf{u}|^2\rangle$, is bounded below for all fields **u** which are admissible as initial fields, that is, for **u** such that $\mathbf{u}|_S=0$ and $\operatorname{div}\mathbf{u}=0$. The existence of a lower bound for this ratio is proved in the "decay constant" lemma:

Let $\theta(x, y, z)$ be any smooth function which vanishes on S. Let l be the smallest distance between two parallel planes which just contain $\mathscr{V}$. Then there exists a constant $\Lambda>2$ such that

$$\frac{l^2}{\Lambda}\langle|\nabla\theta|^2\rangle\geqslant\langle|\theta|^2\rangle\,.^5 \tag{4.1}$$

Similarly, if $\mathbf{u}(x, y, z)$ is any smooth solenoidal vector field vanishing at the boundary, there exists a constant $\hat{\Lambda}>2$ such that

$$\frac{l^2}{\hat{\Lambda}}\langle|\nabla\mathbf{u}|^2\rangle\geqslant\langle|\mathbf{u}|^2\rangle\,. \tag{4.2}$$

Proof: It is enough to prove (4.1). Then (4.2) follows automatically. Let the coordinate z be perpendicular to the boundary of the strip of height l (Fig. 4.1). The domain just fits in this strip. At each (x, y) point, the perpendicular to the strip boundary first enters $\mathscr{V}$ at the point $z_B(x, y)$ and finally leaves at the point

[5] The inequality (4.1) is frequently called "Poincaré's inequality" (see Notes for Appendix B).

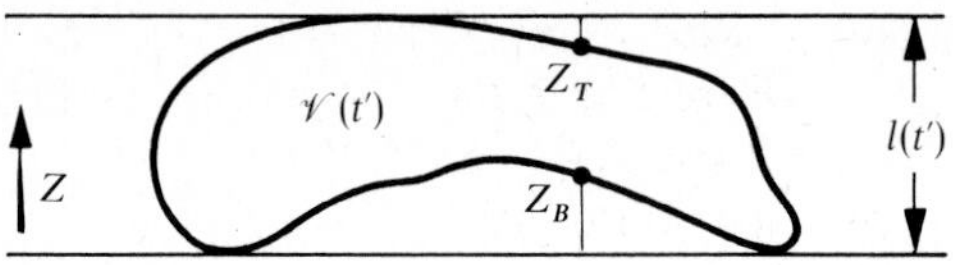

Fig. 4.1: The flow volume is entirely bounded by two parallel planes whose minimum distance apart is $l(t')$

$z_T(x, y)$. Without loss of generality we may take $\theta(x, y, z) \equiv 0$ at points on the outside of $\mathscr{V}$.

Since $\theta(x, y, z_B)=0$ we find, using Schwarz's inequality, that

$$\theta(x, y, z) = \int_{z_B}^{z} \frac{\partial \theta}{\partial z'} dz' \leqslant \sqrt{\int_{z_B}^{z} (1)^2 dz'} \sqrt{\int_{z_B}^{z} \left(\frac{\partial \theta}{\partial z'}\right)^2 dz'} \leqslant \sqrt{z - z_B} \left(\int_{z_B}^{z} |\nabla \theta|^2 dz\right)^{1/2}. \quad (4.3)$$

It follows that

$$\int_{z_B}^{z_T} \theta^2 dz \leqslant \frac{l^2}{2} \int_{z_B}^{z_T} |\nabla \theta|^2 dz \quad (4.4)$$

where z_B and z_T are functions of x and y. Integration of (4.4) over the projection of $\mathscr{V}$ onto the x, y plane leads to the final result:

$$\int_{\mathscr{V}} \theta^2 d\mathscr{V} \leqslant \frac{l^2}{2} \int_{\mathscr{V}} |\nabla \theta|^2 d\mathscr{V}.$$

In the last preliminary to the main theorem we define the kinematically admissible functions. These are scalar fields $\theta(\mathbf{x})$ and vector fields $\mathbf{u}(\mathbf{x})$ such that $\mathbf{u}=0=\theta$ on the boundary S of $\mathscr{V}$ and $\operatorname{div} \mathbf{u}=0$.

The symbols $\mathbf{H}$ and H are used to designate kinematically admissible vectors and functions, respectively. The set of admissible initial conditions $\mathbf{u}$ coincides with the set of kinematically admissible vectors. Solutions of the IBVP (1.4) for $\mathbf{u}$ are kinematically admissible but, of course, not every kinematically admissible vector is a solution.

It is clear from the inequalities (4.1) and (4.2) that the constants Λ and $\hat{\Lambda}$ can be determined from maximum problems posed for kinematically admissible functions; for example,

$$\frac{l^2}{\Lambda} = \max_{\mathbf{H}} \langle |\mathbf{u}|^2 \rangle / \langle |\nabla \mathbf{u}|^2 \rangle. \quad (4.5)$$

A general variational theory for these maximum problems and the values $\nu_{\mathscr{E}}$ of (4.6) below is given in Appendix B.

With the preliminaries aside we can prove the *energy stability theorem I*.[6]

[6] The antecedents for this modification of a theorem of Serrin's (1959B) may be traced back to the work of O. Reynolds (1895) and W. Orr (1907), on the one hand, and to the studies of T.Y. Thomas (1942) and E. Hopf (1941) on the other. The energy theorem I stresses the monotonic decay associated with the criterion $\nu_{\mathscr{E}} < \nu$.

Let $\mathbf{D}[\mathbf{U}(\mathbf{x}, t, \nu)] = \mathbf{D}$ be the stretching tensor for an arbitrary solution of the IBVP (1.1). Let $l(t) = \max l(t')$ for $0 \leqslant t' \leqslant t$. Here $l(t')$ is the minimum distance between parallel planes entirely containing $\mathscr{V}(t')$ (Fig. 4.1).

There exists a positive decay constant $\hat{\Lambda}$ *and a finite stability limit*

$$\nu_{\mathscr{E}}(\nu, t) = \max_{\mathbf{H}} \frac{-\langle \mathbf{u} \cdot \mathbf{D}[\mathbf{U}(\mathbf{x}, t; \nu)] \cdot \mathbf{u} \rangle}{\langle |\nabla \mathbf{u}|^2 \rangle} \tag{4.6}$$

such that when $\nu > \nu_{\mathscr{E}}(\nu, t')$ *for* $0 < t' < t$ *then*

$$\mathscr{E}(t) \leqslant \mathscr{E}(0) \exp\left\{ -\frac{2\hat{\Lambda}\nu}{l^2} \int_0^t \left(1 - \frac{\nu_{\mathscr{E}}}{\nu}\right) dt' \right\}. \tag{4.7a}$$

Suppose further that

$$\nu > \nu_{\mathscr{E}}(\nu) \equiv \sup_{t>0} \nu_{\mathscr{E}}(\nu, t) \tag{4.7b}$$

holds for all t. Then the null solution of (1.4) *is globally and monotonically stable. In contrast, if at the initial instant* $\nu \leqslant \nu_{\mathscr{E}}(\nu, 0)$, *then a kinematically admissible initial condition can be found for which*

$$d\mathscr{E}(0)/dt \geqslant 0. \tag{4.8}$$

We note that the monotonic decay implied by (4.7a) is actually exponential; it applies globally, that is, for $\mathscr{E}(0) < \infty$. The value $\nu = \bar{\nu}_{\mathscr{E}}$ is the smallest value of ν for which (4.7b) holds; $\bar{\nu}_{\mathscr{E}} = \nu_{\mathscr{E}}(\bar{\nu}_{\mathscr{E}})$. The criterion $\nu > \bar{\nu}_{\mathscr{E}}$ is sufficient for stability; Eq. (4.8) shows that it is also necessary for monotonic stability.

Proof: Every solution $\mathbf{u}$ of the IBVP (1.4) satisfies the evolution equation (3.1). Since $\mathbf{u}$ is a solution it is also kinematically admissible, and assuming (4.6) we may rewrite (1.4) as follows:

$$\frac{d\mathscr{E}}{dt'} = \langle |\nabla \mathbf{u}|^2 \rangle \left\{ \frac{-\langle \mathbf{u} \cdot \mathbf{D} \cdot \mathbf{u} \rangle}{\langle |\nabla \mathbf{u}|^2 \rangle} - \nu \right\} \leqslant \langle |\nabla \mathbf{u}|^2 \rangle \{ \nu_{\mathscr{E}}(\nu, t') - \nu \}. \tag{4.9a}$$

Suppose the criterion $\nu_{\mathscr{E}}(\nu, t') < \nu$ holds up to time t. Then, applying (4.5) to (4.9a) we continue the inequality as

$$\frac{d\mathscr{E}}{dt'} \leqslant -2(\hat{\Lambda}/l^2)\{\nu - \nu_{\mathscr{E}}(\nu, t')\} \mathscr{E}. \tag{4.9b}$$

The inequality (4.7a) now follows upon integration of (4.9b) up to time t.

The solution of the problem (4.5) for the decay constant $\hat{\Lambda}$ gives an estimate of the stability limit. To form this estimate one notes that the stretching tensor $\mathbf{D}[\mathbf{U}]$ may be referred to its principal axes at each point $\mathbf{x}$ of $\mathscr{V}$. There at $\mathbf{x}$ we have

$$\mathbf{u} \cdot \mathbf{D} \cdot \mathbf{u} = u_i u_j D_{ij} \geqslant |\mathbf{u}|^2 \min[D_{11}, D_{22}, D_{33}], \tag{4.10}$$

where D_{11}, D_{22}, and D_{33} are the eigenvalues of **D**. Since div $\mathbf{U}$=trace $\mathbf{D}=D_{11}+D_{22}+D_{33}=0$, one or two of these eigenvalues are negative. Now multiply (4.10) by (-1) and note that

$$-\langle \mathbf{u}\cdot\mathbf{D}\cdot\mathbf{u}\rangle \leqslant |\hat{D}_m|\langle|\mathbf{u}|^2\rangle\,, \tag{4.11}$$

where $\hat{D}_m<0$ is the smallest of the three values $[D_{11}, D_{22}, D_{33}]$ over $\mathscr{V}$ up to time t. On taking account of (4.11) and (4.5) we find that

$$\nu_{\mathscr{E}}(\nu, t)=\max_{\mathbf{H}}\frac{-\langle \mathbf{u}\cdot\mathbf{D}\cdot\mathbf{u}\rangle}{\langle|\nabla\mathbf{u}|^2\rangle}\leqslant|\hat{D}_m|\frac{l^2}{\hat{\Lambda}}\,. \tag{4.12}$$

The inequalities (4.2) and (4.12) show that $\bar{\nu}_{\mathscr{E}}$ and $1/\hat{\Lambda}$ are bounded from above. Hence, the functionals in (4.5) and (4.6) have least upper bounds. Suppose $\mathbf{u}=\tilde{\mathbf{v}}$ maximizes (4.6). Let

$$\tilde{\mathbf{v}}=\mathbf{u}_0=\mathbf{u}|_{t=0}\,,$$

that is, choose $\tilde{\mathbf{v}}$ as the initial condition for the disturbance; then, at $t=0$,

$$\frac{d\mathscr{E}}{dt}=\langle|\nabla\tilde{\mathbf{v}}|^2\rangle\left\{\frac{-\langle\tilde{\mathbf{v}}\cdot\mathbf{D}[\mathbf{U}]\cdot\tilde{\mathbf{v}}\rangle}{\langle|\nabla\tilde{\mathbf{v}}|^2\rangle}-\nu\right\}=(\nu_{\mathscr{E}}(\nu,0)-\nu)\langle|\nabla\tilde{\mathbf{v}}|^2\rangle\,. \tag{4.13}$$

It follows that $d\mathscr{E}(0)/dt>0$ when $\nu_{\mathscr{E}}(\nu, 0)>\nu$.

The maximizing field for (4.6) gives the form of the disturbance which is most energetic in the following sense: the energy of this disturbance will increase initially in all of the viscous fluids whose viscosity coefficient is less in value than $\nu_{\mathscr{E}}(\nu, 0)$. For very viscous fluids, those with $\nu>\bar{\nu}_{\mathscr{E}}$, the energy of every disturbance decreases at each and every moment.

The inequality (4.11) for the energy production integral shows that large "strain rates" (large values $|D_m|$ of the stretching tensor D) in the basic flow tend to promote instability.

The inequalities (4.11) and Exercise 4.3 suggest that the basic flow is unstable, not only when the components of $\mathbf{D}[\mathbf{U}]$ are large, but also when the basic flow velocities $\mathbf{U}$ are large relative to the possible rigid motions of the whole flow region $\mathscr{V}$. Two motions which differ by a rigid one have the same tensor $\mathbf{D}$ at each instant and, hence, lead to the same production integral $-\langle\mathbf{u}\cdot\mathbf{D}\cdot\mathbf{u}\rangle$.

It is a remarkable fact that solutions of the Navier-Stokes equations are monotonically and globally stable when $\nu>\bar{\nu}_{\mathscr{E}}$. If the evolution equation for $\mathscr{E}(t)$ were not *homogeneous* of degree two in the disturbances, it would not be possible to obtain a criterion for monotonic stability which is independent of the amplitude of the disturbance. The independence of $d\mathscr{E}/dt$ on the amplitude of $\mathbf{u}$ at $t=0$ stems from the fact that the inertial nonlinearity vanishes in the mean;

$$\langle\mathbf{u}\cdot(\mathbf{u}\cdot\nabla)\mathbf{u}\rangle=0\,.$$

Nonlinearities like those associated with temperature-dependent material parameters (§ 76), with the flow on non-Newtonian fluids (Chapter XIII) or with the effects of surface tension (Chapter XIV) produce energy in the mean. The energy equation is then not homogeneous in the disturbances, and criteria which follow from this equation are conditional.

S. Davis and C. von Kerczek (1973) have proved a different stability theorem which leads to an improved criterion (4.19) for global asymptotic stability of unsteady basic flows in bounded domains. They observe that when the basic flow is unsteady, it should not be necessary for global stability that

$$\frac{1}{\mathscr{E}(t)}\frac{d\mathscr{E}(t)}{dt}=-2[\langle\mathbf{u}\cdot\mathbf{D}[\mathbf{U}(\mathbf{x},t,\nu)]\cdot\mathbf{u}\rangle+\nu\langle|\nabla\mathbf{u}|^2\rangle]/\langle|\mathbf{u}|^2\rangle<0$$

at each instant; for example, if the flow $\mathbf{U}(\mathbf{x},t,\nu)$ is time-periodic, then $\mathscr{E}(t)$ might tend to zero as $t\to\infty$ even though $d\mathscr{E}/dt>0$ for some time in each cycle. As a preliminary to the statement and proof of energy stability theorem II, we define a growth rate functional

$$\mathscr{I}[\mathbf{u},\nu,t,\lambda]=-2[\langle\mathbf{u}\cdot\mathbf{D}[\mathbf{U}(\mathbf{x},t,\nu)]\cdot\mathbf{u}\rangle+\lambda\langle|\nabla\mathbf{u}|^2\rangle]/\langle|\mathbf{u}|^2\rangle$$

where λ is a positive constant. It is easily established, using (4.11), that $\mathscr{I}<2|\hat{D}_m|$ and that

$$\mathscr{I}[\mathbf{u},\nu,t,\lambda]>\mathscr{I}[\mathbf{u},\nu,t,\nu]$$

when $\nu>\lambda$. Let

$$G(\nu,t,\lambda)=\max_{\mathbf{H}}\mathscr{I}[\mathbf{u},\nu,t,\lambda].$$

This maximum problem has a solution and it is a smooth function of the parameters ν, t and λ when $\mathbf{D}[\mathbf{U}]$ is a smooth function of the parameters. Moreover,

$$G(\nu,t,\lambda)>G(\nu,t,\nu) \tag{4.14}$$

when $\nu>\lambda$. Hence, when $\nu\geqslant\lambda$,

$$\frac{d\mathscr{E}(t)}{dt}\leqslant G(\nu,t,\lambda)\mathscr{E}(t). \tag{4.15}$$

Energy stability theorem II: Suppose $\nu\geqslant\lambda$. *Then*

$$\mathscr{E}(t)\leqslant\mathscr{E}(0)\exp\left[\int_0^t G(\nu,t',\lambda)dt'\right]. \tag{4.16}$$

The basic flow is globally and asymptotically stable if

$$\lim_{t\to\infty}\int_0^t G(\nu,t',\lambda)dt'\to-\infty. \tag{4.17}$$

There can be at most one value $\lambda=\bar{\lambda}(v)$ *such that* $G(v,t,\lambda)$ *is integrable*

$$|\int_0^\infty G(v,t,\bar{\lambda})dt|<\infty\,. \tag{4.18}$$

If $\lambda\neq\bar{\lambda}(v)$, *then* $G(v,t,\lambda)$ *is not integrable.*

Proof: The estimate (4.16) follows from (4.15) and if (4.17) holds, (4.16) implies the stability statement. To prove the remaining part of the theorem we note that if $\mathbf{u}_i$ maximizes $\mathscr{I}$ when $\lambda=\lambda_i$, then

$$\mathscr{I}[\mathbf{u},v,t,\lambda]<G(v,t,\lambda_i)=\mathscr{I}[\mathbf{u}_i,v,t,\lambda_i]\qquad(i=1,2)$$

for any $\mathbf{u}\in\mathbf{H}$. It follows that when $\lambda_1>\lambda_2$,

$$\begin{aligned}-2(\lambda_1-\lambda_2)\frac{\langle|\nabla\mathbf{u}_2|^2\rangle}{\langle|\mathbf{u}_2|^2\rangle}&=\mathscr{I}[\mathbf{u}_2,v,t,\lambda_1]-\mathscr{I}[\mathbf{u}_2,v,t,\lambda_2]\\&<G(v,t,\lambda_1)-G(v,t,\lambda_2)\leqslant\mathscr{I}[\mathbf{u}_1,v,t,\lambda_1]-\mathscr{I}[\mathbf{u}_1,v,t,\lambda_2]\\&=-2(\lambda_1-\lambda_2)\frac{\langle|\nabla\mathbf{u}_1|^2\rangle}{\langle|\mathbf{u}_1|^2\rangle}\leqslant-2(\lambda_1-\lambda_2)\hat{\Lambda}/l^2\end{aligned} \tag{4.19}$$

where the last inequality follows from (4.2) when $\lambda_1>\lambda_2$. Passing to the limit $\lambda_1\downarrow\lambda_2$ in (4.19) we find that

$$\frac{\partial G}{\partial\lambda}(v,t,\lambda)=-2\frac{\langle|\nabla\mathbf{u}|^2\rangle}{\langle|\mathbf{u}|^2\rangle}\leqslant-2\frac{\hat{\Lambda}}{l^2} \tag{4.20}$$

so that $\tilde{G}(\lambda)=G(v,t,\lambda)$ is a decreasing function. Returning now to (4.19) with (4.20) we find that the curve $\tilde{G}(\lambda)$ lies below any chord of the curve; that is,

$$(\lambda_1-\lambda_2)\frac{\partial G(v,t,\lambda_2)}{\partial\lambda}\leqslant(\lambda_1-\lambda_2)\frac{\partial G(v,t,\lambda_1)}{\partial\lambda} \tag{4.21}$$

and

$$\begin{aligned}2(\lambda_2-\lambda_1)\int_0^t\langle|\nabla\mathbf{u}_2|^2\rangle/\langle|\mathbf{u}_2|^2\rangle dt'&<\int_0^t[G(v,t',\lambda_1)-G(v,t',\lambda_2)]dt'\\&<2(\lambda_2-\lambda_1)\int_0^t\langle|\nabla\mathbf{u}_1|^2\rangle/\langle|\mathbf{u}_1|^2\rangle dt'\,.\end{aligned} \tag{4.22}$$

Since $\langle|\nabla\mathbf{u}|^2\rangle/\langle|\mathbf{u}|^2\rangle$ is strictly positive, we find that

$$-\infty=\int_0^\infty[G(v,t',\lambda_1)-G(v,t',\lambda_2)]dt' \tag{4.23}$$

when $\lambda_1>\lambda_2$. Using (4.18), the proof of non-integrability follows easily from (4.23), first with $\lambda_2=\bar{\lambda}$ and then with $\lambda_1=\bar{\lambda}$.

We next show that the criterion (4.17) may be associated with a second critical viscosity of energy theory $\nu=\bar{\bar{\nu}}_{\mathscr{E}}$ defining the border $\bar{\bar{\nu}}_{\mathscr{E}}=\bar{\bar{\lambda}}(\bar{\bar{\nu}}_{\mathscr{E}})$ of values $\nu>\bar{\bar{\lambda}}(\nu)$ for which $G(\nu,t,\bar{\bar{\lambda}}(\nu))$ is not integrable. Moreover

$$\lim_{T\to\infty}\frac{1}{T}\int_0^T G(\bar{\bar{\nu}}_{\mathscr{E}},t,\bar{\bar{\nu}}_{\mathscr{E}})\,dt=0 \tag{4.24}$$

If $\nu>\bar{\bar{\nu}}_{\mathscr{E}}$, *then*

$$\lim_{t\to\infty}\int_0^t G(\nu,t',\bar{\bar{\nu}}_{\mathscr{E}})\,dt'=-\infty\,. \tag{4.25}$$

If the basic motion is steady

$$\bar{\nu}_{\mathscr{E}}=\bar{\bar{\nu}}_{\mathscr{E}}\,, \tag{4.26}$$

otherwise

$$\nu_{\mathscr{E}}\geqslant\bar{\bar{\nu}}_{\mathscr{E}} \tag{4.27}$$

Proof: Choose λ_1 and λ_2 in (4.23) so that

$$-\infty=\int_0^\infty\left[G(\nu,t',\bar{\bar{\nu}}_{\mathscr{E}})-G(\nu,t',\bar{\bar{\lambda}}(\nu))\right]dt'$$

and note that $G(\nu,t',\bar{\bar{\lambda}}(\nu))$ is integrable. This proves (4.25). The proof of (4.26) and (4.27) is left as Exercise 4.6.

The energy stability theorem II also has a small technical advantage for the study of steady basic motions. For these $\bar{\nu}_{\mathscr{E}}=\bar{\bar{\nu}}_{\mathscr{E}}$, and the inequality (4.16) reduces to

$$\mathscr{E}(t)\leqslant\mathscr{E}(0)\exp\left[G(\nu,\bar{\nu}_{\mathscr{E}})t\right]. \tag{4.28}$$

It follows that $G(\nu,\bar{\nu}_{\mathscr{E}})$ is a decay constant. Moreover, since

$$G(\nu,\bar{\nu}_{\mathscr{E}})<-\frac{2\hat{\Lambda}}{l^2}(\nu-\bar{\nu}_{\mathscr{E}}) \tag{4.29}$$

when $\nu>\bar{\nu}_{\mathscr{E}}$, the stability theorem II can lead to a larger decay rate in the case of steady flow.

Finally we consider the case when the flow domain $\mathscr{V}=\mathscr{V}_\infty$ is unbounded and cannot be contained between parallel planes. We note, without proof (but see Exercise 3.1 of Appendix B) that

$$\lim_{n\to\infty}\min_{\mathbf{H}(\mathscr{V}_n)}\left\{\frac{\langle|\nabla\mathbf{u}|^2\rangle}{\langle|\mathbf{u}|^2\rangle}\right\}\to 0 \tag{4.30}$$

where, as $n \to \infty$, $\mathscr{V}_n$ are ever larger bounded domains tending to $\mathscr{V}_\infty$. In this limit we lose the inequality (4.2); and (4.9 b) no longer follows from (4.9 a).

We have already noted that the basic motion enters the energy stability theorems only through the tensor $\mathbf{D}[\mathbf{U}]$. In the case of unbounded domains, the components of $\mathbf{D}$ can be expected to bound the values which the ratio

$$-\langle \mathbf{u}\cdot\mathbf{D}\cdot\mathbf{u}\rangle/\langle|\nabla\mathbf{u}|^2\rangle<\infty$$

can take when $\mathbf{u}$ ranges over functions for which

$$\int_{\mathscr{V}_\infty}|\nabla\mathbf{u}|^2 d\mathscr{V}<\infty\,.$$

This will ordinarilly imply that the components of $\mathbf{D}[\mathbf{u}]$ decay with sufficient rapidity as $|\mathbf{x}|\to\infty$ along certain rays. In the case of jets and shear layers, the components of $\mathbf{D}[\mathbf{U}]$ decay exponentially fast with distance from the axis.

Exercise 4.1 (Initially increasing disturbances in a simple case):

$$\frac{d}{dt}\begin{bmatrix}x\\y\end{bmatrix}=\begin{bmatrix}-1 & a\\0 & -2\end{bmatrix}\begin{bmatrix}x\\y\end{bmatrix}.$$

Show that x^2+y^2 must eventually decay exponentially in time. Find initial values for the numbers a, $x(t)$, $y(t)$ such that $d(x^2+y^2)/dt>0$ when $t=0$.

Exercise 4.2: Use the inequality (C.1) (Appendix C) to show that the factors Λ and $\hat{\Lambda}$ in (4.1) and (4.2) satisfy the inequalities $\Lambda>8$ and $\hat{\Lambda}>8$.

Exercise 4.3 (Serrin, 1959A): Prove that

$$\nu\mathbf{u}\cdot\nabla\mathbf{u}\cdot\mathbf{U}\leqslant\tfrac{1}{2}(\nu^2|\nabla\mathbf{u}|^2+|\mathbf{u}|^2|\mathbf{U}|^2)$$

and deduce the inequalities

$$2\frac{d\mathscr{E}}{dt}\leqslant\frac{1}{\nu}\left[\langle|\mathbf{u}|^2|\mathbf{U}|^2\rangle-\nu^2\langle|\nabla\mathbf{u}|^2\rangle\right]$$

and

$$\mathscr{E}(t)\leqslant\mathscr{E}(0)\exp\left[U_m^2-\nu^2\hat{\Lambda}/l^2\right]t/\nu\,,$$

where U_m is the maximum value of $|\mathbf{U}|$.

Hint: For any dyadic $\mathbf{A}$ we have $(\mathbf{A}-\mathbf{u}\mathbf{U}):(\mathbf{A}-\mathbf{u}\mathbf{U})\geqslant 0$.

Exercise 4.4: Show that the rigid body motion of a fluid filling a rigid closed container in arbitrary motion is absolutely, monotonically and globally stable.

Exercise 4.5: Show that $\mathscr{I}\leqslant 2|\hat{D}_m|$. Prove (4.16) and (4.29).

Exercise 4.6: Prove (4.26) and (4.27). Hint: (4.24) shows that there are instants t'' for which $G(\bar{\bar{v}}_{\mathscr{E}},t'',\bar{\bar{v}}_{\mathscr{E}})\geqslant$. Show that $\bar{\bar{v}}_{\mathscr{E}}\leqslant v_{\mathscr{E}}(\bar{\bar{v}}_{\mathscr{E}},t'')\leqslant v_{\mathscr{E}}(\bar{\bar{v}}_{\mathscr{E}})$.

§ 5. Uniqueness

Solutions of the Navier-Stokes problem (1.1) have two fundamental properties of uniqueness:

(1) There is only one solution of (1.1) which starts from an assigned initial field;

(2) When ν is large all solutions of (1.1) tend to a single basic flow.

This section aims at establishing a sharp distinction between these two very different types of uniqueness. These two properties of uniqueness are fundamental to an understanding of fluid mechanics.

Property (1) may be stated as a *uniqueness theorem for the initial boundary value problem* (1.4).

The null solution of (1.4) *is unique.*

Proof: The evolution equation

$$\frac{d\mathscr{E}}{dt} = -\langle \mathbf{u}\cdot\mathbf{D}\cdot\mathbf{u}\rangle - \nu\langle|\nabla\mathbf{u}|^2\rangle \tag{5.1}$$

leads to the inequality

$$\frac{d\mathscr{E}}{dt} \leqslant 2\left(|\hat{D}_m| - \frac{\nu\hat{\Lambda}}{l^2}\right)\mathscr{E} = c\mathscr{E} \tag{5.2}$$

where c is a constant. The estimates leading from (5.1) to (5.2) are (4.11) and (4.5), respectively. The uniqueness theorem follows from the integral of the inequality (5.2) when $\mathscr{E}(0)=0$:

$$\begin{aligned} 0 &\geqslant \int_0^t \left(\frac{d\mathscr{E}}{dt'} - c\mathscr{E}\right)\exp(-ct')dt' \\ &= \int_0^t \frac{d}{dt'}[\mathscr{E}\exp(-ct')]dt' = \mathscr{E}(t)\exp(-ct)\,. \end{aligned} \tag{5.3}$$

If (5.1) holds for all times $t' \leqslant t$ it follows from (5.3) that $\mathscr{E}(t')=0$. Then $\langle|\mathbf{u}|^2\rangle=0$ and $\mathbf{U}^a - \mathbf{U} = \mathbf{u} = 0$ almost everywhere and for all $t' \leqslant t$.

It will be noted that the uniqueness theorem for the initial value problem places no restrictions on the values of the viscosity.

What is the final destiny of all these uniquely determined solutions of the Navier-Stokes equations? For small values of the viscosity the final set of flows which evolve from a given set of initial fields is generally "turbulent", and the resulting flows can perhaps be usefully characterized only in a statistical sense. But when $\nu > \overline{\nu}_{\mathscr{E}}$, the energy stability theorem guarantees that all solutions tend to a single limiting solution which is uniquely determined by the data $\{\mathbf{U}_S(\mathbf{x}, t), \mathbf{F}(\mathbf{x}, t), \mathscr{V}(t)\}$ after the effect of initial conditions has disappeared.

A more precise statement of the uniqueness of limiting solutions when $\nu > \overline{\nu}_{\mathscr{E}}$ can be made when the final form of the given data can be characterized precisely.

We can consider cases in which the data $\{\mathbf{U}_S, \mathbf{F}, \mathscr{V}\}$ is
(a) steady,
(b) periodic in t with period T,
(c) almost periodic in t.

Since constant functions and periodic functions are also almost periodic, it would suffice to consider case (c) alone. It is instructive, however, to consider the cases separately.

Suppose that $\nu > \bar{\nu}_{\mathscr{E}}$. *Then there is at most:*
(a) *one steady solution of* (1.1),
(b) *one periodic solution of* (1.1) *with period* T, *and*
(c) *one almost periodic solution of* (1.1).

Proof: Since $\nu > \bar{\nu}_{\mathscr{E}}$ we have, using (4.7), that

$$\mathscr{E}(t) \leqslant \mathscr{E}(0) e^{-a^2 t}, \qquad a^2 > 0. \tag{5.4}$$

(a) For steady solutions $\mathscr{E}(t) = \mathscr{E}(0)$. This is compatible with (5.4) if and only if $\mathscr{E}(t) = 0$.

(b) For periodic solutions $\mathscr{E}(t) = \mathscr{E}(t+T)$ for any $t \geqslant 0$. Hence, $\mathscr{E}(T) = \mathscr{E}(0) \leqslant \mathscr{E}(0) e^{-a^2 T}$ and $\mathscr{E}(0) = 0$. It follows that $\mathscr{E}(t) = 0$.

(c) For almost periodic solutions we find, using (5.4), that

$$\frac{1}{T} \int_0^T \mathscr{E}^2(t) dt \leqslant \mathscr{E}^2(0) \frac{1 - e^{-2a^2 T}}{2a^2 T}.$$

Hence,

$$\lim_{T \to \infty} \frac{1}{T} \int_0^T \mathscr{E}^2(t) dt = 0.$$

Since $\mathscr{E}(t)$ is almost periodic, we find, using property 8 of Appendix A, that $\mathscr{E}(t) = 0$.

The relation

$$2\mathscr{E}(t) = \langle |\mathbf{u}|^2 \rangle = \langle |\mathbf{U} - \mathbf{U}_a|^2 \rangle = 0 \tag{5.5}$$

implies that $\mathbf{U} = \mathbf{U}^a = 0$ almost everywhere.

We have seen that when $\nu > \bar{\nu}_{\mathscr{E}}$, all flows tend to a single limiting flow which is determined by the data independent of initial values. For smaller values of ν, say in turbulent flow, the limiting solutions can be very sensitive to initial values.

The next problem is to determine what happens when $\nu \leqslant \bar{\nu}_{\mathscr{E}}$. To develop the theory for $\nu \leqslant \bar{\nu}_{\mathscr{E}}$, we shall need to discuss instability.

Exercise 5.1: Prove the uniqueness theorem for limiting solutions when $\nu > \bar{\bar{\nu}}_{\mathscr{E}}$.

Exercise 5.2: Prove the uniqueness theorem for the initial boundary value problem for inviscid fluids. You should assume that the normal component of the velocity is prescribed and equal to the normal velocity of the closed boundary S of $\mathscr{V}$.

Notes for Chapter I

(a) The Reynolds Number

It is conventional, and frequently convenient, to use a Reynolds number rather than the viscosity as the characterizing parameter for stability studies. The Reynolds number is a dimensionless number which is composed as follows: $R=Ul/\nu$ where U and l are constants with the dimensions of velocity and length and ν is the kinematic viscosity. This number is named after Osborne Reynolds whose fundamental 1883 experiments on the transition to turbulence in round pipes revealed that the breakdown of laminar flow to "sinuous motions" occurred at a critical R (defined with the mean laminar velocity and the diameter of the tube) which is the same for different diameter tubes, different velocities and fluids with different viscosities.

Reynolds' observations have not to this day received an adequate theoretical explanation. The instability observed by Reynolds evidently involves the basic nonlinearities in an important way. Reynolds (1895), himself, notes that the abruptness of the transition from laminar flow to an irregular eddying motion "at once suggested the idea that the condition might be one of instability for disturbances of a certain magnitude, and stable for smaller disturbances".

The appearance of a Reynolds number is an immediate consequence of a dimensionless formulation of the IBVP (1.1). This formulation of the problem allows one to study the stability of a "similar" family of flows from a single computation. To characterize the similar family, choose constants U and l^3 which are roughly representative of the fields $\mathbf{U}_s(\mathbf{x},t)$ and $\mathscr{V}(t)$ and, for simplicity, set $\mathbf{F}\equiv 0$. Now, change dimensional variables $\rightarrow$ dimensionless variables:

$$\{t,\mathbf{x},\mathbf{U},\pi\}\rightarrow\{t',\mathbf{x}',\mathbf{U}',\pi'\}=\{tl/U,\mathbf{x}/l,\mathbf{U}/U,\pi_0/U^2\}. \tag{5.6}$$

In the new variables we may write (1.1) as

$$\frac{\partial\mathbf{U}'}{\partial t}+\mathbf{U}'\cdot\nabla\mathbf{U}'-\frac{1}{R}\nabla^2\mathbf{U}'+\nabla\pi'=0, \quad \operatorname{div}\mathbf{U}'=0, \tag{5.7a,b}$$

$$\mathbf{U}'(\mathbf{x}',t')=\mathbf{U}'_{s'}(\mathbf{x}',t'), \quad \mathbf{x}'\in S' \tag{5.7c}$$

and

$$\mathbf{U}'(\mathbf{x}',0)=\mathbf{U}'_0(\mathbf{x}'), \quad \mathbf{x}'\in\mathscr{V}'(0) \tag{5.7d}$$

where

$$R=Ul/\nu=\text{velocity}\times\text{length/kinematic viscosity.}$$

All of the stability concepts which we have defined can be formed for the scaled variables. In particular, the stability properties of the limiting solution $\mathbf{U}'(\mathbf{x}',t',R)$ now depend on R when the data $\{\mathbf{U}'_s(\mathbf{x}',t');\mathscr{V}'(t')\}$ and the attracting radius δ' are given.

The advantage of introducing R is that a value $R=R_c$ is critical not only for one flow but also for a family of flows with different boundary values $U\mathbf{U}'_s(\mathbf{x}',t')$, different domains $l^3\mathscr{V}'(t')$ and different viscosities ν restricted only by the requirement that they share a common value of $R=Ul/\nu$.

(b) Bibliographical Notes

O. Reynolds (1895) was the first to derive an energy equation of the form (3.1). Reynolds' energy equation is not stated as is (3.1) for the basic motion (with stretching tensor $\mathbf{D}[\mathbf{U}]$ and the disturbances $\mathbf{u}$) but applies instead to fluctuations from a mean motion $\mathbf{D}[\mathbf{U}]$. The decomposition of the motion into a basic motion and disturbance is not the same as the decomposition of the same motion into a mean and fluctuations with a zero mean. The governing equations are different though both decompositions lead to an energy equation of the form (3.1).

Reynolds also knew that the equation governing the evolution of energy could be made the basis for a stability criterion. The stability criterion $\nu>\nu_{\mathscr{E}}$ first appears in the works of W. McF. Orr (1907). Orr's study proceeds from a linearization of Reynolds' equation in a manner which is described in the notes for Chapter IV. There I have explained how Orr came to believe that his works applied to nonlinear disturbances. The meaning of the criterion was well understood by Orr who claimed (p. 77) that, "... the numbers I have found are true least values ..., that below them every disturbance must automatically decrease, and that above them it is possible to prescribe a disturbance which will increase for a time."

The actual proof of existence of stability in bounded domains seems to have been given first by T.Y. Thomas (1942). Thomas proved that a fluid motion is monotonically stable in the sense that $d\mathscr{E}/dt<0$ when the components of **D** are sufficiently small. He did not give an algorithm for calculating stability limits in general bounded domains, though he did obtain an explicit value for monotonic stability of flow in an infinitely long round pipe (Hagen-Poiseuille flow). Thomas did not state that this criterion implies exponential decay.

A different proof of stability was given by Hopf (1941, p. 773). Hopf showed that $\mathscr{E}(t)$ decays exponentially when the maximum speed of the basic flow is small enough.

The modern theory of energy dates from the (1959A, B) writings of J. Serrin. Serrin was the first to apply the variational method to the problem which results from decomposing the motion into a basic motion and a disturbance. Serrin proves that the criterion

$$|\hat{D}_m| l^2/\nu < 1/\hat{\Lambda} \tag{5.8}$$

is sufficient to guarantee exponential decay and that the criterion

$$\nu > \nu_{\mathscr{E}}$$

is sufficient to guarantee that $d\mathscr{E}/dt<0$ at every instant. The proof that the criterion $\nu_{\mathscr{E}}<\nu$ implies exponential decay (Eq. 4.7) is given by Joseph and Serrin in the paper of Joseph (1966).

The basic idea for the energy stability theorem II was discovered by Davis and von Kerczek (1973). The second critical viscosity $\bar{\bar{\nu}}_{\mathscr{E}}$ implied by their theorem is introduced first here. Galdi (1975) has exhibited cases in unbounded domains where the energy stability theorem of Davis and von Kerczek fails. In these cases energy theory gives $d\mathscr{E}/dt\leqslant 0$ if and only if $\nu\geqslant\bar{\nu}_{\mathscr{E}}$.

The criterion $\nu>\bar{\nu}_{\mathscr{E}}$ for uniqueness of steady flow is implied by the uniqueness theorem stated by Serrin (1959 A, B). The criterion $\nu>\bar{\nu}_{\mathscr{E}}$ for uniqueness of periodic and almost periodic solutions has not been given elsewhere; the essential idea for the periodic case was given by Serrin (1959 C).

The uniqueness theorem for the initial value problem was found by Foá (1929). The theorem guarantees uniqueness of regular solutions of the initial value problem as long as such solutions exist. These solutions could conceivably develop discontinuities and then Foá's theorem need not hold.

It is generally supposed that discontinuities cannot develop in a viscous fluid but despite great efforts by eminent mathematicians this has never been proved (see Ladyzhenskaya, 1975). Of course, the smoothing of the flow by viscosity could not apply to ideal fluids and Foá's theorem also holds for these. It follows that the problem of uniqueness cannot be separated from existence; it is at least necessary to specify the class of solutions in which uniqueness holds. The most general class of solutions whose existence has been rigourously established are the weak solutions of E. Hopf (1951). Hopf's solutions last for all time; but it is not possible to assert for them the degree of regularity which is required for uniqueness. To get unique solvability various restrictions have to be added. This goes beyond us. Interested readers would do well to consult the papers of Serrin (1963) and Ladyzhenskaya (1975).

Uniqueness and stability theorems for compressible fluids have been given by Serrin (1959 D) and by Hills and Knops (1973). The theory of stability for compressible fluids is not well-developed because the equations governing the flow of compressible fluids are severely nonlinear in the terms which govern the dynamics, in the terms which govern the thermodynamics and in the kinematic equation expressing the conservation of mass.

Chapter II

Instability and Bifurcation

The existence or nonexistence of disturbances whose energy increases initially is determined by a criterion involving the first critical value $\nu_{\mathscr{E}}$. When $\nu>\nu_{\mathscr{E}}$ there is one stable and unique basic flow to which all other flows tend.

§ 6. The Global Stability Limit

When $\nu<\nu_{\mathscr{E}}$ there are disturbances whose energy may increase for a time. The inequality $\nu\leqslant\nu_{\mathscr{E}}$ is not a criterion for instability; flows with $\nu\leqslant\nu_{\mathscr{E}}$ can be globally stable. *The smallest value of* $\nu(\doteq\nu_G)$ *for which global stability can be guaranteed is called the global stability limit.* When $\nu<\nu_G$ the basic flow is unstable to some disturbance though it may be (conditionally) stable to small disturbances (see Fig. 2.1). More than one solution of permanent form may be possible when $\nu\leqslant\nu_G$. We want to know the number of such solutions, their dependence on parameters, their mechanical properties and their stability properties. This is an extremely hard problem for analysis. Some parts of it can be managed through bifurcation theory. This small-amplitude nonlinear theory traces solutions of permanent form which branch from a basic flow when the basic flow loses its stability to an infinitesimally small disturbance. The conditional stability of bifurcating flow is also a topic of interest taken up by bifurcation theory. The starting place for bifurcation theory is the linear theory of stability; the linear theory of stability gives sufficient conditions for instability.

§ 7. The Spectral Problem of Linear Theory

Consider a disturbance of a steady basic motion $\mathbf{U}(\mathbf{x},\nu)$ satisfying (1.4) when the initial values are small. Introducing

$$\mathbf{u}=\varepsilon\mathbf{v}, \qquad p=\varepsilon p^1$$

[1] Footnote: see page 26

into (1.4), we find that

$$\frac{\partial \mathbf{v}}{\partial t}+\mathscr{L}[\mathbf{U},\nu]\mathbf{v}+\varepsilon\mathbf{v}\cdot\nabla\mathbf{v}+\nabla p=0\,,$$

$$\operatorname{div}\mathbf{v}=0,\quad \mathbf{v}|_S=0 \quad\text{and}\quad \mathbf{v}|_{t=0}=\mathbf{v}_0\,, \tag{7.1 a}$$

where

$$\mathscr{L}[0,\nu]=-\nu\nabla^2\,,$$

$$\mathscr{L}[\mathbf{U},\nu]\,a_i\equiv\mathbf{U}\cdot\nabla a_i+\mathbf{a}\cdot\nabla U_i-\nu\nabla^2 a_i \tag{7.1 b,c,d}$$

$$\equiv\mathbf{U}\cdot\nabla a_i+a_j\Omega_{ji}[\mathbf{U}]+a_jD_{ji}[\mathbf{U}]-\nu\nabla^2 a_i=\mathscr{L}[\mathbf{U},0]\,a_i+\mathscr{L}[0,\nu]\,a_i$$

and $\partial_i U_j=D_{ij}+\Omega_{ij}$ is the resolution of the gradient of $\mathbf{U}$ into the stretching tensor and vorticity tensor. The spectral problem of linear theory can be obtained from (7.1) under the assumption of infinitesimal disturbances, $\mathscr{E}(0)<\delta\to 0$. When $\varepsilon=0$, (7.1) is an autonomous linear system and there exist exponential solutions of the form

$$\mathbf{v}(\mathbf{x},t)=e^{-\sigma t}\zeta(\mathbf{x}) \quad\text{and}\quad p(\mathbf{x},t)=e^{-\sigma t}p(\mathbf{x})$$

provided that there exist numbers σ for which the spectral problem

$$-\sigma\zeta+\mathscr{L}[\mathbf{U},\nu]\zeta+\nabla p=0,\qquad \operatorname{div}\zeta=0,\qquad \zeta|_S=0 \tag{7.2}$$

has nontrivial solutions. The special numbers σ are the *eigenvalues* of (7.2) and the nontrivial solutions $(\mathbf{v},p)$ are the *eigenfunctions* belonging to σ.

A system of stability concepts is customarily defined relative to the spectral problem. Here, a flow is called *stable* if there are no eigenvalues such that $\operatorname{re}(\sigma)<0$; *marginally* or *neutrally stable* if there is one eigenvalue with $\operatorname{re}(\sigma)=0$ and $\operatorname{re}(\sigma)>0$ for the other eigenvalues; and *unstable* if at least one eigenvalue has $\operatorname{re}(\sigma)<0$.

---

[1] We are using the same symbol p to denote different functions on the two sides of the equation $p=\varepsilon p$. The pressure in incompressible flow without free surfaces is a passive variable which is to be determined after the velocity field is found. The bifurcation problems could be reformulated with the pressure eliminated. For example, we could consider the governing problems formed from repeated application of the curl operator to (7.1); this procedure is the one followed in numerical computation. Or we could project the equations into a solenoidal subspace in the decomposition of vectors into an orthogonal sum of solenoidal vectors and gradients; this procedure is followed in some mathematical studies. It is therefore unnecessary to rename the pressure in each change of variables. We understand that the p in a certain set of equations is the one that goes with the $\mathbf{u}$ that appears in the same set.

Neutral disturbances are of two kinds. If, when $\mathrm{re}(\sigma)=0$ one also has $\mathrm{im}(\sigma)=0$, then the neutral solution is steady and a *principle of exchange of stability* (PES) is said to hold. The neutral solution is time periodic if $\mathrm{im}(\sigma)\neq 0$ when $\mathrm{re}(\sigma)=0$.

There is a countable set of discrete eigenvalues and a complete set of generalized eigenfunctions for the spectral problem (see Notes for Chapter II). These eigenvalues may be ordered so that $\mathrm{re}(\sigma_1)\leqslant\mathrm{re}(\sigma_2)\leqslant\dots$. The value σ_1 is called the *principal eigenvalue*; the value σ_2 is called the second eigenvalue, etc.

The *critical viscosities of the spectral problem* are the values $\bar{\bar{\nu}}_L$ for which $\mathrm{re}(\sigma_1(\bar{\bar{\nu}}_L))=0$ changes sign as ν is varied across $\bar{\bar{\nu}}_L$. The largest of the critical viscosities, $\bar{\bar{\nu}}_L=\nu_L$, is called the *first critical viscosity of the spectral problem*[2].

A spectral problem,

$$
\begin{aligned}
&-\gamma\zeta+\frac{\partial\zeta}{\partial t}+\mathscr{L}[\mathbf{U},\nu]\zeta+\nabla p=0\,,\\
&\operatorname{div}\zeta=0,\quad \zeta(\mathbf{x},t)=\zeta(\mathbf{x},t+T),\quad \zeta|_S=0\,,
\end{aligned}
\tag{7.3}
$$

for the stability of time-periodic motions $\mathbf{U}(\mathbf{x},t;\nu)=\mathbf{U}(x,t+T;\nu)$ may be derived from the linearization of (7.1) and the Floquet representation

$$
\begin{bmatrix}\mathbf{v}(\mathbf{x},t)\\ p(\mathbf{x},t)\end{bmatrix}=e^{-\gamma t}\begin{bmatrix}\zeta(\mathbf{x},t)\\ p(\mathbf{x},t)\end{bmatrix}=e^{-\gamma t}\begin{bmatrix}\zeta(\mathbf{x},t+T)\\ p(\mathbf{x},t+T)\end{bmatrix}.
\tag{7.4}
$$

The eigenvalues γ of (7.3) are the *Floquet exponents*. Asymptotic stability in the spectral theory is guaranteed when $\mathrm{re}\,\gamma>0$. At critically $\mathrm{re}\,\gamma(\nu_L)=0$ and

$$
\gamma(\nu_L)\equiv\gamma_L=i\omega_L\,.
$$

If $\omega_L=2n\pi/T$, where n is any integer, then $e^{-\gamma_L t}\zeta(x,t)$ is T-periodic. If $\omega_L=\pm\pi/T$, then $e^{-\gamma_L t}\zeta$ is $2T$-periodic. If $\omega_L=\alpha\pi/T$, $|\alpha|\neq 1$, then $e^{-\gamma_L t}\zeta$ is quasi-periodic with periods T and $2T/\alpha$.

A brief explanation of the methods of Floquet for systems of ordinary differential equations is given below. The extension of these methods to partial differential equations can be found in the papers of Yudovich (1970A,B) and Iooss (1972).

The stability of a time-periodic solution $\mathbf{x}(t)=\mathbf{x}(t+T)$ of a system of autonomous ordinary differential equations

$$
\dot{\mathbf{x}}=\mathbf{F}(\mathbf{x})
\tag{7.5}
$$

[2] The largest critical viscosity corresponds to the smallest critical Reynolds number R_L. A second critical Reynolds number $R=\bar{\bar{R}}_L$ occurs in some problems; for example, the spectral problem for the stability of Poiseuille flow, represented in Fig. 34.1, gives, for each fixed wave number $\alpha<\alpha_m$, two values of R, R_L and $\bar{\bar{R}}_L>R_L$, such that $\xi=\mathrm{re}\,\sigma_1(R)=0$. Poiseuille flow loses conditional stability when R is increased past R_L and regains conditional stability when R is increased past $\bar{\bar{R}}_L$.

can be determined by computing the *Floquet exponents* for the linearized equations

$$\mathring{\mathbf{y}} = \mathbf{F}'(\mathbf{x})\cdot\mathbf{y} \qquad \left(\mathring{y}_i = \frac{\partial F_i}{\partial x_j} y_j;\ i,j=1,2,\ldots,n\right). \tag{7.6}$$

The system (7.6) is derived by linearizing (7.5) for small disturbances $\delta\mathbf{y}$ of $\mathbf{x}(t)$, $(\delta\to 0)$.
Eq. (7.6) may be written as

$$\mathring{\mathbf{y}} = \mathbf{A}(t)\cdot\mathbf{y}, \quad (\mathbf{F}'(\mathbf{x}) = \mathbf{A}(t) = \mathbf{A}(t+T)). \tag{7.7}$$

There are n linearly independent solution vectors for (7.7), $\mathbf{y}_1, \mathbf{y}_2, \ldots, \mathbf{y}_n$. The columns of an $n\times n$ fundamental solution matrix for (7.7) are any set of linearly independent solution vectors:

$$\mathbf{Y} = [\mathbf{y}_1, \mathbf{y}_2, \ldots, \mathbf{y}_n] \quad (Y_{ij} = y_{ij})$$

and

$$\mathring{\mathbf{Y}} = \mathbf{A}(t)\cdot\mathbf{Y} \quad (\mathring{\mathbf{Y}}_{ij} = A_{ik} Y_{kj}). \tag{7.8}$$

Since the $\mathbf{y}_i$ are independent, $\mathbf{Y}$ is non-singular.

Floquet theory is developed from the observation that if $\mathbf{Y}(t)$ is a solution matrix satisfying (7.8), then

$$\mathring{\mathbf{Y}}(t+T) = \mathbf{A}(t+T)\cdot\mathbf{Y}(t+T) = \mathbf{A}(t)\cdot\mathbf{Y}(t+T),$$

so that $\mathbf{Y}(t+T)$ is also a solution matrix. Since every solution of (7.7) can be expressed as a linear combination of the $\mathbf{y}_i$, there is a constant non-singular matrix $C(T)$, independent of t, such that

$$\mathbf{Y}(t+T) = \mathbf{Y}(t)\cdot\mathbf{C}(T). \tag{7.9}$$

The matrix $\mathbf{C}(T)$ is called the *monodromy matrix*. Its eigenvalues are called *Floquet multipliers* $\mu(T)$. The monodromy matrix arises from the initial value problem

$$\mathring{\boldsymbol{\Phi}} = \mathbf{A}(t)\cdot\boldsymbol{\Phi} \qquad \boldsymbol{\Phi}(0) = \mathbf{1}. \tag{7.10}$$

Replace $\mathbf{Y}$ with $\boldsymbol{\Phi}$ and set $t=0$ in (7.9) to show that

$$\boldsymbol{\Phi}(T) = \boldsymbol{\Phi}(0)\cdot\mathbf{C}(T) = \mathbf{C}(T). \tag{7.11}$$

So the multipliers are eigenvalues of $\boldsymbol{\Phi}(T)$:

$$\boldsymbol{\Phi}(T)\cdot\boldsymbol{\psi} = \mu(T)\boldsymbol{\psi}, \quad (\Phi_{ij}\psi_j = \mu\psi_i) \tag{7.12}$$

and

$$\boldsymbol{\Phi}^n(T)\cdot\boldsymbol{\psi} = \mu^n(T)\boldsymbol{\psi}. \tag{7.13}$$

Moreover, since $\boldsymbol{\Phi}(t)$ is a fundamental solution matrix, we may write (7.9) as

$$\boldsymbol{\Phi}(t+T) = \boldsymbol{\Phi}(T)\cdot\boldsymbol{\Phi}(t). \tag{7.14}$$

The relation

$$\boldsymbol{\Phi}(nT) = \boldsymbol{\Phi}^n(T) \tag{7.15}$$

follows from (7.14) and

$$\boldsymbol{\Phi}(nT)\cdot\boldsymbol{\psi} = \boldsymbol{\Phi}^n(T)\cdot\boldsymbol{\psi} = \mu^n(T)\boldsymbol{\psi} \tag{7.16}$$

follows from (7.14) and (7.13). (7.16) shows that $\mu^n(T) = \mu(nT)$ are eigenvalues of the matrix $\boldsymbol{\Phi}(nT)$.

Hence, we may put

$$\mu(T)=e^{-\gamma T}. \tag{7.17}$$

The values γ are the *Floquet exponents.*

The matrix $\boldsymbol{\Phi}(T)$ may be diagonalized when the multipliers $\mu(T)$ are all simple. If there are multipliers $\mu(T)$ of higher multiplicity the matrix $\boldsymbol{\Phi}$ may be reduced to a normal Jordan form (see Coddington and Levinson, 1955, p. 76). *Floquet exponents* are not uniquely determined by the *multiplier*; the same multiplier

$$\mu=e^{-\gamma T}=e^{-\gamma T+2i\pi m}$$

is associated with each and every integer m.

The Floquet exponents may be determined from the initial-value problem (7.7) with $\mathbf{y}(0)=\boldsymbol{\psi}$. (7.9) and (7.11) imply that each solution $\mathbf{y}$ of (7.7) satisfies

$$\mathbf{y}(t+T)=\boldsymbol{\Phi}(T)\cdot\mathbf{y}(t), \quad (y_i(t+T)=\Phi_{ij}(T)\,y_j(t)). \tag{7.18}$$

The vector

$$\mathbf{w}(t)=e^{\gamma t}\mathbf{y}(t), \quad (w_i=e^{\gamma t}y_i) \tag{7.19}$$

then satisfies the differential equation

$$-\gamma\mathbf{w}+\mathring{\mathbf{w}}-\mathbf{A}(t)\cdot\mathbf{w}=0$$

and periodicity conditions $\mathbf{w}(T)=\mathbf{w}(0)$;

$$\mathbf{w}(T)=e^{\gamma T}\mathbf{y}(T)=e^{\gamma T}\boldsymbol{\Phi}(T)\cdot\boldsymbol{\psi}=\boldsymbol{\psi}=\mathbf{y}(0)=\mathbf{w}(0).$$

It follows that $\mathbf{w}(t)$ satisfies

$$\mathring{\mathbf{w}}(t)=\mathbf{B}(t)\cdot\mathbf{w}, \quad \mathbf{w}(0)=\mathbf{w}(T) \tag{7.20 a, b}$$

where $\mathbf{B}(t)=[\gamma\mathbf{1}+\mathbf{A}(t)]=\mathbf{B}(t+T)$. Eq. (7.9) applies to $\mathbf{w}$ satisfying (7.20 a, b) and implies that, for all t,

$$\mathbf{w}(t+T)=\mathbf{C}(T)\cdot\mathbf{w}(t),$$

and using (7.20 b),

$$\mathbf{w}(T)=\mathbf{C}(T)\cdot\mathbf{w}(0)=\mathbf{w}(0).$$

Hence, the monodromy matrix for solutions of (7.20 a, b) is the unit matrix $\mathbf{C}=\mathbf{1}$, and

$$\mathbf{w}(t+T)=\mathbf{1}\cdot\mathbf{w}(t)=\mathbf{w}(t)$$

is a T-periodic function. The Floquet exponents for the stability of T-periodic solutions $\mathbf{x}(t)$ may therefore be determined as eigenvalues of

$$-\gamma\mathbf{w}+\mathring{\mathbf{w}}-\mathbf{A}(t)\cdot\mathbf{w}=0, \quad \mathbf{w}(t)=\mathbf{w}(t+T). \tag{7.21 a, b}$$

The vector $\mathring{\mathbf{x}}$ is always an eigenfunction of (7.21 a, b) with eigenvalue $\gamma=0$. The eigenfunctions $\mathbf{w}(t)+k\mathring{\mathbf{x}}(t)$ of (7.21 a, b) are of indeterminate phase depending on the constant k. The conditional stability of $\mathbf{x}(t)$ solving (7.5) (Coddington and Levinson, 1955, p. 323) therefore gives asymptotic stability not of a single solution but of a set of solutions $\mathbf{x}(t+\alpha)$ depending on the phase α (see 9.6). If small disturbances are attracted to this set, the set of periodic solutions is said to have conditional, asymptotic, orbital stability. An interesting application of Floquet theory is given in Exercise 16.1.

Exercise 7.1: Show that all solutions of the functional equation $\mu(nT)=\mu^n(T)$ may be represented as $\mu(T)=e^{\gamma T}$ for some complex constant γ. *Hint:* Differentiate the functional equation with respect to n, fiddle with the result, extend the definition of n to complex numbers and then integrate.

§ 8. The Spectral Problem and Nonlinear Stability

It is instructive to note that there is no way in which an attracting radius δ can appear in the spectral problem. Hence, the relation of the spectral problem (7.2) to the nonlinear problem (7.1) can only be inferred by comparison of the two problems. A *theorem of comparison of critical viscosities* establishes one such relation.

The first critical viscosity of the spectral problem (7.2) *is not larger than the first critical viscosity of energy theory; that is,*

$$\nu_{\mathscr{E}} \geqslant \nu_L . \tag{8.1}$$

Proof: From (7.2) we obtain

$$-\sigma\langle\zeta\cdot\bar{\zeta}\rangle+\langle\mathscr{L}\zeta\cdot\bar{\zeta}\rangle=0$$

where ζ is the complex conjugate of $\bar{\zeta}$. Using the symmetry $D_{ij}=D_{ji}$ and the antisymmetry $\Omega_{ij}=-\Omega_{ji}$, we find that

$$\langle\bar{\zeta}\cdot\mathscr{L}\zeta+\zeta\cdot\mathscr{L}\bar{\zeta}\rangle=2\langle\zeta\cdot\mathbf{D}[\mathbf{U}]\cdot\bar{\zeta}\rangle+2\nu\langle|\nabla\zeta|^2\rangle . \tag{8.2}$$

Hence,

$$\mathrm{re}\,\sigma\langle|\zeta|^2\rangle=\langle\zeta\cdot\mathbf{D}[\mathbf{U}]\cdot\bar{\zeta}\rangle+\nu\langle|\nabla\zeta|^2\rangle . \tag{8.3}$$

At criticality, $\mathrm{re}\,\sigma(\nu_L)=0$. Hence

$$\nu_L=\frac{-\langle\zeta\cdot\mathbf{D}[\mathbf{U}]\cdot\bar{\zeta}\rangle}{\langle|\nabla\zeta|^2\rangle}=\frac{-\langle\mathbf{a}\cdot\mathbf{D}[\mathbf{U}]\cdot\mathbf{a}+\mathbf{b}\cdot\mathbf{D}[\mathbf{U}]\cdot\mathbf{b}\rangle}{\langle|\nabla\mathbf{a}|^2+|\nabla\mathbf{b}|^2\rangle}\leqslant \max_H\frac{-\langle\mathbf{a}\cdot\mathbf{D}[\mathbf{U}]\cdot\mathbf{a}\rangle}{\langle|\nabla\mathbf{a}|^2\rangle}=\nu_{\mathscr{E}} \tag{8.4}$$

where $\zeta=\mathbf{a}+i\mathbf{b}$. In proving (8.4) we have used the following inequality:

Let a_n and b_n be any real numbers and suppose $b_n>0$. Then

$$\sum a_n/\sum b_n=\sum(b_n a_n/b_n)/\sum b_n\leqslant \max_n a_n/b_n . \tag{8.5}$$

A second theorem comparing the nonlinear problem (7.1) with the spectral problem can be framed as a *theorem of conditional stability*. This theorem is sometimes called *the principle of linearized stability:*

Suppose that $\text{re}(\sigma_1)>0$. *Then there exists an attracting radius* $\delta(\nu)>0$ *such that when* $\mathscr{E}(0)<\delta$, $\mathscr{E}(t)/\mathscr{E}(0)\to 0$ *as* $t\to\infty$. *In contrast, if* $\text{re}(\sigma_1(\nu))<0$, *then* $\mathbf{U}(\mathbf{x},\nu)$ *is unstable.*

For a proof of this theorem we refer the reader to the paper of Sattinger (1970). A brief discussion of this theorem and further references are given in the notes for Chapter II.

The principle of linear stability says that if the eigenvalues of the spectral problem are such that $\text{re}(\sigma(\nu))>0$ then the null solution will be asymptotically stable, at least to small disturbances with $\mathscr{E}(0)<\delta(\nu)$. However, this principle does not give an algorithm for the actual computation of values δ and only guarantees existence of "sufficiently small" $\delta>0$.

The principle of linearized stability does not give the most useful stability result since stability is guaranteed only when the disturbances are indefinitely small. It follows that the criteria of linear theory do not insure global stability and one must turn to nonlinear theory or experiment for the ultimate justification of the stability criteria of linear theory. In fact, this justification is so frequently lacking that one cannot rely on the stability predictions of linear theory.

The main value of the principle of linearized stability is that it gives the guarantee that if $\text{re}(\sigma_1)<0$ then there is instability[3].

The linear theory and the energy theory have a complementary character which allows one to deduce sufficient conditions for the stability and instability of basic steady flows. It is of interest that the first critical values $\nu_{\mathscr{E}}$ and ν_L may be computed as eigenvalues of *linear* partial differential equations.

A principle of linearized stability for the orbital asymptotic stability of time-periodic flows is expressed in Theorem 2.2, p. 323, of Coddington and Levinson (1955). This principle has been extended to the Navier-Stokes equations by Yudovich (1970 A, B) and Iooss (1972).

Exercise 8.1: Show that $\nu_{\mathscr{E}}\geqslant\nu_G\geqslant\nu_L$.

Exercise 8.2 (Lasalle and Lefschetz, 1961): Consider the pair of ordinary differential equations

$$\frac{dx}{dt}=y-xf(x,y),\qquad \frac{dy}{dt}=-x-yf(x,y)$$

where f is a convergent power series and $f(0,0)=0$. Show that solutions of the linearized equations corresponding to this pair of equations are "neutrally stable". Form an evolution equation for x^2+y^2 and show that the stability of the solution $x=y=0$ is controlled by the nonlinear term f.

Exercise 8.3 (A simple example of stability concepts): Consider the stability of the null solution $x(t)=0$ of the equation

$$\mathring{x}=x-\nu x f(x) \tag{8.6}$$

where $f(x)$ has a convergent power series and $f(0)=f_0>0$.

---

[3] No statement of stability can be made relative to (7.1) when $\nu=\bar{\bar{\nu}}_L$ is a critical viscosity of the spectral problem. In this case im $\sigma_1(\bar{\bar{\nu}}_L)=0$ and growth or decay is controlled by the nonlinear terms in (7.1) (see Exercise 8.2).

(a) Show that the spectral problem for the null solution of (8.6) is

$$-\sigma x=(1-\nu f_0)x$$

and that at criticality $\nu=1/f_0$. Conclude that $x(t)=0$ is stable to small disturbances when $\nu>1/f_0$ and is unstable when $\nu<1/f_0$.

(b) Suppose that $f(x)\geqslant f(0)$ for all real numbers x with equality only when $x=0$. Show that $x(t)=0$ is globally and monotonically stable when $\nu>1/f_0$.

Exercise 8.4 (Conditional stability): Suppose that the function $f(x)=f_0-Ax^2$ with $A>0$. Show that $x(t)=0$ is conditionally stable when $\nu>1/f_0$ and the initial disturbance $x(0)$ satisfies

$$x^2(0)<|(1-\nu f_0)/A\nu|=2\delta(\nu)$$

where $\delta(\nu)$ is the attracting radius. Show that $x(t)=0$ is unstable when

$$x^2(0)>2\delta(\nu)\,.$$

Exercise 8.5 (A simple example of stability concepts when the eigenvalues of the spectral problem are complex):

Consider the stability of the null solution $x(t)$ and $y(t)=0$ of the equations

$$\begin{aligned}\mathring{x}&=(1-\nu f(x,y))x+y\\ \mathring{y}&=(1-\nu f(x,y))y-x\end{aligned} \tag{8.7}$$

where $\mu>0$ is a constant, $f(x,y)$ has convergent power series, $f(0,0)=f_{00}>0$. Show that the spectral problem for the null solution of (8.7) is

$$\begin{bmatrix}\sigma+(1-\nu f_{00}) & 1\\ -1 & \sigma+(1-\nu f_{00})\end{bmatrix}\begin{bmatrix}x\\ y\end{bmatrix}=0\,, \tag{8.8}$$

that the eigenvalues of (8.8) are complex, that at critically $f_{00}=1/\nu$ and $\sigma(f_{00})=\pm i$. Conclude that $x(t)=y(t)=0$ is stable to small disturbances when $\nu>1/f_{00}$ and is unstable when $\nu<1/f_{00}$. Suppose that $f(x,y)\geqslant f_{00}$ for x and y with equality only when $x=y=0$. Show that the null solution is globally and monotonically stable when $\nu>1/f_{00}$.

Exercise 8.6 (Conditional stability): Suppose that $f(x,y)=f_{00}-A(x^2+y^2)$ with $A>0$. Show that $x(t)=y(t)=0$ is conditionally stable when $\nu>1/f_{00}$ and the initial disturbance satisfies

$$x^2(0)+y^2(0)<|(1-\nu f_{00})/A\nu|=2\delta(\nu)$$

where $\delta(\nu)$ is the attracting radius. Show that $x(t)=y(t)=0$ is unstable when $x^2(0)+y^2(0)>2\delta(\nu)$.

§ 9. Bifurcating Solutions

In the past, the stability of flow was judged only by the criteria of linear theory. In some cases the predictions of the spectral problem of linearized theory would lead to good agreement between theory and experiments and in other cases the agreement between theory and experiment was less good or no good.

Now it is understood that a flow which is stable by the criteria of the linearized theory need not be stable at all. Understanding the main physical features of the instability of these flows requires analysis of the nonlinear problem. Though a fully general nonlinear theory is not known, we can learn something about the full problem by considering the kinds of solutions which can develop as a result of the instability.

Consider a steady basic flow $\mathbf{U}(x, \nu)$. Such flows are globally stable when the viscosity is large, and, if the disturbances are kept small, $\mathbf{U}$ will lose stability only if the viscosity is reduced past the level of the first critical viscosity ν_L of linearized theory.

Out of the instability of $\mathbf{U}$ new solutions $\mathbf{U}+\mathbf{u}$ develop. Bifurcation theory is concerned with the new solutions $\mathbf{u}(\mathbf{x}, t, \nu)$ of "permanent form" which branch off the basic flow in a continuous fashion; that is, $\mathbf{u}\to 0$ as $|\nu-\nu_L|\to 0$. The "permanent form" of the bifurcating solutions is related to the properties of the principal eigenvalue $\sigma_1(\nu)$ at criticality: $\sigma_1(\nu_L)=i\omega_0$. We assume that $\sigma_1(\nu_L)$ is an isolated simple eigenvalue of $\mathscr{L}[U, \nu_L]$,[2] that there are no other eigenvalues of the form $\pm in\omega_0$ $(n\neq\pm 1)$ of $\mathscr{L}$ at criticality (see Exercise 11.5) and that the loss of stability is strict, $\sigma_\nu(\nu_L)\neq 0$. The consequences of these assumptions are summarized in Fig. 15.1.[4]

(A) If $\omega_0=0$ then the linearized problem (7.1 with $\varepsilon=0$) has a single steady solution of undetermined amplitude. This solution is a limit $(\varepsilon\to 0)$ of a one parameter (ε) family of steady bifurcating solutions which depend analytically on ε.

(B) If $\omega_0\neq 0$ then the linearized problem has complex-valued solutions $e^{i\omega_0 t}\mathbf{v}(\mathbf{x};\nu)$ which are periodic in t. In this case a unique one parameter family of time periodic solutions of the nonlinear problem bifurcates from the solution $\mathbf{U}$. These solutions are analytic functions of the amplitude ε.

It is possible to define the amplitude ε in various ways and it is best, as in Chapters IV and XI, to base the choice on physics. In this chapter: (A) $\varepsilon^2=\mathscr{E}(t)=\mathscr{E}(0)$ is the kinetic energy of a steady bifurcating solution and (B) $\varepsilon^2=T^{-1}\int_0^T\mathscr{E}(t)\,dt$ is the average energy per cycle. In both cases we set

$$\mathbf{u}=\varepsilon\mathbf{v}, \qquad p=\varepsilon p\,. \tag{9.1}$$ [5]

In the time-periodic case it will be convenient to introduce the scalar product

$$[\mathbf{a}, \mathbf{b}]=\frac{1}{T}\int_0^T\langle\mathbf{a}\cdot\overline{\mathbf{b}}\rangle\,dt \tag{9.2}$$

---

[4] A simple eigenvalue (an eigenvalue of multiplicity one) is associated with a single eigenfunction unique to within an arbitrary multiplicative constant. A more precise statement of our basic assumption is that the geometric and algebraic multiplicity of $\sigma_1(\nu_L)$ are the same and equal to one. It is possible for $\mathscr{L}[\mathbf{U}, \nu]$ to have more than one imaginary eigenvalue at criticality; for example, $\operatorname{re}\sigma_1=\operatorname{re}\sigma_2=0$ and $\operatorname{im}\sigma_1\neq\operatorname{im}\sigma_2$. To determine the number of bifurcating solutions and their analytic properties we must consider how various linear combinations of the critical eigenfunctions perturb with ε (cf. Chapter X). We shall not study this further complication here and shall assume that the eigenvalues σ_1 and $\overline{\sigma}_1$ are the only critical eigenvalues of the operator $\mathscr{L}$ of the form $\pm$ in ω_0 $(n=0, 2, 3, \ldots)$ (see Exercise 11.5).

[5] Other definitions of ε which are equivalent to this one when ε is small are useful in computations (Exercise 12.3 and 12.4).

where the overbar signifies "complex conjugate". In the notation of (9.1) we may write

$$2\varepsilon^2=[\mathbf{u},\mathbf{u}]=\varepsilon^2[\mathbf{v},\mathbf{v}] \tag{9.3}$$

where we have assumed that **u**, **v** and ε have only real values.

With these preliminaries aside we may form the boundary value problems for the bifurcating solutions:

$$\frac{\partial \mathbf{v}}{\partial t}+\mathscr{L}\mathbf{v}+\varepsilon\mathbf{v}\cdot\nabla\mathbf{v}+\nabla p=0, \qquad \operatorname{div}\mathbf{v}=0, \qquad \mathbf{v}|_S=0. \tag{9.4a, b, c}$$

For steady bifurcating solutions

$$\tfrac{1}{2}\langle|\mathbf{v}|^2\rangle=1\,. \tag{9.4d}$$

For time-periodic bifurcating solutions

$$\mathbf{v}(\mathbf{x},t)=\mathbf{v}(\mathbf{x},t+T)\,, \tag{9.4e}$$

$$\tfrac{1}{2}[|\mathbf{v}|^2]=1\,. \tag{9.4f}$$

The most important results of the bifurcation analysis concern the "direction" and stability of the bifurcating solutions. The explanation of these results requires some further analysis.

Exercise 9.1 (Simple examples of bifurcation):

(a) Show that bifurcating steady solutions of (8.6) lie on a curve through the origin of the $(1/\nu-f_0,x)$ plane given by

$$x=f^{-1}(1/\nu)$$

where f^{-1} is the inverse of the function f.

(b) Consider the time-periodic solutions which bifurcate from the null solution of (8.7) when $\nu=1/f_{00}$. Suppose that $f(x,y)=g(x^2+y^2)$ and show that the time-periodic solution

$$\begin{aligned} x&=\varepsilon\sin(t+\alpha)\,,\\ y&=\varepsilon\cos(t+\alpha) \end{aligned} \tag{9.6}$$

where

$$\frac{1}{\nu}=g(\varepsilon^2)$$

bifurcates from the null solution when $\operatorname{re}\sigma(1/g(0))=0$.

§ 10. Series Solutions of the Bifurcation Problem

The parameter ε appears analytically in the coefficients of (9.4a) and analytic solutions of the problems (9.4) exist. To obtain properties of these solutions, we shall develop them in a series of powers of the parameter ε. It will be convenient here to consider the somewhat more involved time-periodic problem (B); the steady problem (A) can then be obtained from (B) by a few simple adjustments.

Since a periodic solution of (9.4a, b, c) is a solution of a special type, we would not expect such solutions to exist for all values of ε. Indeed, when $\varepsilon=0$, by assumption, there is only the conjugate pair of solutions belonging to the principal eigenvalues $\sigma_1 = \pm i\omega_0$, and there are no other solutions corresponding to small values of $|v - v_L|$ with $\varepsilon=0$. For this reason we seek a family of solutions parameterized by ε and such that for each small value ε, there is at least one value of the viscosity $v=v(\varepsilon)$ for which a periodic solution exists. We call the graph of $v(\varepsilon)$ a *bifurcation curve.*

Moreover, experience with other nonlinear problems suggests that the period should depend on the amplitude of the solution; we take note of this possibility by mapping the solution with the period $T(\varepsilon)$ into a fixed period by means of the substitution

$$s=\omega(\varepsilon)t, \qquad T(\varepsilon)=2\pi/\omega(\varepsilon) \tag{10.1}$$

where $\omega(\varepsilon)$ is the frequency which depends on amplitude. It is also convenient to introduce the space $P_{2\pi}$ of 2π-periodic vectors

$$P_{2\pi}=\{\boldsymbol{\phi}(\mathbf{x},s)\colon \operatorname{div}\boldsymbol{\phi}=0,\ \boldsymbol{\phi}|_S=0,\ \boldsymbol{\phi}(\mathbf{x},s)=\boldsymbol{\phi}(\mathbf{x},s+2\pi)\} \tag{10.2}$$

and the operator

$$J=\omega\frac{\partial}{\partial s}+\mathscr{L}[\mathbf{U},v] \tag{10.3}$$

where, for notational simplicity, we write $\mathbf{U}(\mathbf{x},v)=\mathbf{U}$. The operator $\mathscr{L}$ is defined under (7.1a).

A compact statement of the bifurcation problem follows from the insertion of (10.1, 2, 3) into (9.4a):

$$J\mathbf{v}+\varepsilon\mathbf{v}\cdot\nabla\mathbf{v}+\nabla p=0, \qquad \mathbf{v}\in P_{2\pi}, \qquad [\mathbf{v},\mathbf{v}]=2 \tag{10.4}$$

where

$$[\mathbf{v},\mathbf{v}]=\frac{1}{2\pi}\int_0^{2\pi}\langle\mathbf{v}\cdot\mathbf{v}\rangle\,ds\,.$$

The solution of the bifurcation problem may be formed in a series of powers of ε:

$$\begin{bmatrix}\mathbf{v}(\mathbf{x},s;\varepsilon)\\ p(\mathbf{x},s;\varepsilon)\\ v(\varepsilon)\\ \omega(\varepsilon)\end{bmatrix}=\sum_{n=0}\varepsilon^n\begin{bmatrix}\mathbf{v}_n(\mathbf{x},s)\\ p_n(\mathbf{x},s)\\ v_n\\ \omega_n\end{bmatrix}. \tag{10.5}$$

To obtain the equations satisfied by the quantities on the right side of (10.3) we must expand the operator J

$$J=J_0+\sum \varepsilon^l J_l\,,$$

$$J_0=\omega_0\frac{\partial}{\partial s}+\mathscr{L}_0\,,\qquad \mathscr{L}_0=\mathscr{L}[\mathbf{U}_0,v_0]\,,$$

$$J_1=\omega_1\frac{\partial}{\partial s}+v_1 L_1\,,$$

$$J_l=\omega_l\frac{\partial}{\partial s}+v_l L_1+L_l\,,\qquad l>1\,,$$

$$L_1=\mathscr{L}[\mathbf{U}_1,1]\,,$$

$$L_l=\mathscr{L}[\mathbf{G}_l,0]\,,\qquad l>1$$

where

$$\mathbf{U}(\mathbf{x},v)=\sum_{m=0}\mathbf{U}_m(v-v_0)^m=\mathbf{U}_0+\mathbf{U}_1\sum_{l=1}\varepsilon^l v_l+\sum_{l=2}\varepsilon^l\mathbf{G}_l$$

and

$$\mathbf{G}_l=\sum_{m=2}^{l}\mathbf{U}_m\sum_{l_1+l_2+\cdots+l_m=l}v_{l_1}v_{l_2}\ldots v_{l_m}\,,\qquad 1\leqslant l_i<l.$$

Inserting (10.5) into (10.4), we find that

$$J_0\mathbf{v}_0+\nabla p_0=0\,,\qquad \mathbf{v}_0\in P_{2\pi}\,,\qquad [\mathbf{v}_0,\mathbf{v}_0]=2 \tag{10.6a}$$

and, for $m>0$,

$$\begin{aligned}&J_0\mathbf{v}_m+\omega_m\dot{\mathbf{v}}_0+v_m L_1\mathbf{v}_0+\mathbf{F}_m+\nabla p_m=0\,,\\ &\mathbf{v}_m\in P_{2\pi}\,,\qquad [\mathbf{v},\mathbf{v}]_m=0\end{aligned} \tag{10.6b}$$

where

$$\dot{\mathbf{v}}_0\equiv\partial\mathbf{v}_0/\partial s\,,$$

$$\mathbf{F}_m=L_m\mathbf{v}_0+\sum_{\substack{l+j=m\\ l,j>0}}J_l\mathbf{v}_j+\sum_{l+j=m-1}(\mathbf{v}_l\cdot\nabla)\mathbf{v}_j\,,\qquad m>1\,,$$

and

$$\mathbf{F}_1=\mathbf{v}_0\cdot\nabla\mathbf{v}_0\,.$$

The unknowns of (10.6) at order m are

$$\mathbf{v}_m,\ p_m,\ v_m\quad\text{and}\quad\omega_m\,.$$

The vector $\mathbf{F}_m$ contains only terms of order $l<m$. The perturbation problems (10.6) may be solved sequentially for the functions $\mathbf{v}_m$ and p_m provided that the

numbers ω_m and v_m can be selected appropriately. To explain this method of selection, we shall need to introduce the adjoint spectral problem.

Exercise 10.1: Expand the function $e^{is}=e^{i\omega(\varepsilon)t}$ as a power series in ε when $\omega(\varepsilon)$ is analytic. Does the power series converge? Does it converge uniformly in t?

Suppose (10.5) holds. Show that

$$\mathbf{v}_n(\mathbf{x},s)=\sum_{l=-N(n)}^{N(n)}\mathbf{v}_{n_l}(\mathbf{x})e^{-ils}.$$

Consider the expansion of the function

$$\mathbf{v}_n(\mathbf{x},\omega(\varepsilon)t)$$

into a series of powers of ε. Show that this power series leads to secular terms which become unbounded for large times. Explain why it is better to introduce the Poincaré-Lindstedt variable s into the bifurcation problem for periodic solutions.

§ 11. The Adjoint Problem of the Spectral Theory

Consider arbitrary solenoidal vector fields **a** and **b** which vanish on S. Recalling that

$$\langle\mathbf{a},\mathbf{b}\rangle=\frac{1}{\mathscr{M}}\int_{\mathscr{V}}\mathbf{a}\cdot\overline{\mathbf{b}},$$

we define the adjoint operator $\mathscr{L}^*[\mathbf{U},v]$ through the relation

$$\langle\mathbf{a},\mathscr{L}\mathbf{b}\rangle=\langle\mathscr{L}^*\mathbf{a},\mathbf{b}\rangle \tag{11.1}$$

and find that

$$(\mathscr{L}^*\mathbf{a})_i=-\mathbf{U}\cdot\nabla a_i-a_j\Omega_{ji}[\mathbf{U}]+a_jD_{ji}[\mathbf{U}]-v\nabla^2a_i.$$

The adjoint eigenvalue problem is defined by

$$\langle\mathscr{L}^*\zeta^*,\zeta\rangle=\langle\zeta^*,\mathscr{L}\zeta\rangle=\langle\zeta^*,\sigma\zeta\rangle=\overline{\sigma}\langle\zeta^*,\zeta\rangle.$$

Hence,

$$-\overline{\sigma}\zeta^*+\mathscr{L}^*\zeta^*+\nabla p^*=0 \tag{11.2}$$

where $\zeta^*(\mathbf{x})$ is solenoidal and vanishes on the boundary.

Now consider arbitrary solenoidal vector fields **a** and **b** which vanish on S and are 2π-periodic in s. We seek the adjoint of the operator

$$J_0=\omega_0\frac{\partial}{\partial s}+\mathscr{L}_0 \tag{11.3}$$

relative to the scalar product

$$[\mathbf{a},\mathbf{b}]=\frac{1}{2\pi}\int_0^{2\pi}\langle\mathbf{a},\mathbf{b}\rangle ds\,. \tag{11.4}$$

We require that

$$[J_0\mathbf{a},\mathbf{b}]=[\mathbf{a},J_0^*\mathbf{b}] \quad \text{for all } \mathbf{a},\mathbf{b}\in P_{2\pi}\,,$$

and find that

$$J_0^*=-\omega_0\frac{\partial}{\partial s}+\mathscr{L}_0^*\,. \tag{11.5}$$

A consequence of the uniqueness assumption for the principal eigenvalue, $\sigma_1(v_L)=i\omega_0$ at criticality, is that apart from multiplicative constants there are just two solutions of

$$J_0\mathbf{u}+\nabla p=0, \qquad \mathbf{u}\in P_{2\pi}\,. \tag{11.6}$$

These are (see Exercise 11.5)

$$\mathbf{Z}_1=e^{-is}\zeta \quad \text{and} \quad \mathbf{Z}_2=\overline{\mathbf{Z}}_1\,. \tag{11.7}$$

In the same way, there are just two solutions of the adjoint problem:

$$\mathbf{Z}_1^*=e^{-is}\zeta^* \quad \text{and} \quad \mathbf{Z}_2^*=\overline{\mathbf{Z}}_1^*\,. \tag{11.8}$$

Finally, for a later application, we shall prove that

$$-\sigma_v\langle\zeta,\zeta^*\rangle+\langle L_1\zeta,\zeta^*\rangle=0\,. \tag{11.9}$$

To prove (11.9) differentiate (7.2) with respect to v at $v=v_0$,

$$(-\sigma+\mathscr{L})\zeta_v-\sigma_v\zeta+L_1\zeta+\nabla p_v=0\,, \qquad \operatorname{div}\zeta_v=0\,, \qquad \zeta_v|_S=0\,,$$

and note that

$$\langle(-\sigma+\mathscr{L})\zeta_v,\zeta^*\rangle=\langle\zeta_v,(-\bar{\sigma}+\mathscr{L}^*)\zeta^*\rangle=0\,.$$

Exercise 11.1: Derive (11.5).

Exercise 11.2: Verify that (11.7) satisfies (11.6).
Verify that (11.8) satisfies the adjoint problem.

Exercise 11.3: Show that

$$\langle\zeta,\bar{\zeta}^*\rangle=0\,, \tag{11.10}$$

and

$$[\mathbf{Z}_i, \mathbf{Z}_j^*] = \delta_{ij}\langle\zeta, \zeta^*\rangle . \tag{11.11}$$

Exercise 11.4: Show that the condition

$$\operatorname{re}\sigma_\nu = \operatorname{re}\frac{d\sigma(\bar{\nu}_L)}{d\nu} \neq 0 \tag{11.12}$$

implies that there is a change of stability for $\mathbf{U}(\mathbf{x}, \nu)$ as ν crosses $\bar{\nu}_L$. Show that

$$\operatorname{re}\sigma_\nu(\bar{\nu}_L) > 0 \tag{11.13}$$

implies that $U(\mathbf{x}, \nu)$ loses stability as ν crosses $\bar{\nu}_L$ from above. Show that

$$\sigma_\nu\langle\zeta, \zeta^*\rangle = \langle\nabla\zeta, \nabla\zeta^*\rangle \tag{11.14}$$

if $\mathbf{U}(\mathbf{x})$ is independent of ν.

Exercise 11.5: Assume that $\mathscr{L}_0$ has no eigenvalues of the form $\pm in\omega_0$, $n = 0, 2, 3, \ldots$. Show $\mathbf{Z}_1$ and $\mathbf{Z}_2$ are the only solutions of (11.6). *Hint:* Expand $\mathbf{u}$ and p in a Fourier series.

§ 12. Solvability Conditions

The following solvability lemma is important in the construction of the series solution of the bifurcation problem.

Suppose $\mathbf{b}(\mathbf{x}, s)$ *is an arbitrary vector field which is* 2π*-periodic in* s. *Then the problem*

$$J_0\mathbf{u} + \nabla p = \mathbf{b}, \qquad \mathbf{u} \in P_{2\pi}, \tag{12.1}$$

has a solution if and only if

$$[\mathbf{b}, \mathbf{Z}_i^*] = 0, \qquad i = 1, 2. \tag{12.2}$$

Proof: Eq. (12.2) follows from the relation

$$[J_0\mathbf{u}, \mathbf{Z}_i^*] = [\mathbf{u}, J_0^*\mathbf{Z}_i^*] = 0$$

and Eq. (12.1). The proof that (12.2) is also a sufficient condition for solvability is given in the (1972) paper of Joseph and Sattinger.

The solvability lemma may be simplified when, as in the perturbation problem (10.4), the inhomogeneous terms $\mathbf{b}$ are real-valued. Then, instead of the two conditions (12.2) we may apply a single (complex-valued) condition

$$[\mathbf{b}] \equiv [\mathbf{b}, \mathbf{Z}_1^*] = [\mathbf{b}, \bar{\mathbf{Z}}_2^*] = \overline{[\mathbf{b}, \mathbf{Z}_2^*]} = 0 . \tag{12.3}$$

The reader should note that (12.3) defines a shorthand notation for $[\mathbf{b}, \mathbf{Z}_1^*]$.

Now we can complete the algorithm for computing the perturbation series. First, consider (10.6a). The general solution of this problem is a linear combination of $\mathbf{Z}_1$ and $\mathbf{Z}_2$ which, without loss of generality (see Exercise 12.1), we may take as

$$\mathbf{v}_0 = 2\,\mathrm{re}\,\mathbf{Z}_1\,. \tag{12.4}$$

The condition $[\mathbf{v}_0, \mathbf{v}_0] = 2$ fixes the scale of $\mathbf{Z}_1$ (see Exercise 12.2).

The solvability lemma in the form (12.3) assures that we may solve (10.6b) provided that

$$\omega_m \left[\frac{\partial \mathbf{v}_0}{\partial s}\right] + \nu_m [L_1 \mathbf{v}_0] + [\mathbf{F}_m] = 0\,. \tag{12.5}$$

The real and imaginary parts of (12.5) determine the values ω_m and ν_m.

When $m=1$, using (12.10) and (12.11) we find that

$$\omega_1 \left[\frac{\partial \mathbf{v}_0}{\partial s}\right] + \nu_1 [L_1 \mathbf{v}_0] + [\mathbf{F}_1] = (-i\omega_1 + \sigma_\nu \nu_1)\langle \zeta, \zeta^* \rangle + [\mathbf{v}_0 \cdot \nabla \mathbf{v}_0] = 0\,.$$

The time behavior of $\mathbf{v}_0$ and $\mathbf{Z}_1^*$ is such that

$$[\mathbf{v}_0 \cdot \nabla \mathbf{v}_0] = 0\,.$$

Hence, assuming that $\mathrm{re}\,\sigma_\nu \neq 0$, we have

$$\omega_1 = \nu_1 = 0\,. \tag{12.6}$$

In fact, we may prove by induction (Exercise 12.6) that

$$\omega_{2l+1} = \nu_{2l+1} = 0 \quad \text{for} \quad l = 1, 2, \ldots\,. \tag{12.7}$$

Hence,

$$\omega = \omega(\varepsilon^2) \quad \textit{and} \quad \nu = \nu(\varepsilon^2)$$

are even functions of ε.

Exercise 12.1: Show that any real-valued linear combination of $\mathbf{Z}_1$ and $\mathbf{Z}_2$ may be reduced to (12.4) by translating the origin of the time s.

Exercise 12.2: Show that the condition $[\mathbf{v}_0, \mathbf{v}_0] = 2$ fixes the scale of the function $\zeta(\mathbf{x})$ which is otherwise determined only up to a multiplicative constant. Show that the solution (10.5) of (9.4) is independent of the scale of $\zeta^*(\mathbf{x})$; we can choose ζ^* so that $\langle \zeta, \zeta^* \rangle = 1$.

Exercise 12.3: Show that a solution of (10.6b) for some fixed $m > 0$ may always be found in the form

$$\mathbf{v}_m + a\mathbf{v}_0 \tag{12.8}$$

for an arbitrary constant a. Show that $[\mathbf{v}, \mathbf{v}]_m = 0$ fixes the value of a uniquely.

Exercise 12.4: Investigate the consequences of replacing the definition (9.3) of ε with

$$[\mathbf{u}]=\varepsilon=\varepsilon[\mathbf{v}], \quad [\mathbf{v}_m]=0 \quad (m\geqslant 1).$$

Show that this choice of norm implies that the constant a in Eq. (12.8) vanishes. Show that the solution of (9.4) with a prescribed normalizing condition

$$[\mathbf{u}]=\varepsilon a(\varepsilon)=\varepsilon[\mathbf{v}]$$

may always be obtained in the form of an orthogonal sum

$$\mathbf{v}(x,s;\varepsilon)=a(\varepsilon)\mathbf{v}_0+\tilde{\mathbf{v}}(\mathbf{x},s,\varepsilon) \tag{12.9}$$

where $[\tilde{\mathbf{v}}]=0$.

Exercise 12.5: Show that

$$\left[\frac{\partial \mathbf{v}_0}{\partial s}\right]=-i[\mathbf{v}_0]=-i\langle\zeta,\zeta^*\rangle, \tag{12.10}$$

and using (11.9) show that

$$v_m[L_1\mathbf{v}_0]=v_m\sigma_v\langle\zeta,\zeta^*\rangle \tag{12.11}$$

Exercise 12.6: Prove Eq. 12.7. *Hint:* Assume that $\mathbf{v}_{2l}$ is an even polynomial and $\mathbf{v}_{2l+1}$ is an odd polynomial in e^{is} and e^{-is} and that $\lambda_{2l+1}=\omega_{2l+1}=0$ when $l<m$. Prove that the assumptions hold when $m\geqslant l$.

Exercise 12.7[6]: Show that

$$\mathbf{F}_2=\mathbf{v}_1\cdot\nabla\mathbf{v}_0+\mathbf{v}_0\cdot\nabla\mathbf{v}_1+L_2\mathbf{v}_0 \tag{12.12}$$

and

$$i\omega_2-v_2\sigma_v=[F_2]/\langle\zeta,\zeta^*\rangle. \tag{12.13}$$

§ 13. Subcritical and Supercritical Bifurcation

One of the more important results to be obtained from bifurcation analysis is concerned with the direction of the bifurcation. To understand this, recall that $\mathbf{U}(\mathbf{x},v)$ is stable by the criterion of linear theory if $v>v_L$. Hence, according to this criterion, the energy of all sufficiently small disturbances decays to zero. However, if $v(\varepsilon)>v_L$, there is a solution $\mathbf{U}(\mathbf{x},v)+\mathbf{u}(\mathbf{x};t,\varepsilon)$ of permanent form whose energy does not decay even though the stability criterion of linear theory is satisfied.

---

[6] It is convenient to choose ζ^* so that

$$\langle\zeta,\zeta^*\rangle=1; \tag{12.14}$$

then $[F_2]/\langle\zeta,\zeta^*\rangle=[F_2]$.

Bifurcating solutions which exist for values of ν deemed stable by the linear theory are called *subcritical.* Bifurcating solutions which exist for values of ν deemed unstable by linear theory are called *supercritical*[7].

The basic flow $\mathbf{U}(\mathbf{x}, \nu)$ is unstable near the first critical value of linear theory if $\nu - \nu_L < 0$. Since $\nu_1 = 0$, we find that

$$\nu(\varepsilon) - \nu_L = \nu_2 \varepsilon^2 + 0(\varepsilon^4)$$

where

$$\nu_2 = \tfrac{1}{2} \frac{d^2 \nu(0)}{d\varepsilon^2} = -\operatorname{re}\{\mathbf{F}_2]/\langle \zeta, \zeta^* \rangle\}/\operatorname{re} \sigma_\nu(\nu_L).$$

When ε is small, we find that time-periodic bifurcating solutions are subcritical if $\nu_2 > 0$ and supercritical if $\nu_2 < 0$.

The notion of subcritical and supercritical bifurcation can be generalized to situations in which ε is not small. In this generalization we think of the direction of bifurcation as fundamental. There are three possible directions of bifurcation: $\nu'(\varepsilon) > 0$, $\nu'(\varepsilon) = 0$ and $\nu'(\varepsilon) < 0$. We regard the case in which $\nu'(\varepsilon) = 0$ on an arc as unusual; then $\nu'(\varepsilon) = 0$ at critical points $\varepsilon = \varepsilon_*$ of the bifurcation curve. $\varepsilon_* = 0$ is a critical point. When $\nu'(\varepsilon) < 0$, the amplitude of the bifurcating solution increases with $1/\nu(\varepsilon)$, the "Reynolds number". In the other, less natural, case, the amplitude of the bifurcating solution decreases as the "Reynolds number" is increased. The bifurcating curve in physical problems is properly regarded as a response curve generated by a response functional evaluated on solutions. For example, the response curve for pipe flow, treated in Chapter IV, relates the pressure gradient $P'(\varepsilon)$ to the mass flux discrepancy ε. Here, ε is the difference between the mass flux delivered by the time-periodic bifurcating flow and the mass flux delivered by laminar flow with the same pressure gradient. The physically unrealistic bifurcating solutions are those for which the mass flux discrepancy decreases as the pressure gradient is increased. In Chapter IV we shall see that the physically unrealistic solutions are the only solutions to bifurcate (subcritically) when ε is small. It is natural to expect that these bifurcating solutions are unstable, not only when ε is small, but also for all values ε for which $P'(\varepsilon) < 0$.

---

[7] These definitions are often framed in terms of critical Reynolds numbers R_c of linear theory (see Notes for Chapter I). Since $R < R_c = Ul/\nu_L$ for stability, the relation $R(\varepsilon) < R_c$ implies bifurcation under the critical value of linearized theory; hence, subcritical bifurcation. This definition is inadequate because of the existence of more than one critical viscosity. For example, in the Poiseuille flow example treated in Chapter IV, the criterion $R < R_c = Ul/\bar{\bar{\nu}}_L$ where $\bar{\bar{\nu}}_L$ is a second critical viscosity implies instability and by our definition is supercritical. (The second critical viscosity for Poiseuille flow is associated with parameters on the upper branch (CE') of Fig. 34.1 of the curve giving the wave numbers and viscosities required for spatially periodic solutions of the bifurcation problem with $\varepsilon = 0$. This curve is called a neutral curve.)

Exercise 13.1: Show that near any critical viscosity $\bar{\nu}_L$ of linearized theory, there is subcritical bifurcation if

$$\nu_2 \operatorname{re}(\sigma_\nu(\bar{\nu}_L)) > 0 \tag{13.3}$$

and supercritical bifurcation when the expression in (13.3) is negative.

Exercise 13.2: Specify conditions on $f(x)$ of Eq. (8.6) and on $f(x,y)$ of Eq. (8.7) which make the bifurcating solutions bifurcate subcritically.

§ 14. Stability of the Bifurcating Periodic Solution

We are now going to construct a mathematical theory of bifurcation in which the intuitive idea about the instability of unrealistic bifurcating solutions may be given a precise form[8].

In trying to determine the stability of the bifurcating time-periodic motion $\mathbf{U}(\mathbf{x};\nu(\varepsilon))+\mathbf{u}(\mathbf{x},s;\varepsilon)$ one is led to consider the linearized equations

$$\frac{\partial \mathbf{q}}{\partial t}+\bar{\bar{\mathscr{L}}}\mathbf{q}+\nabla p=0\,,\qquad \operatorname{div}\mathbf{q}=0 \quad\text{and}\quad \mathbf{q}|_S=0 \tag{14.1}$$

for a small disturbance $\mathbf{q}(x,t)$ of $\mathbf{U}+\mathbf{u}$. Here,

$$\bar{\bar{\mathscr{L}}}\mathbf{q}=\mathscr{L}[\mathbf{U}+\mathbf{u};\nu]\mathbf{q}=(\mathbf{U}+\mathbf{u})\cdot\nabla\mathbf{q}+\mathbf{q}\cdot\nabla(\mathbf{U}+\mathbf{u})-\nu(\varepsilon)\nabla^2\mathbf{q} \tag{14.2}$$

where $\mathbf{U}(\mathbf{x};\nu(\varepsilon))$, $\mathbf{u}(\mathbf{x},s;\varepsilon)$ and $\nu(\varepsilon)$ are evaluated on the bifurcation solution for a fixed value of ε. The coefficients of (14.1) are periodic in t with period $2\pi/\omega(\varepsilon)$; $\mathbf{u}(\mathbf{x},s;\varepsilon)$ is 2π-periodic in s. Hence, Floquet theory applies and, without loss of generality, we look for solutions of (14.1) in the form

$$\mathbf{q}=e^{-\gamma t}\boldsymbol{\Gamma}(\mathbf{x},s)\,,\qquad p'=e^{-\gamma t}p(\mathbf{x},s) \tag{14.3}$$

where γ is the Floquet exponent and $\boldsymbol{\Gamma}$ and p are 2π-periodic in $s=\omega t$.

Substitution of (14.3) into (14.1) leads to

$$-\gamma\boldsymbol{\Gamma}+\mathscr{J}\boldsymbol{\Gamma}+\nabla p=0\,,\qquad \boldsymbol{\Gamma}\in\mathbf{P}_{2\pi} \tag{14.4}$$

where

$$\mathscr{J}(\cdot)=\omega\frac{\partial(\cdot)}{\partial s}+\bar{\bar{\mathscr{L}}}(\cdot)\,.$$

When $\varepsilon=0$, we have $\omega=\omega_0$, $\nu=\nu_L$, $\mathbf{u}\equiv 0$, $\bar{\bar{\mathscr{L}}}=\mathscr{L}_0$, $\mathscr{J}=J_0$ and

$$-\gamma\boldsymbol{\Gamma}+J_0\boldsymbol{\Gamma}+\nabla p=0\,,\qquad \boldsymbol{\Gamma}\in\mathbf{P}_{2\pi}\,.$$

---

[8] The ultimate goal of our stability analysis may perhaps be most easily understood through study of the simple example leading to the bifurcation diagrams shown in Figs. 14.1 and 15.1.

It follows that $\boldsymbol{\Gamma}(\mathbf{x},s]$ may be developed in a Fourier series of terms $e^{iks}\boldsymbol{\Gamma}_k(\mathbf{x})$ where the Fourier coefficients $\boldsymbol{\Gamma}_k(\mathbf{x})$ are solenoidal, vanish on S and satisfy

$$(-\gamma+ik\omega_0)\boldsymbol{\Gamma}_k+\mathscr{L}_0\boldsymbol{\Gamma}_k+\nabla p_k=0. \tag{14.5}$$

At criticality the eigenvalues of $\mathscr{L}_0$ are σ_j. Then all the solutions of (14.5) are in the form

$$(-\gamma+ik\omega_0)\boldsymbol{\Gamma}_k+\sigma_j\boldsymbol{\Gamma}_k+\nabla p_k=0.$$

We are assuming that at criticality $\sigma_1=\pm i\omega_0$; apart from these two values $(k=\pm 1)$

$$\operatorname{re}\gamma=\operatorname{re}\sigma_j, \quad j=2,3,\ldots,$$

and all eigenvalues have positive real parts leading to stability. We can, therefore, restrict our attention to the Floquet exponents $\gamma=i(k+1)\omega_0$, $k=0, \pm 1, +2,\ldots$. Each and every one of these exponents corresponds to one and the same Floquet multiplier $\mu=e^{i(k\pm 1)2\pi}=1$. Without losing generality we may restrict our attention to the exponent γ which is zero when $\varepsilon=0$. If, say, $\gamma=i(k+1)\omega_1$ for some definite k when $\varepsilon=0$, then

$$\boldsymbol{\Gamma}=e^{i(k+1)s}\hat{\boldsymbol{\Gamma}} \quad \text{and} \quad p=e^{i(k+1)s}\hat{p}$$

where $\hat{\boldsymbol{\Gamma}}$ and $\hat{p}$ are 2π-periodic in s and satisfy the eigenvalue problem

$$\tilde{\gamma}\hat{\boldsymbol{\Gamma}}+\mathscr{J}\hat{\boldsymbol{\Gamma}}+\nabla\hat{p}=0, \quad \hat{\boldsymbol{\Gamma}}\in P_{2\pi}$$

where $\tilde{\gamma}=\gamma-i(k+1)\omega_0$ vanishes when $\varepsilon=0$.

We next note that

$$\boldsymbol{\Gamma}=\mathring{\mathbf{u}}(\equiv\partial\mathbf{u}/\partial s), \quad \gamma=0 \tag{14.6}$$

is always a solution of (14.4), even when $\varepsilon\neq 0$. To check this, replace $\varepsilon\mathbf{v}$ with $\mathbf{u}$ in (7.1a), and with respect to s and compare the resulting equation with (14.1).

It will be recalled that $\pm i\omega_0$ are simple eigenvalues of the operator $\mathscr{L}_0$. These two eigenvalues both correspond to a zero eigenvalue of the operator J_0 on the space $P_{2\pi}$. A basis for the two-dimensional null space of J_0 is the set of vectors $\mathbf{Z}_1$ and $\mathbf{Z}_2=\overline{\mathbf{Z}}_1$. We are going to show that this double eigenvalue splits, at criticality, into two branches. On one branch, $\gamma=0$ is an eigenvalue of $\mathscr{J}$ (with eigenfunction $\mathring{\mathbf{u}}$) and $\mathscr{J}\to J_0$ as $\varepsilon\to 0$. On the other branch $\gamma=\gamma(\varepsilon)$; this branch controls the stability of $\mathbf{U}+\mathbf{u}$.

We seek solutions on the second branch in the form

$$\boldsymbol{\Gamma}(\mathbf{x},s;\varepsilon)=a(\varepsilon)\mathring{\mathbf{u}}+\boldsymbol{\phi}(\mathbf{x},s;\varepsilon) \tag{14.7}$$

where $a(\varepsilon)$ is to be determined. The decomposition (14.7) splits $\boldsymbol{\Gamma}$ into parts: the contribution $a(\varepsilon)\mathring{\mathbf{u}}(\mathbf{x},s;\varepsilon)$ lies in the null space of the operator $\mathscr{J}$; this contribution varies with ε.

Substituting (14.7) into (14.4), we find that

$$\mathscr{J}\boldsymbol{\phi}-\gamma\boldsymbol{\phi}+\tau\mathring{\mathbf{u}}+\nabla p=0\,,\qquad \boldsymbol{\phi}\in P_{2\pi} \tag{14.8}$$

where $\tau=-\gamma a(\varepsilon)$. $\boldsymbol{\phi}$ may be normalized arbitrarily.

It is possible at this point to construct the solution $\boldsymbol{\phi}(\mathbf{x},s;\varepsilon)$, $\gamma(\varepsilon)$ and $\tau(\varepsilon)$ as a Taylor series in powers of ε. It is much better, however, to introduce the *fundamental factorization*[9]:

$$\begin{aligned}
&\boldsymbol{\phi}(\mathbf{x},s;\varepsilon)=\mathbf{u}_\varepsilon(\mathbf{x},s;\varepsilon)+v_\varepsilon(\varepsilon)\hat{\boldsymbol{\phi}}(\mathbf{x},s;\varepsilon)\,,\\
&\tau(\varepsilon)=\omega_\varepsilon(\varepsilon)+v_\varepsilon(\varepsilon)\hat{\tau}(\varepsilon)\,,\\
&\gamma(\varepsilon)=v_\varepsilon(\varepsilon)\hat{\gamma}(\varepsilon)\,.
\end{aligned}\tag{14.9}$$

The subscript ε denotes differentiation with respect to ε. Substituting (14.9) into (14.8) we find that

$$\mathscr{J}\mathbf{u}_\varepsilon+v_\varepsilon\mathscr{J}\hat{\boldsymbol{\phi}}+(\omega_\varepsilon+v_\varepsilon\hat{\tau})\mathring{\mathbf{u}}-v_\varepsilon\hat{\gamma}(\mathbf{u}_\varepsilon+v_\varepsilon\hat{\boldsymbol{\phi}})+\nabla p=0\,. \tag{14.10}$$

Reverting to (9.4a) expressed in terms of $\mathbf{u}=\varepsilon\mathbf{v}$, $p=\varepsilon p$, we find, after differentiating with respect to ε, that

$$\mathscr{J}\mathbf{u}_\varepsilon+\omega_\varepsilon\mathring{\mathbf{u}}+v_\varepsilon\mathscr{L}[\mathbf{U}_v,1]\mathbf{u}+\nabla p_\varepsilon=0 \tag{14.11}$$

where $\mathbf{U}_v=\partial\mathbf{U}/\partial v$. When $\varepsilon=0$, $\mathbf{U}_v=\mathbf{U}_1$ and $\mathscr{L}[\mathbf{U}_1,1]=L_1$. When Eq. (14.11) is subtracted from (14.10), $\mathscr{J}\mathbf{u}_\varepsilon+\omega_\varepsilon\mathring{\mathbf{u}}$ cancels and v_ε may be factored from each of the remaining terms. It follows that

$$\mathscr{J}\hat{\boldsymbol{\phi}}+\hat{\tau}\mathring{\mathbf{u}}-H\mathbf{u}-\hat{\gamma}(\mathbf{u}_\varepsilon+v_\varepsilon\hat{\boldsymbol{\phi}})+\nabla\hat{p}=0\,,\qquad \hat{\boldsymbol{\phi}}\in P_{2\pi} \tag{14.12}$$

where

$$H=\mathscr{L}[\mathbf{U}_v,1]$$

and $\hat{p}$ is a new "pressure".

It remains to show that (14.12) may be solved for analytic functions

$$\begin{bmatrix}\hat{\boldsymbol{\phi}}(\mathbf{x},s;\varepsilon)\\ \hat{p}(\mathbf{x},s;\varepsilon)\\ \hat{\tau}(\varepsilon)-\hat{\tau}_0\\ \hat{\gamma}(\varepsilon)\end{bmatrix}=\sum_{l=1}\varepsilon^l\begin{bmatrix}\hat{\boldsymbol{\phi}}_l(\mathbf{x},s)\\ \hat{p}_l(\mathbf{x},s)\\ \hat{\tau}_l\\ \hat{\gamma}_l\end{bmatrix}. \tag{14.13}$$

[9] This factorization and the theorems which follow from it are due to Joseph (see Joseph and Nield, 1975).

To obtain the equations satisfied by the quantities on the right of (14.12), we must also expand the operators $\mathscr{J}$ and H and the function $\mathbf{u}(\mathbf{x},s;\varepsilon)$. Thus,

$$\mathbf{u}(\mathbf{x},s;\varepsilon)=\sum_{l=1}\mathbf{u}_l\varepsilon^l=\sum\mathbf{v}_{l-1}\varepsilon^l\,,\qquad \mathbf{u}_l=\mathbf{v}_{l-1}\,;$$
$$H=H_0+\sum_{l=1}\varepsilon^l H_l\,,\qquad H_0=L_1\,;$$
$$H_l=\mathscr{L}[K_l,0]\,,$$
$$K_l=\sum_{m=1}^{l}(m+1)\,U_{m+1}\sum v_{l_1}v_{l_2}\dots v_{l_m}\,,\qquad 1\leqslant l_i<l\,;$$

where the second summation is over integers l_i, such that $l_1+l_2+\dots+l_m=l$,

$$\mathscr{J}=J_0+\sum_{l=1}\varepsilon^l\mathscr{J}_l\,,$$

and

$$\mathscr{J}_l(\cdot)=J_l(\cdot)+(\cdot)\cdot\nabla\mathbf{u}_l+\mathbf{u}_l\cdot\nabla(\cdot)\,.$$

The series representations are inserted into (14.12). The coefficients of separate powers of ε are zero if

$$J_0\hat{\boldsymbol{\phi}}_1+\hat{\tau}_0\mathring{\mathbf{u}}_1-L_1\mathbf{u}_1-\hat{\gamma}_1\mathbf{u}_1+\nabla\hat{p}_1=0,\qquad \hat{\boldsymbol{\phi}}_1\in P_{2\pi}\,, \tag{14.14a}$$

$$J_0\hat{\boldsymbol{\phi}}_l+\hat{\tau}_0\mathring{\mathbf{u}}_l-L_1\mathbf{u}_l-\hat{\gamma}_l\mathbf{u}_1+\sum_{\substack{n+m=l\\ n,m>0}}\mathbf{R}_{nm}-\sum_{\substack{n+m+r=l\\ n,m,r>0}}T_{nmr}+\nabla p_l=0\,, \tag{14.14b}$$

$$\hat{\boldsymbol{\phi}}_l\in P_{2\pi}\,,\qquad l\geqslant 2$$

where

$$\mathbf{R}_{nm}=-H_n\mathbf{u}_m+\hat{\tau}_n\mathring{\mathbf{u}}_m-(m+1)\mathbf{u}_{m+1}\hat{\gamma}_n+\mathscr{J}_n\hat{\boldsymbol{\phi}}_m$$

and

$$T_{nmr}=(n+1)\,v_{n+1}\hat{\gamma}_m\hat{\boldsymbol{\phi}}_r\,,\qquad (T_{nmr}=0 \text{ when } l=2)\,.$$

Eqs. (14.14), subject to any convenient normalization for the function $\hat{\boldsymbol{\phi}}$, are uniquely solvable, and have bounded inverses, if the inhomogeneous terms satisfy (12.2).

Using (12.3), and noting, from (12.11), and (12.14), that $[L_1\mathbf{u}_1]=\sigma_v$, we find that

$$\sigma_v+i\hat{\tau}_0+\hat{\gamma}_1=0\,, \tag{14.15a}$$

$$i\hat{\tau}_{m-1}+\hat{\gamma}_m+[\xi_m]=0\,,\qquad m>1 \tag{14.15b}$$

where ξ_m is a composition of terms involving $\hat{\boldsymbol{\phi}}_l$, $\hat{\gamma}_l$ and $\hat{\tau}_{l-1}$, $l<m$. It follows that Eqs. (14.14) and (14.15) may be solved sequentially for the quantities

$$\hat{\boldsymbol{\phi}}_l(\mathbf{x},s),\qquad \hat{p}_l(\mathbf{x},s),\qquad \hat{\tau}_{l-1},\qquad \hat{\gamma}_l$$

defining the right side of (14.13).

We leave the proof that

$$\hat{\tau}_{2l-1}=\hat{\gamma}_{2l}=0, \quad l\geqslant 1 \tag{14.16}$$

as Exercise (14.2); the method of proof imitates the induction leading to (12.7).

The implications of the results achieved so far may be collected in the form of a theorem concerning the fundamental factorization (see Fig. 15.1).

Suppose $\mathbf{u}(\mathbf{x},s;\varepsilon)$, $\omega(\varepsilon)$ *and* $v(\varepsilon)$ *are real analytic functions on an open interval* I_1 *containing the point* $\varepsilon=0$. *Then,*

$$\boldsymbol{\phi}(\mathbf{x},s;\varepsilon)=\mathbf{u}_\varepsilon(\mathbf{x},s;\varepsilon)+v_\varepsilon(\varepsilon)\hat{\boldsymbol{\phi}}(\mathbf{x},s;\varepsilon),$$

$$\tau(\varepsilon)=\omega_\varepsilon(\varepsilon)+v_\varepsilon(\varepsilon)\hat{\tau}(\varepsilon)$$

and

$$\gamma(\varepsilon)=v_\varepsilon(\varepsilon)\hat{\gamma}(\varepsilon)$$

where $\hat{\boldsymbol{\phi}}(\mathbf{x},s;\varepsilon)$, $\hat{\tau}(\varepsilon)$ *and* $\hat{\gamma}(\varepsilon)$ *are real analytic functions on an interval* $I_2\subset I_1$ *containing the point* $\varepsilon=0$. *Moreover,* $\hat{\tau}(\varepsilon)$ *and* $\hat{\gamma}(\varepsilon)/\varepsilon$ *are even functions of* ε *and such that*

$$\hat{\gamma}_1=-\operatorname{re}\sigma_v, \qquad \tau_0=-\operatorname{im}\sigma_v.$$

The functions mentioned in the theorem are known (Joseph and Nield, 1975) to be analytic in a complex neighborhood of $\varepsilon=0$. Suppose, as in the hypotheses of the theorem, that the bifurcating solution may be analytically continued onto the interval I_1. In the same way the roofed functions may be continued as real analytic functions. The interval $I_2\subset I_1$ because, by hypothesis, each term of Eq. (14.9) defining the roof functions, loses analyticity on the points of closure of the interval I_1.

The representation $\gamma(\varepsilon)=v_\varepsilon(\varepsilon)\hat{\gamma}(\varepsilon)$ shows that the Floquet exponent for the bifurcating solution $\mathbf{u}(\mathbf{x},s;\varepsilon)$ vanishes at critical points $\varepsilon=\varepsilon_*$, where $v_\varepsilon(\varepsilon_*)=0$, of the bifurcation curve. The sign of the Floquet exponent is the same as the slope of the "Reynolds number", $(1/v)_\varepsilon$, wherever $\hat{\gamma}(\varepsilon)<0$. We shall see that $\hat{\gamma}(\varepsilon)<0$ when ε is small. Whenever $\hat{\gamma}(\varepsilon)<0$, whether or not ε be small, bifurcating solutions whose amplitude ε increases as the Reynolds number decreases,

$$\frac{d}{d\varepsilon}\left(\frac{1}{v(\varepsilon)}\right)<0,$$

are unstable. This implies, for example, that time-periodic bifurcating pipe flows are unstable when the mass flux decreases as the pressure gradient is increased.

Unfortunately, the sign of $\hat{\gamma}(\varepsilon)$ is not known when ε is large. For small ε, the following *theorem of stability of time-periodic bifurcating solutions* holds:

When ε *is small,*

$$\gamma(\varepsilon)=-v_\varepsilon\varepsilon\operatorname{re}\sigma_v+v_\varepsilon O(\varepsilon^3).$$

Subcritical solutions are unstable and supercritical solutions are stable.

This theorem depends in a fairly strong way on the assumption that σ_1 and $\bar{\sigma}_1$ are simple eigenvalues of $\mathscr{L}_0$ at criticality. Though such an assumption is likely to hold for almost all bounded domains, it can fail in particular domains. The instability of the subcritical bifurcating solutions, however, may not depend so strongly on the assumed simplicity. If the principal eigenvalue of $\mathscr{L}_0$ has a multiplicity greater than one, then more solutions are available to destabilize an otherwise stable motion.

Exercise 14.1 (Stability of periodic solutions on ascending branches of the bifurcation curve): Show that the time periodic bifurcating solutions (9.6) of problem (8.7) with $f(x,y)=g(x^2+y^2)>0$ have a constant radius $\varepsilon=(x^2+y^2)^{1/2}$. Consider the stability of the bifurcating solution at a fixed value of $v=1/g(\varepsilon^2)$. At the given v show that any solution $x=R\cos\theta$, $y=R\sin\theta$ of (8.7) with $f=g$ satisfies

$$\frac{1}{2}\frac{d(R^2)}{dt}-\left(1-\frac{g(R^2)}{g(\varepsilon^2)}\right)R^2=0\,. \tag{14.18}$$

Show that $R^2=\varepsilon^2$ is conditionally stable whenever

$$g'(\varepsilon^2)>0$$

and is unstable whenever

$$g'(\varepsilon^2)<0\,.$$

Show that supercritical bifurcating solutions are stable and subcritical bifurcating solutions are unstable when ε is sufficiently small (see Fig. 14.1).

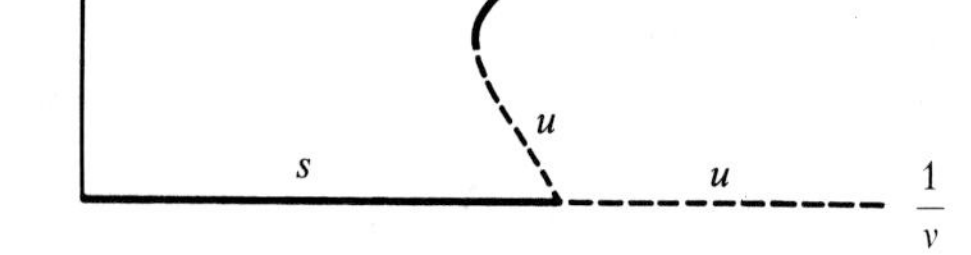

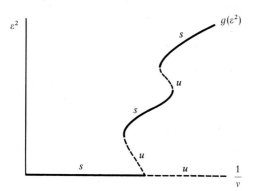

Fig. 14.1: A sample bifurcation curve which exhibits the instability of the subcritical solution (9.6) bifurcating from the null solution $\varepsilon=0$ and the recovery of stability on ascending branches $(g'(\varepsilon^2)>0)$ of the bifurcation curve

Exercise 14.2: Find the form of the quantity ξ_m appearing in (14.15b). Prove (14.16).

§ 15. Bifurcating Steady Solutions; Instability and Recovery of Stability of Subcritical Solutions

When $\omega_0=0$ (case A of § 9) the solution of the linearized problem (7.1, with $\varepsilon=0$) is steady. In this case we set $\omega(\varepsilon)=0$ and seek a steady solution of (9.4a, b, c, d). The perturbation series for this problem is just (10.5) with $\omega_n=0$.

Replacing the normalizing condition used in (10.6) is the relation

$$\langle|\mathbf{v}|^2\rangle_m = \begin{cases} 2 & \text{for } m=0 \\ 0 & \text{for } m>0 \end{cases}. \tag{15.1}$$

At zeroth order we may take

$$\mathbf{v}_0=\mathbf{u}_1=\boldsymbol{\zeta}, \quad \langle|\boldsymbol{\zeta}|^2\rangle=2 \tag{15.2}$$

where $\boldsymbol{\zeta}$ is the principal eigenfunction of the spectral problem belonging to the eigenvalue $\sigma=0$. The adjoint problem for the spectral theory is defined through (11.1) and leads to (11.2) with $\bar{\sigma}=0$. Eq. (11.9) holds in the steady case.

To solve the perturbation problems which arise in the steady case, we consider problems of the form

$$\mathscr{L}_0\mathbf{u}+\nabla p=\mathbf{b}(\mathbf{x}), \quad \operatorname{div}\mathbf{u}=0, \quad \mathbf{u}|_S=0.$$

These are solvable if and only if

$$\langle\mathbf{b},\boldsymbol{\zeta}^*\rangle=0.$$

It should be noted that when $\omega_0=0$, we may assume that $\boldsymbol{\zeta}^*(\mathbf{x})$ and the other fields are real-valued.

At first order we find, computing as in § 12, that

$$v_1\langle L_1\mathbf{v}_0,\boldsymbol{\zeta}^*\rangle+\langle(\mathbf{v}_0\cdot\nabla)\mathbf{v}_0,\boldsymbol{\zeta}^*\rangle=0.$$

Hence, using (11.9)

$$v_1=-\langle(\mathbf{v}_0\cdot\nabla)\mathbf{v}_0,\boldsymbol{\zeta}^*\rangle/\sigma_v\langle\boldsymbol{\zeta},\boldsymbol{\zeta}^*\rangle. \tag{15.3}$$

The scale of $\boldsymbol{\zeta}^*$ is arbitrary and may be chosen so that $\langle\boldsymbol{\zeta},\boldsymbol{\zeta}^*\rangle=1$.

In general, v_1 does not vanish, though it can vanish in special problems (see Exercise 15.2).

The main difference between steady and time-periodic bifurcating solutions is implied by (15.3), which shows that

A: $v=v(\varepsilon)$, for steady bifurcating solutions, (15.4)

whereas

B: $v=v(\varepsilon^2)$ for periodic bifurcating solutions. (15.5)

It follows from (15.4) and (15.5) that

A: Two-sided bifurcation is possible. Since

$$v(\varepsilon)-v_L=v_1\varepsilon+O(\varepsilon^2), \tag{15.6}$$

the bifurcation is subcritical when $v_1\varepsilon$ is positive and supercritical when $v_1\varepsilon$ is negative[10].

B: Only one-sided bifurcation is possible,

$$v(\varepsilon^2)-v_L=v_2\varepsilon^2+O(\varepsilon^4)\,. \tag{15.7}$$

In case B, $\varepsilon=\varepsilon_*=0$ is a critical point of the bifurcation curve.

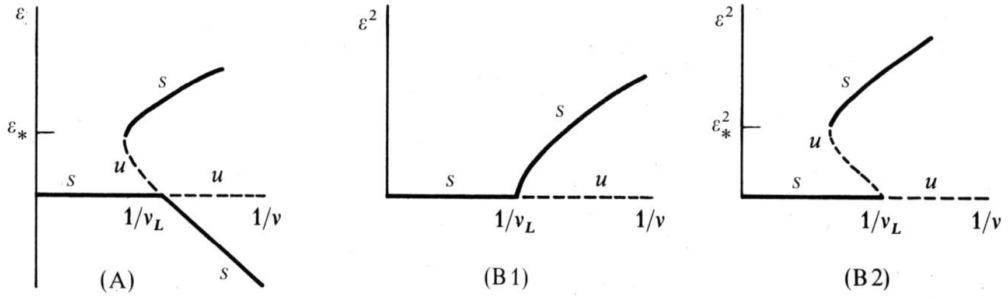

Fig. 15.1: Bifurcation from a simple eigenvalue. The solid lines represent stable bifurcating solutions. The dotted lines represent unstable bifurcating solutions. In the steady case (A), a two-sided bifurcation can occur (see (15.6)). In the time-periodic case (B), $v=v(\varepsilon^2)$ and the bifurcating solution is either supercritical (B1) or subcritical (B2)

The study of the stability of steady bifurcating solutions is like the one given in § 14, only easier. The bifurcating solution satisfies

$$\mathscr{L}\mathbf{u}+\mathbf{u}\cdot\nabla\mathbf{u}+\nabla p=0\,,\qquad \operatorname{div}\mathbf{u}=0\,,\qquad \mathbf{u}|_S=0\,,\qquad \langle|\mathbf{u}|^2\rangle=2\varepsilon^2\,. \tag{15.8}$$

The stability of $\mathbf{U}+\mathbf{u}$ to small disturbances $\boldsymbol{\phi}e^{-\gamma t}$ is determined by the spectral problem

$$-\gamma\boldsymbol{\phi}+\overline{\overline{\mathscr{L}}}\boldsymbol{\phi}+\nabla p'=0\,,\qquad \operatorname{div}\boldsymbol{\phi}=0\,,\qquad \boldsymbol{\phi}|_S=0\,. \tag{15.9}$$

$\boldsymbol{\phi}$ may be normalized arbitrarily.

As in the study of periodic solutions the study of the stability of steady bifurcating solutions is greatly simplified by the *fundamental factorization*:

$$\boldsymbol{\phi}(\mathbf{x};\varepsilon)=\mathbf{u}_\varepsilon(\mathbf{x},\varepsilon)+v_\varepsilon\hat{\boldsymbol{\phi}}(\mathbf{x},\varepsilon)\,,\qquad \gamma(\varepsilon)=v_\varepsilon\hat{\gamma}(\varepsilon)\,. \tag{15.10}$$

The factorization (15.10), (15.9) and the equation for $\mathbf{u}_\varepsilon$ which arises from differentiating (15.8),

$$\overline{\overline{\mathscr{L}}}\mathbf{u}_\varepsilon+v_\varepsilon\mathscr{L}[\mathbf{U}_v,1]\mathbf{u}+\nabla p_\varepsilon=0\,,$$

[10] Changing the sign of ε at small ε is equivalent to reversing the direction of motion. At larger $|\varepsilon|$ the bifurcating motions would be physically distinct.

can be combined so as to cancel $\bar{\bar{\mathscr{L}}}\mathbf{u}_\varepsilon$ and to leave v_ε as a factor in each of the remaining terms. After cancellation of this factor,

$$\bar{\bar{\mathscr{L}}}\hat{\boldsymbol{\phi}} - \mathscr{L}[\mathbf{U}_v, 1]\mathbf{u} - \hat{\gamma}(\mathbf{u}_\varepsilon + v_\varepsilon \hat{\boldsymbol{\phi}}) + \nabla p = 0, \quad \operatorname{div}\hat{\boldsymbol{\phi}} = 0, \quad \hat{\boldsymbol{\phi}}|_S = 0. \tag{15.11}$$

The solution of the bifurcation problem (15.11) is nearly the same as the solution of (14.12) except that the functions do not depend on s and $\tau = \hat{\tau} = 0$. With slight modifications, Eqs. (14.13), (14.14) and (14.15) also hold for steady bifurcating solutions; in particular

$$\hat{\gamma}_1 = -\sigma_v .$$

Hence, when ε is small

$$\gamma(\varepsilon) = -v_\varepsilon \sigma_v \varepsilon + v_\varepsilon O(\varepsilon^2). \tag{15.12}$$

Eq. (15.12) shows that when ε is small the most dangerous small disturbances of the bifurcating solution are of the form

$$\boldsymbol{\phi}(\mathbf{x}, \varepsilon) \exp\{\varepsilon v_\varepsilon t[\sigma_v + O(\varepsilon)]\}.$$

If $\mathbf{U}(\mathbf{x}; v)$ loses stability strictly at criticality, then $\sigma_v > 0$ and supercritical solutions are those for which $1/v(\varepsilon)$ increases when the amplitude $|\varepsilon|$ increases. Supercritical solutions have $v_\varepsilon \varepsilon < 0$. Hence, (15.12) shows that subcritical solutions are unstable and supercritical solutions are stable when ε is small. The physical interpretation of the factorization for steady bifurcating flow is similar to the one already given for time-periodic bifurcating flow. Solutions for which

$$\frac{d}{d\varepsilon}\left(\frac{1}{v(\varepsilon)}\right) < 0$$

are unstable whenever $\varepsilon\hat{\gamma}(\varepsilon) > 0$, whether or not ε be small. When applied to bifurcation of heat conduction into convection (see Chapter X) we verify the expected result—convection is unstable if less heat is transported when the temperature differences are increased.

Unfortunately, not much is known about $\hat{\gamma}(\varepsilon)$ when ε is large. It is of particular interest to know the sign of $\hat{\gamma}(\varepsilon)$ at critical points $\varepsilon = \varepsilon_*$. At a critical point $v_\varepsilon = 0$ and, differentiating (15.8) with respect to ε at $\varepsilon = \varepsilon_*$, we find that

$$\bar{\bar{\mathscr{L}}}\mathbf{u}_\varepsilon + \nabla p_\varepsilon = 0.$$

Hence zero is an eigenvalue of $\bar{\bar{\mathscr{L}}}$ when $\varepsilon = \varepsilon_*$. The adjoint operator $\bar{\bar{\mathscr{L}}}^*$ may be defined by the requirement that

$$\langle \bar{\bar{\mathscr{L}}}\mathbf{a}, \mathbf{b}\rangle = \langle \mathbf{a}, \bar{\bar{\mathscr{L}}}^*\mathbf{b}\rangle$$

for all solenoidal vectors $\mathbf{a}$ and $\mathbf{b}$ which vanish on S. There is a solenoidal vector $\mathbf{u}^*$, vanishing on S, and such that

$$\bar{\bar{\mathscr{L}}}^*\mathbf{u}^* + \nabla p^* = 0.$$

If we assume that zero is a simple eigenvalue of the operator $\bar{\bar{\mathscr{L}}}$ when $\varepsilon=\varepsilon_*$, then zero is also a simple eigenvalue of $\bar{\bar{\mathscr{L}}}^*$ and inhomogeneous problems of the form

$$\bar{\bar{\mathscr{L}}}\boldsymbol{\phi}-\gamma\boldsymbol{\phi}+\nabla p=0\,,\qquad \operatorname{div}\boldsymbol{\phi}=0\,,\qquad \boldsymbol{\phi}|_S=0$$

are boundedly invertible if and only if

$$\langle\boldsymbol{\phi},\mathbf{u}^*\rangle=0\,. \tag{15.13}$$

Applying the solvability condition (15.13) to (15.11) we find that when $\varepsilon=\varepsilon_*$

$$-\langle\mathscr{L}[\mathbf{U}_\nu,1]\,\mathbf{u},\mathbf{u}^*\rangle=\hat{\gamma}(\varepsilon_*)\langle\mathbf{u}_\varepsilon,\mathbf{u}^*\rangle. \tag{15.14}$$

Eq. (15.14) gives a formula for $\hat{\gamma}(\varepsilon_*)$ at critical points which holds whenever zero is an isolated, simple eigenvalue of the operator $\bar{\bar{\mathscr{L}}}$. Under this same hypothesis, it is possible to construct $\gamma(\varepsilon)$, $\boldsymbol{\phi}(\mathbf{x};\varepsilon)$ and $p'(\mathbf{x};\varepsilon)$ as a convergent series of powers of $\varepsilon-\varepsilon_*$ (see Joseph and Nield, 1975). Since all of the quantities appearing in (15.14) are unknown, (15.14) does not give the sign of $\hat{\gamma}(\varepsilon_*)$.

In the theory of bifurcation the function $\hat{\gamma}(\varepsilon)$ is one of the more interesting objects needing study.

Exercise 15.1: Find a formula for ν_2.

Exercise 15.2: Show that $\nu_1=0$ when $\mathscr{L}=\mathscr{L}^*$ is self-adjoint.

Exercise 15.3: Carry out the stability analysis of § 14 for steady bifurcating solutions with small amplitudes.

Exercise 15.4: Show that solutions which bifurcate at lower critical viscosities $\nu=\nu_n$ associated with higher eigenvalues $\sigma_n(\nu_n)=0$ $(n>1)$ are unstable.

§ 16. Transition to Turbulence by Repeated Supercritical Bifurcation

It is almost always true that when the viscosity of the fluid is small or, what amounts to the same thing, when the Reynolds number R is high, the flows which are observed are complicated, fluctuating and irregular. Such irregular flows, for want of an adequate theoretical description of their dynamic properties, are called turbulent.

To a degree, the bifurcation results given in this chapter are relevant to a description of the transition to turbulence. The bifurcation analysis separates flows which bifurcate subcritically from those which bifurcate supercritically. The subcritical bifurcating solutions are unstable and the supercritical bifurcating solutions are stable. This suggests that the transition may be regarded as:

(I) A continuous process of transition between flows involving repeated instability and bifurcation of supercritical solutions as ν is decreased.

(II) A discontinuous process of transition between flows involving instabilities which "snap through" subcritical bifurcating solutions to other "larger" solutions.

A continuous process of transition between flows involving repeated supercritical bifurcations could be described as follows: first there is a steady basic flow with $\nu > \nu_L$. At $\nu = \nu_L$ this flow loses stability and is replaced by a second more complicated flow which is stable for $\nu < \nu_L$ and which tends to zero as $\nu_L - \nu \to 0$. The second flow again loses stability as ν is decreased further. The second bifurcating flow then loses its stability at a second point of bifurcation $\nu_2 < \nu_L$ and a third, still more complicated flow, bifurcates supercritically, and so on. The transition between flows is continuous when the bifurcation is supercritical because there is no abrupt change in the flow as the point of bifurcation is crossed.

Subcritical bifurcations cannot lead to a continuous transition between flows because the bifurcating flow is unstable from the start; it has no domain of attraction and an arbitrary initial disturbance which escapes the domain of attraction of the basic flow "snaps through" the unstable bifurcating flow to some flow (or set of flows) with a larger amplitude. In this case an abrupt change in the flow would be observed as ν crosses ν_G from above.

The existence of subcritical solutions and the "snap through" description of their instability appears to be typical for many physical problems. The physical importance of the snap through phenomena, however, is related to the extent of subcriticality. For example, in steady convection extreme conditions are often required to induce a modest degree of subcriticality (see Chapter XI). On the other hand, in Poiseuille flow down an annular pipe, the extent of subcriticality is always deep (see Figs. 27.1, 27.2 and 35.1).

Experimentally, the bottom of the bifurcation curve (Fig. 35.1) is $\dfrac{U_m(b-a)}{\nu} \equiv R = R_G \cong 2000$. However, calculations show that R_L varies monotonically from a minimum of about 11,600 as the radius of the inner pipe is reduced. In the round pipe $R_L = \infty$; hence, Poiseuille flow is stable in the linearized theory for all R and any other motions (the observed ones) are of the subcritical type.

We note that the Landau-Hopf conjecture about the continuous transition to turbulence through repeated branching of solutions which was mentioned in the introduction and is discussed in the notes for this chapter is not consistent with the discontinuous type of transition which occurs in the subcritical case.

Exercise 16.1 (Iooss, 1973—74; Quasi-periodic bifurcation in a simple case): Consider the initial-value problem for the vector $U = (u, v)$:

$$\frac{dU}{dt} = \mathscr{A}_\lambda(t)\, U + M(U), \tag{16.1}$$

$$\mathscr{A}_\lambda(t) = \begin{bmatrix} \lambda - 1 + \cos t & \alpha + \sin t \\ -(\alpha + \sin t) & \lambda - 1 + \cos t \end{bmatrix},$$

$$M(U) = \begin{bmatrix} -u(u^2 + v^2) \\ -v(u^2 + v^2) \end{bmatrix}.$$

The null solution $U=(0,0)$ of these equations is analogous to a basic flow which is time-periodic with a period $T=2\pi$.

(1) Illustration of concepts from Floquet theory. Let $[u_1(t), v_1(t)]$ and $[u_2(t), v_2(t)]$ be the independent solutions of the linearization of (16.1) which have initial values $[u_1(0), v_1(0)]=[1,0]$ and $[u_2(0), v_2(0)]=[0,1]$. Show that the fundamental solution matrix for (7.10) is

$$\Phi_{(t)}=e^{(\lambda-1)+\sin t}\begin{bmatrix} \cos(\alpha t-\cos t+1) & \sin(\alpha t-\cos t+1) \\ -\sin(\alpha t-\cos t+1) & \cos(\alpha t-\cos t+1) \end{bmatrix}$$

Show that the eigenvalue of $\Phi(2\pi)$ satisfy the eigenvalue equation

$$\mu^2-\mu[e^{2\pi i\alpha}+e^{-2\pi i\alpha}]e^{2(\lambda-1)\pi}+e^{4(\lambda-1)\pi}=0$$

and the Floquet multipliers are

$$\mu=e^{2\pi(\lambda-1)}e^{\pm 2\pi i\alpha}.$$

What are the values of the Floquet exponent? *Hint:* Consider the equation satisfied by the complex variable $z=u(t)+iv(t)$.

(2) Show that the evolution of any solution $(u,v)\equiv(r\cos\theta, r\sin\theta)$ of (16.1) is governed by

$$\theta=\theta_0-\alpha t+\cos t-1,$$

$$\frac{1}{r^2}=\frac{1}{r_0^2}\exp(-2[(\lambda-1)t+\sin t])$$
$$+2\int_0^t\exp(-2[(\lambda-1)t+\sin t])\exp(2[(\lambda-1)\tau+\sin\tau])\,d\tau$$

where θ_0 and r_0 are the values of θ and r at $t=0$.

(3) Show that the null solution is globally and monotonically stable when $\lambda<1$. Infer that subcritical bifurcation is not possible.

(4) Show that a solution of permanent form whose radius $r(t)$ is periodic with period 2π bifurcates supercritically when $\lambda>1$. Show that the bifurcation curve is given by

$$\lambda=1+\varepsilon^2$$

where $\varepsilon^2=r^2(0)=r^2(2\pi n)$, $n=\pm 1, \pm 2, \ldots$. Show that the phase $\theta(t)$ of this solution is 2π-periodic if α is an integer and is $2\pi n$-periodic for $(n=\pm 2, \pm 3, \ldots)$ if $\alpha=k/n$ for some integer k. Suppose α is irrational and show that the bifurcating solution is quasi-periodic with two frequencies $(\omega_1,\omega_2)=(1,\alpha)$.

Notes for Chapter II

§ **7.** Two important mathematical properties of the spectral problem in a bounded domain (or in domains which can be made bounded by assuming disturbances with fixed spatial period) are that the eigenvalues of (7.2) are discrete and the generalized eigenfunctions are complete in the underlying Hilbert space (see S. Krein (1953), Browder (1953), Schensted (1960), DiPrima and Habetler (1969) and Sattinger (1970)).

Generally applicable necessary and sufficient conditions for the eigenvalues of the spectral problem to be real-valued are unknown. Of course, the spectrum of self-adjoint problems is necessarily real-valued. Problems which can be reduced to ordinary differential equations through the assumptions of periodicity can sometimes be shown to have a principle of exchange of stability even when they are nonself-adjoint. (Davis, 1969 together with DiPrima and Habetler, 1969; Yih, 1972A, B.)

The term "exchange of stability" can be traced back to the paper of Poincaré (1885) on the figures of equilibrium of rotating liquid masses. This classical problem treats the bifurcation of equilibrium figures in which the stability of one equilibrium figure is "exchanged" with the stability of another figure.

The now conventional understanding of the principle of exchange of stability, which implies that $\sigma=0$ is the principal eigenvalue of (7.2), seems to have been introduced by Jeffreys (1926). This understanding has only an oblique correspondence to the earlier understanding. Loss of stability of the basic flow when $\sigma=0$ portends the bifurcation of the basic flow into a secondary steady motion which may itself be stable or unstable, depending on conditions.

Poincaré's concept is an intuitive one which has no essential connection with steadiness. The "exchange of stability" between different equilibrium figures could be generalized to "transfer of stability" in a more general context. However, Jeffreys' use of these words, which is now conventional, is both inaccurate and confusing.

Linear stability theory is applied to interesting special problems in the following books and general reviews: Chandrasekhar (1961) treats many different kinds of special problems; C.C. Lin (1955) also considers different problems but he emphasizes the Orr-Sommerfeld equation; Synge (1938), Stuart (1963), Shen (1964), Monin and Yaglom (1971) give general reviews; Reid (1965) and the computer-oriented book of Betchov and Criminale (1967) confine their analysis to the Orr-Sommerfeld theory; Drazin and Howard (1966) consider the stability of inviscid flow; Greenspan (1969) gives results for the stability of rotating flow; and Yih (1965) deals with the stability of stratified flows. A wider range of applications of interest to chemical engineers can be found in the book of Denn (1975). The books of Monin and Yaglom and Denn also treat nonlinear problems.

Finally, we note that the energy method of Stuart (1958) is also strongly connected with the linear stability theory. This is an approximate method which assumes that the spatial form (shape) of the nonlinear disturbances is the same as the shape of marginal disturbances of the linearized theory, but with an unknown amplitude A (see Segel (1966) for details and further references). This energy method yields interesting nonlinear results but does not yield sufficient conditions for stability or the form of the disturbance which increases initially at the largest viscosity.

§ 8. The conditional stability theorem for steady flows was proved by G. Prodi (1962). The companion instability theorem was proved in (1968) by Kirchgässner and Sorger. General theorems of instability and conditional stability were stated earlier by Yudovich (1965) but without proofs. A full statement and proof of these theorems was given by Sattinger (1970). Further results and a thorough discussion of the principle of linearized stability can be found in the articles of Iooss (1971), Kirchgässner and Kielhöfer (1972) and in the monographs of Sattinger (1973) and Iooss (1973—1974).

§ 9. Poincaré, in his 1885 study of the figures of equilibrium of rotating masses, describes the junction where the Maclaurin spheroids exchange stability with the Jacobi ellipsoids as a *point of bifurcation*.

Bifurcation theory now has status as a branch of analysis which is closely allied to the theory of nonlinear eigenvalue problems which arise in fluid mechanics and in other branches of science and engineering. An introduction to the basic features of bifurcation theory, its principal applications and an extensive bibliography is given by Stakgold (1971). The monographs of Sattinger (1973) and Iooss (1973—74) treat bifurcation problems in a Banach space. Their studies include the bifurcation problem for the Navier-Stokes equations as a special case.

In the theory of nonlinear eigenvalue problems with analytic nonlinearities, the "eigenvalue" and the solutions are continuous, usually analytic, in a measure of the amplitude of the solution. For this reason bifurcation theory and the theory of nonlinear eigenvalue problems are conveniently treated for small amplitudes by perturbation methods.

§ 10. The method of power series in the amplitude appears to have started with the astronomical studies of Lindstedt (1882) and Poincaré (1892). Poincaré's work was extended to differential equations governing non-conservative systems depending on two parameters (our ν and ε) by Hopf (1942). The periodic solutions given by Hopf represent the simplest limiting case of Poincaré's periodic solutions of the second type (see Poincaré, Vol. III, Chapters XVIII and XXX). Hopf's fundamental contributions have only recently become well known. He considers bifurcation with a real eigenvalue (steady bifurcation) and bifurcation with a complex eigenvalue (time-periodic bifurcation). He was the first to prove that subcritical bifurcating solutions are unstable and that supercritical bifurcating solutions are stable. Hopf felt that his results might extend to partial differential equations and he speaks repeatedly of the Taylor problem and other famous problems of hydrodynamic stability. Hopf's results and Hopf's methods were extended to partial differential equations by Joseph and Sattinger (1972) following earlier studies by Joseph (1971) of the existence and stability of steady bifurcating solutions and of existence (without stability) by Sattinger (1971A) of time-periodic bifurcating solutions. Other authors have obtained Hopf's results using other methods (Iooss, 1972; Yudovich, 1971).

The method of power series was applied to steady bifurcation problems in fluid mechanics by Gor'kov (1957) and independently by Malkus and Veronis (1958). These authors consider a stability and bifurcation problem (the Bénard problem) in the theory of thermal convection. Two simplifying features of this problem are that it is steady ($\omega(\varepsilon)\equiv 0$) and that the linearized problem with $\varepsilon=0$ is self-adjoint. More extensive use of these power series in the amplitude were made by Lortz (1961), Busse (1962) and Schlüter, Lortz and Busse (1965).

Closely related to the method of power series is the method of Schmidt (1908) and Liapounov (1906) (see Stakgold (1971) and Vainberg and Trenogin (1962)). This method has been applied to bifurcation problems in fluid mechanics by Sorokin (1961), Kirchgässner and Sorger (1969), and Schwiderski (1972). Yudovich (1971) has used the method of Liapounov and Schmidt to prove that a time-periodic solution bifurcates from a stationary solution when the eigenvalue σ of the spectral problem is complex. Yudovich uses the same method to establish stability of supercritical bifurcations and instability of subcritical bifurcations.

To compare the method of Liapounov and Schmidt with the method of power series in the amplitude we first recall that in some problems (steady bifurcation, see § 15)

$$\varepsilon=\langle \mathbf{u},\zeta^*\rangle \quad \text{and} \quad \nu-\nu_L=f(\varepsilon) \tag{*}$$

whereas in other problems (time-periodic bifurcation, see Exercise 12.4)

$$\varepsilon=[\mathbf{u}] \quad \text{and} \quad \nu-\nu_L=g(\varepsilon^2). \tag{**}$$

In the method of Liapounov and Schmidt the solution is always expanded in powers of the discrepancy $\nu-\nu_L$; corresponding to (*) the method of Liapounov and Schmidt gives

$$\mathbf{u}(\mathbf{x})=\sum \mathbf{u}_n(\mathbf{x})(\nu-\nu_L)^n$$

and corresponding to (**)

$$\mathbf{u}(\mathbf{x},t)=\sum \mathbf{u}_n(\mathbf{x},t)(\nu-\nu_L)^{n/2}.$$

Combining these different representations of the same solution

$$\varepsilon=\sum \langle \mathbf{u}_n,\zeta^*\rangle(\nu-\nu_L)^n \quad \text{and} \quad \nu-\nu_L=f(\varepsilon) \tag{*}$$

and

$$\varepsilon=\sum [\mathbf{u}_n](\nu-\nu_L)^{n/2} \quad \text{and} \quad \nu-\nu_L=g(\varepsilon^2). \tag{**}$$

It follows that the series in powers of the discrepancy invert the bifurcation curves $f(\varepsilon)$ and $g(\varepsilon^2)$.

We note that if $df/d\varepsilon$ or $dg/d\varepsilon^2$ is not zero we may invert the amplitude expansions into the series of Liapounov and Schmidt. In the case of subcritical bifurcation there are always critical points, (like those at the nose of the bifurcation curve in A of Fig. 15.1) at which the bifurcating curve is not invertible. The functions $f(\varepsilon)$ and $g(\varepsilon^2)$ are good single-valued functions; for subcritical bifurcation, the inverse of these functions, given by Schmidt-Liapounov series, cannot generally be single-valued and two or more values of ε can correspond to a single value of $\nu-\nu_L$. For this reason, others apart, the method of power series is better than the Liapounov-Schmidt method. The two methods, of course, are entirely equivalent when the bifurcation curve is invertible, for example, in the limit $\varepsilon\to 0$.

A third method deals with growth of perturbations and replaces ε with a function of time $A(t)$. In this method it is assumed that

$$\frac{dA}{dt}=a_0A+a_1A^3+a_2A^5+\dots. \tag{***}$$

This method, at lowest order (A^3), leads to an amplitude equation of type conjectured by Landau (Eq. (****)). The method associated with (***) was suggested by Stuart (1960) and Watson (1960), while Palm (1960) simultaneously introduced related ideas in thermal convection. This method is very closely related to the Liapounov-Schmidt method (compare Reynolds and Potter, 1967A with Andreichikov and Yudovich, (1972).

A fourth method, which is related to the third, is given in the monograph of Eckhaus (1965). This method is based on "Fourier-series" in biorthogonal functions with time-dependent Fourier coefficients.

A partial justification for the third and fourth method at lowest order can be found in the work of Guiraud and Iooss (1968) and Iooss (1971). At present, there is no strict justification for the full series (***).

A fifth method utilizes the fact that in some bifurcation problems it is possible to define two different time scales, a fast time and slow time. One application of the method of two times to bifurcation problems is as a theory of transients between solutions of permanent form. The theory of two times is basically a perturbation theory which, when valid, holds for small amplitudes. It can lead to statements about the stability of bifurcating solutions; these also follow from standard methods. The method of two times goes beyond the standard method of studying the stability of the bifurcating solution because it treats the initial value problem for disturbances. However, no amount of fussing with time scales can relax the assumption that the amplitude is small. Therefore the method of two times does not go much beyond linear theory of stability of the bifurcating solution; it does not give global regions of attraction but only leads to statements about "sufficiently small" disturbances. Some of the superficially greater generality of the method of two times is apparent and not real. Like its counterpart in "les affaires d'amour", "two timing" need not lead to better results. When carefully worked, the method of two times can be used to study the initial-value problem for small disturbances of the bifurcating solution.

Newell and Whitehead (1969) have given an interesting analysis of bifurcation into motion and the stability of the motion in a layer of fluid heated from below. They use the method of multiple scales in time and space. Their method will be discussed in § 82. When the wave number spectrum is assumed to be discrete, Newell and Whitehead's amplitude equation reduces to an equation of the Landau type.

$$\frac{dA}{dt} = a_0 A + a_1 A^3 \qquad (****)$$

The same equation was obtained by Matkowsky (1970) as a first approximation to an asymptotic expansion of initial value problems of the parabolic type. Higher approximations are obtained recursively. Habetler and Matkowsky (1974) and Hoppensteadt and Gordon (1975) have demonstrated asymptotic convergence, uniform in time, for solutions to certain parabolic problems exhibiting steady one-sided, supercritical bifurcation. Matkowsky's method has been applied to study supercritical transients between unstable and stable solutions in the Taylor and Bénard problems by Kogleman and Keller (1971). A method of multiple scales has been developed by Stewartson and Stuart (1971) to derive an equation governing the evolution of spatially-localized, small-amplitude disturbances of plane Poiseuille flow. The Stewartson and Stuart study may be regarded as a formal extension of the method of multiple scales to cases in which the bifurcating solution is time-periodic.

A sixth method, used by Iooss, could be called the method of variation of constants in function space. Iooss uses abstract semigroup theory to invert the evolution equation in Banach space which is equivalent to the NS equations. The final form of the inverted equations is analogous to the elementary variation of constants formula used to solve systems of ordinary differential equations. Iooss' method is the only one which inverts the initial value problem and he has used it to prove many important results (see Iooss, 1972, 1973—1974, 1975, 1976).

Another approach to problems of stability and bifurcation utilizes the theory of invariant manifolds for dynamical systems. The theory of invariants is discussed, for example, by Hale (1971). This approach to stability and bifurcation has been developed by Ruelle and Takens (1971A, B). A detailed account of this approach, with worked examples, can be found in the monograph of Marsden and McCracken (1976).

Other methods which have been used to study bifurcation theory are successive approximations (Fife, 1970; Rabinowitz, 1968) and the topological method of Leray-Schauder degrees of a mapping.

Leray-Schauder degree refers to a powerful abstract theory which leads to definite statements about the existence and number of bifurcating solutions (Ukhovskii and Yudovich, 1963; Velte, 1964, 1966; Yudovich 1966B), the global properties of bifurcating solutions (Rabinowitz, 1971) and the stability of bifurcating solutions (Sattinger, 1971B; McLeod and Sattinger, 1973; Benjamin, 1974). It is not yet known how to apply arguments about topological degree to problems with complex eigenvalues (time-periodic bifurcations). The topological method is not constructive—it does not give explicit representations of the solutions. Results which are obtained by topological methods can often be obtained by elementary perturbation methods (see Joseph, 1971 and Crandall and Rabinowitz, 1973); this is an example of Weinberger's theorem: if it can be proved in one way, it can be proved in another way. The opposite of this theorem also holds and it is even easier to prove.

For a review of some of these methods and further references see Segel (1966), Görtler and Velte (1967), Kirchgässner and Kielhöffer (1972), Monin and Yaglom (1971), Stuart (1971), Sattinger (1973), Iooss (1973—1974), Fife (1974), Palm (1975) and Ladyzhenskaya (1975).

§§ 11, 12. The material in these sections is taken from the work of Joseph and Sattinger (1972).

§§ 13, 14. The results concerning the stability of bifurcating solutions of arbitrary amplitude follow from the fundamental factorization. Though the sign of $\hat{\gamma}(\varepsilon)$ is not generally known when ε is large, the numerical study of the nonlinear stability of Poiseuille flow (Zahn, Toomre, Spiegel and Gough, 1974) suggests that $\hat{\gamma}(\varepsilon)$ is positive. In their study, the subcritical bifurcating time-periodic solution gains stability at the critical point of the bifurcating flow. When ε is small the fundamental factorization shows that subcritical solutions are unstable and supercritical solutions are stable. This result was achieved earlier by Joseph and Sattinger (1972) and, for ordinary differential equations, by Hopf (1942). They assumed that $\nu_2 \neq 0$; this assumption is now shown to be unnecessary (see Joseph and Nield, 1975).

§ 15. The bifurcation results of this section are taken from Joseph (1971) following earlier results of Fife and Joseph (1969) and Busse (1962). The stability results following from the fundamental factorization are taken from the paper of Joseph and Nield (1975). Similar results for general problems with non-analytic nonlinearities have been proved by Crandall and Rabinowitz (1973). Related results for the Navier-Stokes equations are also implied by Benjamin's theorem (1974) "Except possibly at isolated values of $1/\nu$ whenever the steady-flow problem does not have a unique solution then one of its solutions is an unstable flow". Benjamin's theorem follows from a careful analysis of topological degree using the theory of Leray and Schauder.

§ 16. Landau and Hopf both conjecture that transition to turbulence can be described by repeated supercritical bifurcation of manifolds of quasi-periodic solutions into similar manifolds of higher dimension[9].

Three ideas have a prominent place in Landau's (1944) essay (these are repeated in the book by Landau and Lifschitz (1959)):

(1) The initial exponential growth of small disturbances when $\nu < \nu_L (R > R_L)$ will actually be limited by nonlinear interactions governed, when $|A|$ is small, by the amplitude equation (****).

(2) The terminal state for a disturbance of steady flow when ν is slightly smaller than ν_L is a time-periodic motion.

(3) As ν is decreased further, the time-periodic solution loses its stability to a quasi-periodic solution with two frequencies; then the two-frequency solution gives up its stability to a solution with three frequencies. In this way more and more frequencies are introduced into the ever larger manifolds of stable solutions.

---

[9] A quasi-periodic function of n variables is a function $f(\omega_1 t, \omega_2 t, \ldots, \omega_n t)$ containing a finite number n of rationally independent frequencies $\omega_1, \omega_2, \ldots, \omega_n$, which is periodic with period 2π in each of its variables. For example, the function $f(\omega_1 t, \omega_2 t) = g(t) = \cos t \cos \pi t$ is a quasi-periodic function with frequencies $\omega_1 = 2\pi$ and $\omega_2 = 2$. The value $g(t) = 1$ occurs when $t = 0$ but not again; though $g(t) < 1$ when $t \neq 0$, there is always $t(\varepsilon) > 0$ such that $|g(t) - g(0)| < \varepsilon$ for preassigned $\varepsilon > 0$.

"In the course of the further increase of the Reynolds number new periods appear in succession, and the motion assumes an involved character typical of turbulence ... so a turbulent motion is to a certain extent a quasi-periodical motion." (Landau and Lifschitz, p. 105.)

Unfortunately, this very plausible description of transition to turbulence encounters two difficulties when one tries to make it precise. First, supercritical bifurcating solutions may not be quasi-periodic, and second, the transition to turbulence may occur discontinuously by a snap-through instability.

Explicit examples which lead to the bifurcation of solutions with more than one frequency are not presently known. However, Joseph (1973) has considered the problem of bifurcation of time-periodic basic flows with a fixed frequency into quasi-periodic flows with two frequencies. Using a method of "two times" and a suggestion of Landau, he gives an algorithm for the construction of a series solution in powers of ε for this problem. Nearly all of the results which can be obtained rigorously in the steady case can be obtained formally from the construction. Again, subcritical bifurcating solutions are unstable, etc. However, the construction of the bifurcating solution generates a problem of small divisors and the formal computation of stability of the bifurcating solution relies on perturbing an eigenvalue which is not isolated.

The Landau-Hopf conjecture has been studied for systems of ordinary differential equations by D. Ruelle and F. Takens (1971A, B). They use concepts which arise out of the modern qualitative theory of differential equations. Among other things they prove that, given conditions, an unstable time-periodic solution bifurcates into an invariant torus. The periodic solution is a closed orbit in phase space; the torus can be visualized as a closed surface which contains the orbit of the periodic solution. The radius of the torus is related to the amplitude of the bifurcating solution but the character of the solutions on the torus, their "periodicity" or "quasi-periodicity" etc., is left vague. Their analysis uses a very powerful theorem, the central manifold theorem, which allows one to reduce the original problem which is set in a Banach space of n dimensions to $n=2$. The paper by O. Lanford (1973) gives a clearly written account of this bifurcation result which also includes a proof of the central manifold theorem.

G. Iooss (1975) has proved that the bifurcation results of Ruelle and Takens can be extended to the case in which a torus of solutions bifurcate from a forced periodic solution of fixed frequency of the Navier-Stokes equations. As in the work of Ruelle and Takens the supercritical torus is stable and the subcritical torus is unstable. Iooss's results, like those of Ruelle and Takens, are incomplete because the analytic properties of the solutions on the bifurcating torus are not completely specified. Though no complete statement can be made about the quasi-periodicity of solutions on the torus, it is possible to complete the leading term in the expression giving the torus in powers of $(\nu_c-\nu)^{1/2}$. The computation of Iooss shows that the leading term is quasi-periodic and, in fact, coincides with the leading term of the formal series proposed by Joseph (1973).

Ruelle and Takens (1971A) also discuss successive supercritical bifurcation into tori of increasing dimensions. Higher-dimensional tori are topological objects which are not easy to describe in an intuitive way. For example, a torus in k-dimensional space, $k\geqslant 4$, does not divide the space into an inside and an outside. The image in phase space of the quasi-periodic vector field

$$\mathbf{x}=\mathbf{f}(\omega_1 t, \omega_2 t, \ldots, \omega_k t)$$

is a special case of a k-dimensional torus T^k. Ruelle and Takens discussion of bifurcating tori rest on some restrictive assumptions. They assume that systems of differential equations treated by them are "generic" in the sense of residual sets[1] and they require that their "viscosity" parameter should be bounded away from critical values. They prove that near a quasi-periodic solution with k frequencies ($k\geqslant 4$) on T^k there is, on T^k, a "strange attractor" (technically, this is one of Smale's strange axiom A attractors).

To understand the significance of the work of Ruelle and Takens in the description of transition to turbulence it is useful to replace the word "strange" with "non-periodic". There is then an at-

[1] A property P of a dynamical system, $\dot{x}=X(x)$, $x\in B$ where B is a Banach space, is generic if it is satisfied on a residual subset of B (a subset which contains a countable intersection of dense open sets). P is non-generic if it is not generic. The status of the Navier-Stokes equations with regard to "genericity" is presently unknown.

tractor on the $k \geqslant 4$ torus and it is non-periodic[1]. There are examples of non-periodic attractors which are not strange axiom A attractors. An interesting asymptotic analysis of one of these has been given by Baker, Moore and Spiegel (1971). The non-periodic attractor of Lorenz (1963) is of more than ordinary importance in the present discussion since it arises from a problem of convection—albeit a badly mutilated problem in which the partial differential equations are reduced to three nonlinear ordinary differential equations in the Fourier components of the solution. A similar set of equations, which contain the Lorenz equations as a special case, arises in the Robbin (1975) study of disk dynamos. An important property of these non-periodic attractors is that they "phase mix". That is, unlike quasi-periodic functions which do not phase mix (see property 10 of Appendix A) the Lorenz and strange axiom A attractors have an autocorrelation function which decays rapidly in time. This property of the autocorrelation function is in strict agreement with experiments. A second property shared by observed turbulence and non-periodic attractors is a sensitive dependence on initial conditions. Though average values in turbulent flows with steady data are steady and therefore independent of initial conditions, the detailed structure of each of the realizations of turbulent flow is different and this difference may presumably be associated with different initial conditions. This high sensitivity to initial conditions is never shared by quasi-periodic attractors; on these, a condition which occurred once will occur, nearly, again and again if you have the patience to wait long enough.

The main idea suggested by the work of Lorenz (1963) and of Ruelle and Takens (1971A) is that transition to turbulence in the supercritical case occurs as a definite higher bifurcation. Instead of the bifurcation into a sequence of quasi-periodic manifolds of increasing dimension along the lines of Landau and Hopf we get a few bifurcations of the Landau-Hopf type followed by bifurcation into a non-periodic attractor with phase mixing. This bifurcation would be observed in experiments as a sudden transition to turbulence. Something like this is sometimes observed in experiments (see the discussion of the experiments of Gollub and Swinney (1975) given at the end of § 39, and the theoretical papers of McLaughlin and Martin (1974) in which a large number of experiments are cited).

It is perhaps necessary to add that the Ruelle-Takens theory needs justification at many points. Conclusions about supercritical bifurcation into strange attractors for $k \geqslant 4$ assumes the existence of T^k, of a quasi-periodic solution on T^k and that properties of Navier-Stokes equations are generic in the sense of residual sets. Moreover, a catalogue of attractors for ordinary differential equations in $\mathbb{R}^n$ is not known—there are many types of attractors which still elude analysis. Our knowledge of attractors for solutions of the Navier-Stokes equations is certainly no better. Despite this, it is a step forward to think of non-periodic, phase-mixing attractors in the description of transition to turbulence through repeated supercritical branching.

The theory of non-periodic attractors does not lead to a description of the spatial structure of turbulent fields. Interesting notions about the spatial structure of turbulent fields arise in the variational theory of turbulence.

This theory is based on the discovery of L. Howard (1963) that it is possible to define bounded functionals of statistically stationary turbulence. F. Busse (1969 B, 1970 A) suggested that the variational problem posed by Howard might be solved by a new type of solution—the multi-α solution and he gave a Galerkin type of approximation for these solutions in an asymptotic limit. Some necessary preliminaries for a rigorous theory of the multi-α solutions were given by Busse and Joseph (1972) and a rigorous bifurcation theory was established by Joseph (1974A). The branching theory establishes the multi-α solutions as a model of repeated supercritical branching of solutions[2].

Even if the Landau-Hopf conjecture were true in the supercritical case, it could not hold in the subcritical case. As we have seen, the instability of the subcritical bifurcating solutions precludes continuous branching.

Landau (but not Hopf) did consider the possibility that the loss of stability could occur under subcritical conditions but he discounts the importance of this possibility. Landau speaks of the two values of R_{cr} (we have called these values R_L and R_G) in the quote given below (Landau and Lifschitz, p. 107).

---

[1] An attractor for a differential equation is a compact set Λ such that all points sufficiently close to Λ tend to Λ under time evolution when time tends to $+\infty$.

[2] A detailed account of the variational theory of turbulence is given in Chapter IV and especially in Chapter XII. However, the existence and bifurcation theory of Joseph (1974A) is not included in this book.

"We have introduced the concept of the critical Reynolds number as being the value of R at which instability of steady flow, in the sense described above, first occurs. The critical Reynolds number can, however, be regarded from a somewhat different point of view. For $R<R_{cr}$ there are no stable, non-steady solutions of the equations of motion that are not damped in time. After the critical value has been reached, a stable non-steady solution appears, which will actually occur in a moving fluid.

"As far as experimental investigations of the flow past ordinary finite bodies are concerned, the two definitions of R_{cr} seem to be the same. Logically, however, this need not be so, and cases could in principle occur where there are two different critical values: one above which non-steady flow can occur without being damped, and another above which steady flow becomes absolutely unstable. The second must obviously be greater than the first. However, since there is at present no indication that such cases of instability actually exist, we shall not pause to investigate them more closely".

In the footnote to this last paragraph, he says "We are not here concerned with (e.g.) flow in a pipe, where the loss of stability has unusual properties." (The temptation to compare the statement that "no ... such cases ... exist ..." with the footnote just quoted is irresistible. The unusual properties of instability of flow in pipes are described by Landau (p. 114, read $R'_{cr}=R_G$) as follows:

"However, the experimental results also show that there is another critical Reynolds number (which we denote by R'_{cr}); this determines the limit beyond which stable non-steady flow can exist ... If, in any section of the pipe, turbulent flow occurs, then for $R<R'_{cr}$ the turbulent region will be carried downstream and will diminish in size until it disappears altogether; if, on the other hand, $R>R'_{cr}$, the turbulent region will extend in the course of time to include more and more of the flow. If perturbations of the flow occur continually at the entrance to the pipe, then for $R<R'_{cr}$ they will be damped out at some distance down the pipe, no matter how strong they are initially. If, on the other hand, $R>R'_{cr}$, the flow becomes turbulent throughout the pipe, and this can be achieved by perturbations which are the weaker, the greater R. Thus laminar flow in a pipe with $R>R'_{cr}$ is metastable, being unstable with respect to perturbations of finite intensity; the necessary intensity is the smaller, the greater R.

"... non-steady flow arising by the disruption of metastable laminar flow is already fully-developed turbulence. In this sense the appearance of turbulence in a pipe is essentially different from the appearance of turbulence owing to the absolute instability of steady flow past finite bodies. In the latter case non-steady flow seems to appear in a continuous manner as we pass through R_{cr}, the number of degrees of freedom increasing gradually ... For flow in a pipe, however, turbulence appears discontinuously. This difference causes, in particular, the different dependence of the drag on the Reynolds number in the two cases. For example, if we consider the motion of any body in a fluid, the drag force F on it is continuous at $R=R_{cr}$, where steady flow becomes absolutely unstable. At this point the curve $F=F(R)$ can have only a bend corresponding to the change in the nature of the flow. For flow in a pipe, on the other hand, there are essentially two different laws of drag for $R\geqslant R_{cr}$: one for steady flow, and the other for turbulent flow. The drag is discontinuous for whatever value of R marks the transition from one type of flow to the other."

Even if it were granted that the flow around bodies is more "typical" than the flow in pipes, the facts involved do not support the notion of "continuous" transition to turbulence in the manner envisaged by Landau. In fact, experiments suggests that there is a sense in which the drag force is a discontinuous function of R.

A very beautiful set of elementary experiments of A. Shapiro (1961) illustrated this discontinuity in the drag law. In these experiments a three-inch diameter sphere is immersed in an air stream.

"... Then we measure the drag for a constantly-rising series of air speeds. At 80 miles per hour the force is about 1.5 units on the scale. At 100 mph the drag has increased to about 2.4 units. Up a little more to 115 mph and the drag has climbed to 3.0 units. All this is quite regular, and just what you might expect. But as the speed goes above 115 mph to our astonishment the drag begins to fall, and at 140 mph it is only about 2.3 units. After this it begins to rise again and at 155 mph it has once more reached 3.1 units. Subsequently, the drag continues to rise as the speed increases, with no further irregularities.

"These observations are summarized graphically in (our Fig. 16.1), which shows drag force on the vertical scale and wind speed on the horizontal scale. As the speed goes up, starting from zero, the drag also increases, just as it does when you put your hand out of an accelerating automobile. But after the speed reaches a certain critical value, a further increase of speed causes the drag to decrease. At this stage, too, the pointer on the drag scale oscillates quite strongly, indicating some

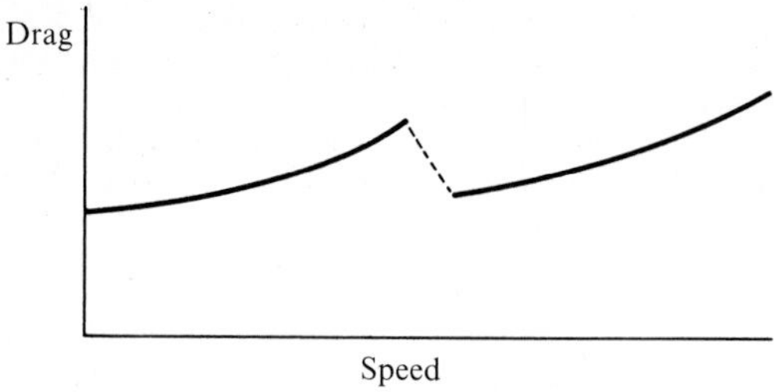

Fig. 16.1: Schematic summary of the experiment of A. Shapiro (1961) giving the drag on a sphere as the speed of the air-stream is varied

sort of instability. Finally we reach a speed at which the drag once again increases smoothly with speed, but along a curve different from that for the first smoothly-ascending section.

"Thus there seem to be two distinct types of flow with a rapid transition between the two, occurring in a certain range of speeds. The drag falls sharply when a quite moderate increase of speed causes the flow to change from one pattern to another." (Shapiro, pp. 29—30.)

In a second set of experiments, Shapiro compares the drag on a slightly roughened ball with the drag on a smooth ball of the same size. At low speeds the drag is greater on the rough ball but the drag on the smooth ball is greater when the speed is larger than 125 mph.

"While it is surprising that surface roughness sometimes decreases drag, it is even more astonishing to discover from quantitative measurements with the weighing scale that at some speeds the drag of the rough ball is only about one-fifth the drag of the smooth ball! We like to think that in the world of natural phenomena small changes in circumstances produce correspondingly small changes in behavior. How can such a very small cause as a roughening barely perceptible to the touch produce such a very large effect as a five-fold reduction in drag?" (Shapiro, p. 37.)

The sudden decrease of the drag as the Reynolds number is increased and the surprising fact that the drag on a rough sphere can be much smaller than on a smooth sphere can be explained, following Prandtl, as follows: In potential flow around a sphere, the pressure is largest at the stagnation points at the front and back of the sphere and is smallest at the top and bottom of the sphere where the speed is greatest. Viscosity alters this symmetry and a boundary layer develops on the sphere. The fluid in the boundary layer is retarded by friction. This retardation by friction together with the adverse pressure gradient on the rear of the sphere finally set up a return flow against the stream in the boundary layer. This return flow leads to the separation of the boundary layer and the formation of a wake. In the wake the pressure is nearly constant and equal to the value of the pressure at the point of separation. It follows that the boundary layer separation prevents the pressure recovery which would otherwise occur and this increases the drag on the sphere.

The development of the separated boundary layer is a laminar flow problem which is continuous in the Reynolds number (the upper branch of Fig. 16.1). The sudden decrease in drag, the drag crisis as it is sometimes called, is connected with the instability of the laminar boundary layer at a critical Reynolds number. This critical Reynolds number is smaller when the disturbances are larger. Roughening the sphere will reduce the critical Reynolds number and:

"It may be mentioned that the degree of turbulence in the main stream affects the drag crisis; the greater the incident turbulence, the sooner the boundary layer become turbulent (i.e., the smaller is R when this happens). The decrease in the drag coefficient therefore begins at a smaller Reynolds number, and extends over a wider range of R." (Landau and Lifschitz, p. 170.)

The instability of the laminar flow has a large effect on the drag because it moves the point of separation of the boundary layer toward the rear of the sphere[1] and allows the flow to more nearly recover the pressure it has at the forward stagnation point. This increased pressure recovery in the wake is responsible for the sharp decrease in drag.

In a given experiment it is possible to find a continuous drag curve with a rapid but continuous transition (cf. Landau, quoted earlier) between two very different values of the drag (see Figs. 16.2, 3). The disturbance level of the flow is fixed by the experiment. If larger disturbances are introduced at

[1] The turbulent fluctuations bring fluid with high forward momentum close to the boundary on the back of the sphere. If it were not for this transport of momentum, the adverse pressure gradient would produce a strong deceleration and early separation (see G. Batchelor (1967), pp. 337—343, for a fuller explanation).

a fixed R, the value of the drag may be greatly reduced. Thus, the transition between values of the drag at a fixed R could be regarded as a "discontinuous" transition.

The stability of separated boundary layers which is central to the problem of the drag on a sphere is not, as yet, well understood. Until this problem is better understood, observations about the drag on a sphere do not constitute a firm basis for a general hypothesis about the transfer of stability between solutions.

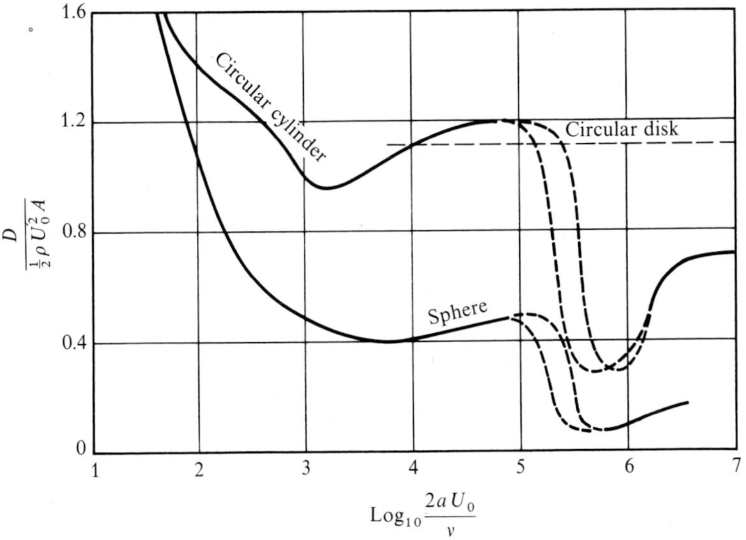

Fig. 16.2: The measured drag on unit axial length of a cylinder ($A=2a$), on a sphere ($A=\pi a^2$), and on a circular disk normal to the stream ($A=\pi a^2$), all of radius a. The broken curves represent results obtained in different wind tunnels (G.K. Batchelor, 1967)

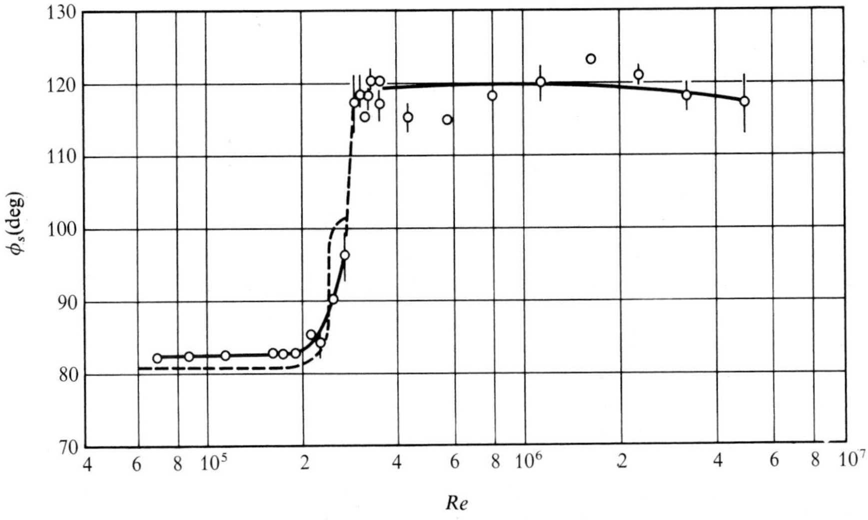

Fig. 16.3: Position of boundary-layer separation for smooth spheres (E. Achenbach, 1972)

Chapter III

Poiseuille Flow: The Form of the Disturbance whose Energy Increases Initially at the Largest Value of ν

The problem to be considered in the next two chapters is the stability and mechanics of flow driven down a straight pipe of annular cross section by a pressure gradient. We want to know what new regimes of flow are created as ν is decreased.

When the viscosity is large, there is only one possible flow corresponding to the given geometry and operating conditions. This uniquely determined motion is the basic flow; it is called laminar Poiseuille flow and it is described by Eqs. (17.4) and (17.5). Hagen-Poiseuille flow is Poiseuille flow in a round tube in which there is no inner cylinder (Eq. 17.6, see Exercise 17.1). Plane Poiseuille flow is flow down a channel with plane parallel walls, and can be obtained as a limiting case when the gap size in the annulus is small relative ro the inner cylinder radius (Eq. 17.7). More flows become possible as various critical values of ν are passed[1]. In this chapter we will consider the first critical value.

To make clear the nature of our mathematical idealization of the physical problem, we shall start with a physical motion which is generated by the apparatus shown in Fig. 17.1.

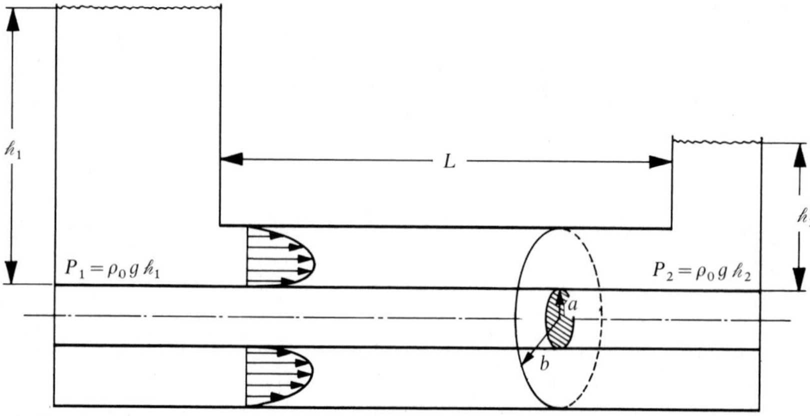

Fig. 17.1: Poiseuille flow down an annular pipe

[1] It will be convenient to retain the name Poiseuille flow for flows which replace laminar Poiseuille flow as ν is increased; we can, for example, consider time-periodic Poiseuille flow or turbulent Poiseuille flow.

The head heights h_1 and h_2 are to be imagined as large relative to the outer pipe diameter b. Then the pressure at the annular entrance and exit of the pipe is almost constant over the cross section. Spillways at the top of each tank are employed to keep the heads at a constant level so that the pressure drop across the pipe of length L is held at a fixed constant value $p_1 - p_2$.

The regimes of flow which are possible in the apparatus correlate strongly with the value of the Reynolds number, here defined as

$$R = \frac{U_m(b-a)}{\nu}$$

where U_m is the maximum velocity of the laminar Poiseuille flow. The Reynolds number arises in the dimensionless formulation of the initial value problem in which the length is measured in units of $b-a$ and the velocity in units of U_m (see Notes to Chapter I). The dimensionless formulation is useful in correlating data. Of course, for a fixed pressure drop and tube, varying R is entirely equivalent to varying $1/\nu$.

§ 17. Laminar Poiseuille Flow

One can make the following two observations about the pipe flow when the viscosity is large: (a) the pressure gradient, when measured at various stations across the length of the pipe, is constant except in the region near the entrance and exit of the pipe; and (b) the trace of a dye streak down the pipe is a straight line. It is possible to find a steady solution of the Navier-Stokes equations for a problem which coincides with the one which we have just described on the region over which the pressure gradient is a constant. This solution satisfies the Eqs. (1.1) for a steady basic flow down an infinite pipe. The velocity components are

$$W=0\,, \qquad V=0\,, \qquad U=U(r)\,, \tag{17.1}$$

where (r, θ, x) are cylindrical polar coordinates and the x axis is at the pipe center, $U(a)=U(b)=0$ and

$$-\frac{d\pi}{dx}+\nu\left(\frac{d^2U}{dr^2}+\frac{1}{r}\frac{dU}{dr}\right)=0\,. \tag{17.2}$$

Here, we must have

$$\frac{d\pi}{dx} = -\hat{P} = \text{const}\,, \tag{17.3}$$

where $\pi = p/\rho_0$ and the constant coincides with the pressure gradient observed in the experiment. If we suppose that

$$-\frac{1}{\nu}\frac{d\pi}{dx} = \frac{p_1 - p_2}{\nu\rho_0 L} \tag{17.4}$$

is a constant, then

$$\frac{U}{U_m}=\frac{a^2-r^2+(b^2-a^2)[\ln(r/a)]/\ln(b/a)}{a^2-r_c^2+2r_c^2\ln(r_c/a)}, \tag{17.5}$$

where

$$r_c^2=\tfrac{1}{2}(b^2-a^2)/\ln(b/a)$$

is the value of r at which $U(r)$ has its maximum value $U(r_c)=U_m$.

It will be noted that the family of laminar flows (17.5) exists for all values of ν. For small ν, (17.5) will be unstable and will be replaced by an eddying motion.

Exercise 17.1: Consider the limit $a\to 0$ when $r>a$ is fixed. Show that (17.1) reduces to Hagen-Poiseuille flow; that is, to

$$U(r)=U_m(1-r^2/b^2) \tag{17.6}$$

where $U_m=\hat{P}b^2/4\nu$ is the centerline velocity. Show that the force per unit length which is exerted on the fluid by the inner cylinder tends to zero with a.

Consider the limit $a\to\infty$ when $(b-a)=l$ is fixed. Show that (17.5) reduces uniformly to plane Poiseuille flow

$$U=U_m(1-4z^2/l^2), \tag{17.7}$$

where $U_m=\hat{P}l^2/8\nu$ is the centerline velocity and $z=0$ is the centerline of the channel. (The reduction requires an expansion of two terms: $\ln b/a=\ln(1+l/a)=\frac{l}{a}-\frac{l^2}{2a^2}+0\left(\frac{l^3}{a^3}\right)$.)

Exercise 17.2: Show that

$$\frac{U}{\langle U\rangle}=2\frac{b^2-r^2+2r_c^2\ln(r/b)}{b^2+a^2-2r_c^2}\equiv U_0(r) \tag{17.8}$$

where

$$\langle U\rangle=\frac{2}{b^2-a^2}\int_a^b rU\,dr$$

is the mean velocity. What is the ratio of the maximum velocity to the mean velocity? Show that $U_m/\langle U\rangle=2$ for Hagen-Poiseuille flow and that $U_m/\langle U\rangle=\frac{3}{2}$ for plane Poiseuille flow.

§ 18. The Disturbance Flow

The laboratory flow is now disturbed by agitation of the entrance region, by vibrations of the apparatus, etc. It is observed that if ν is very large, these disturbances are immediately puffed out. For smaller ν the disturbances may grow for a distance and then decay. For still smaller values of ν, it is possible to maintain a "sinuous and eddying" motion other than laminar Poiseuille flow.

It is clear that all flows, like the laboratory flow, occupy a limited region of space; for this reason the bounded domain is physically of greatest interest.

But the mathematical problem is not always best set in a bounded domain. When the bounded region is not closed by a material surface on which one can pose reasonable prescribed boundary conditions, it may be better to extend the problem onto an infinite domain and to restrict the search for solutions to a class in which one hopes to represent the most important properties of the true solution on the bounded domain.

Speaking precisely, laminar Poiseuille flow is not the true laboratory flow but is the extension of the laboratory flow onto the infinite cylinder with $-\infty < x < \infty$. We are, on this account, moved to search for a similar extension for the disturbances. In other words, we confront the problem of how to extend disturbances onto the whole cylinder so that they will coincide with true disturbances over the region of constant pressure gradient.

Given a smooth disturbance on a finite domain, it is always possible to find a smooth extension to the infinite domain as (a) a periodic function, (b) an almost periodic function or (c) a function of a Fourier transform class. A natural choice for the extensions would appear as the one which leaves unaltered the dominant structures of the observed motions.

We have chosen to set problems in the class of almost periodic disturbances (see Appendix A). An almost periodic function $f(x)$ can oscillate arbitrarily provided that its value at any point x repeats itself, nearly, at some distant point $x+\tau$. Every periodic function is also almost periodic but the converse is not true; e.g., the almost periodic function $\cos x + \cos \pi x$ is definitely not periodic though it is a sum of periodic functions. Its value is 2 at $x=0$ and not elsewhere. In a pipe with axis x, it is possible to consider almost periodic functions of x.

The disturbance equations are again formed by considering the difference between the basic flow $(U(r), d\pi/dx)$ given by (17.4) and (17.5) and another flow $(\mathbf{U}^a(\mathbf{x}), \nabla\pi^a)$ which is also an almost periodic solution of the IBVP but has an initial velocity and pressure different from (17.4) and (17.5). In addition to the usual conditions, we want to maintain the same average pressure drop, $(p_1 - p_2)/L\rho_0 = \hat{P}$, along the pipe.

In the extended problem on the infinitely long pipe,

$$\pi(x) = -\hat{P}x + \text{constant}$$

and

$$\pi^a(x, y, z, t) = p(x, y, z, t) + \pi(x)$$

are both unbounded but their difference $p = \pi^a - \pi$ must be almost periodic, hence bounded.

We call the difference variables p and

$$\mathbf{u} = \mathbf{U}^a - \mathbf{e}_x U(r)$$

"disturbances". They are assumed to be almost periodic in x (or periodic with period L) and they must satisfy (1.4a, b, d) and

$$\mathbf{u} = 0 \quad \text{at} \quad r = a, b. \tag{18.1}$$

The equations which govern the difference between Poiseuille flow and any other almost periodic motion in the annulus which has the same constant component of the pressure gradient $\hat{P}$ are

$$\frac{\partial \mathbf{u}}{\partial t}+\mathbf{U}\cdot\nabla\mathbf{u}+\mathbf{u}\cdot\nabla\mathbf{U}+\mathbf{u}\cdot\nabla\mathbf{u}=-\nabla p+\nu\nabla^2\mathbf{u}, \qquad \operatorname{div}\mathbf{u}=0 \tag{18.2}$$

and (18.1).

§ 19. Evolution of the Disturbance Energy

The average energy of a disturbance $\mathbf{u}$ of Poiseuille flow in an annular cylinder of length $2L_x$ containing constant mass is given by

$$2\mathscr{E}(t)=\langle|\mathbf{u}|^2\rangle=\frac{1}{2L_x}\int_{X_1}^{X_2}\langle|\mathbf{u}|^2\rangle_A\,dx \tag{19.1}$$

where

$$\langle\cdot\rangle_A=\frac{1}{\mathscr{M}(A)}\int\int\cdot\,dy\,dz=\frac{1}{\pi(b^2-a^2)}\int_a^b\int_0^{2\pi}\cdot\,r\,dr\,d\theta \tag{19.2}$$

and

$$2L_x=X_2(t)-X_1(t).$$

Since mass is constant, $dL_x/dt=0$. Calculation shows that

$$2\frac{d\mathscr{E}}{dt}=2\left\langle\mathbf{u}\cdot\frac{d\mathbf{u}}{dt}\right\rangle+\frac{1}{2L_x}\langle[|\mathbf{u}|^2]\rangle_A\frac{dX_1}{dt} \tag{19.3}$$

where $[g]=g(x_2)-g(x_1)$ gives the jump in g.

The average of any periodic function of period $2L_x$ vanishes. The average jump of any almost periodic function vanishes; when g is almost periodic, we average over the entire layer, $L_x\to\infty$, and since $|g|<\infty$, $[|g|]<\infty$ and $\lim_{L_x\to\infty}\frac{1}{L_x}[g]=0$. Hence, in (19.3) we have $d\mathscr{E}/dt=\left\langle\mathbf{u}\cdot\frac{d\mathbf{u}}{dt}\right\rangle$; moreover,

$$\langle\mathbf{v}\cdot\nabla\phi\rangle=\langle\nabla\cdot\phi\mathbf{v}\rangle=\frac{1}{2L_x}\langle[\phi\mathbf{v}]\rangle_A=0 \tag{19.4}$$

for any ϕ and any solenoidal $\mathbf{v}$ having a zero normal component at $r=a,b$. As in the bounded domain, we find that

$$\tfrac{1}{2}\frac{d}{dt}\langle|\mathbf{u}|^2\rangle=\left\langle\mathbf{u}\cdot\frac{d\mathbf{u}}{dt}\right\rangle=-\langle\mathbf{u}\cdot\nabla\mathbf{U}\cdot\mathbf{u}\rangle-\tfrac{1}{2}\langle\mathbf{u}\cdot\nabla|\mathbf{u}|^2\rangle-\langle\mathbf{u}\cdot\nabla p\rangle+\nu\langle\mathbf{u}\cdot\nabla^2\mathbf{u}\rangle. \tag{19.5}$$

Here we have set

$$\frac{d\mathbf{u}}{dt}=\frac{\partial\mathbf{u}}{\partial t}+\mathbf{U}\cdot\nabla\mathbf{u}$$

as in § 3.

For the basic Poiseuille flow (17.1), we have

$$\mathbf{u}\cdot\mathbf{D}\cdot\mathbf{u}=uD_{xr}w+wD_{rx}u=uw\,dU/dr$$

where D_{xr} is the only nonvanishing component of the stretching tensor $\mathbf{D}$ and $U(r)$ is given by (17.5).

Integration by parts of (19.5) leads us to

$$\frac{d\mathscr{E}}{dt}=-\left\langle wu\frac{dU}{dr}\right\rangle-\nu\langle|\nabla\mathbf{u}|^2\rangle=\mathscr{I}[\mathbf{u}]-\nu\mathscr{D}[\mathbf{u}] \tag{19.6}$$

where

$$-\left\langle wu\frac{dU}{dr}\right\rangle=\frac{2U_m\left\langle\left(r-\frac{r_c^2}{r}\right)wu\right\rangle}{a^2-r_c^2+2r_c^2\ln(r_c/a)}=\mathscr{I}\quad\text{and}\quad\langle|\nabla\mathbf{u}|^2\rangle=\mathscr{D}\,. \tag{19.7}$$

The coordinate form of $\langle|\nabla\mathbf{u}|^2\rangle$ is given under (37.2).

Exercise 19.1: Show that (19.6) holds for all flows having the same mass flux but different pressure gradients.

§ 20. The Form of the Most Energetic Initial Field in the Annulus

The disturbance $\tilde{\mathbf{u}}_0(\mathbf{x})$ of Poiseuille flow which makes the time derivative of the disturbance energy

$$\frac{d\mathscr{E}}{dt}=\mathscr{I}[\tilde{\mathbf{u}}_0]-\nu\mathscr{D}[\tilde{\mathbf{u}}_0] \tag{20.1}$$

increase initially at the largest value of ν is the field $\tilde{\mathbf{u}}_0(\mathbf{x})$ which maximizes $\mathscr{I}/\mathscr{D}$; that is

$$\nu_{\mathscr{E}}=\max_{\mathbf{H}_x}\mathscr{I}/\mathscr{D}=\mathscr{I}[\tilde{\mathbf{u}}_0]/\mathscr{D}[\tilde{\mathbf{u}}_0] \tag{20.2}$$

where

$$\mathbf{H}_x=\{\mathbf{u}(r,\theta,x)\colon \operatorname{div}\mathbf{u}=0,\ \mathbf{u}|_{a,b}=0,\ \mathbf{u}\in AP(x)\}$$

(see the energy stability theorem of § 4).

Two problems are to be distinguished:

(1) Find the form of the most energetic initial disturbance. (Solve 20.2; see §§ 21—23.)

(2) Given the form of the most energetic initial disturbance, find the properties of its subsequent evolution. (See §§ 24—26.)

Both problems can be made simpler by first using the Fourier series for almost periodic functions to simplify (20.2). To facilitate this simplification, we introduce index notation $(x_1, x_2, x_3) = (x, y, z)$, $\mathbf{u} = (u_1, u_2, u_3) = (u, u_2, u_3)$ and $\mathbf{U} = (U(x_2, x_3), 0, 0)$ where $U(x_2, x_3)$ is the laminar Poiseuille flow (17.5).

We may write (20.2) as

$$\nu_{\mathscr{E}} = \max_{\mathbf{H}_x} \frac{-\langle u_1 u_\alpha \partial_\alpha U \rangle}{\langle |\partial_\beta u_i|^2 + |\partial_1 u_i|^2 \rangle} \tag{20.4}$$

where the summation convention is to hold; repeated indices are summed over their range; Latin scripts have range 1, 2, 3; Greek scripts have range 2,3. For example,

$$\partial_i u_i = \partial_1 u_1 + \partial_2 u_2 + \partial_3 u_3 ,$$

but

$$\partial_\alpha u_\alpha = \partial_2 u_2 + \partial_3 u_3 .$$

The following "normal mode" reduction of the problem (20.4) is a consequence of the fact that almost periodic functions have a Fourier series with mean square convergence. We shall prove that

$$\nu_{\mathscr{E}} = \max_\alpha \max_{\mathbf{H}_A} \mathscr{F}[\mathbf{u}; \alpha] = \max_\alpha \nu(\alpha) \tag{20.5 a}$$

where

$$\mathscr{F}[\mathbf{u}; \alpha] = \hat{\mathscr{I}}[\mathbf{u}]/\hat{\mathscr{D}}[\mathbf{u}; \alpha] = \frac{-\langle (u_1 \bar{u}_\beta + \bar{u}_1 u_\beta) \partial_\beta U \rangle_A}{2\langle |\partial_\beta u_i|^2 + \alpha^2 |u_i|^2 \rangle_A} , \tag{20.5 b}$$

α is any real number, the angle brackets with subscript A are defined by (19.2), and $\mathbf{H}_A$ is the set of complex-valued vector fields,

$$\mathbf{H}_A \equiv \{\mathbf{u} = \mathbf{u}(\mathbf{x}_2, \mathbf{x}_3) \colon \mathbf{u}|_{a,b} = 0, \ \partial_\beta u_\beta + i\alpha u_1 = 0\} .$$

To prove (20.5) we replace u_i in (20.4) with its Fourier series (Appendix A).

$$\mathbf{u} \sim \textstyle\sum_{-\infty}^{\infty} \mathbf{u}_l(x_2, x_3; \alpha_l) e^{i\alpha_l x_1} .$$

Since $\mathbf{u}(x_1, x_2, x_3)$ is real,

$$\bar{\mathbf{u}}_l = \mathbf{u}_{-l} , \qquad \alpha_l = -\alpha_{-l}$$

where the overbar designates the complex conjugate. This substitution leads to

$$\frac{-\langle u_1 u_\beta \partial_\beta U\rangle}{\langle|\partial_\beta u_i|^2+|\partial_1 u_i|^2\rangle}=\frac{\sum(\hat{\mathscr{I}}[\mathbf{u}_l]\hat{\mathscr{D}}[\mathbf{u}_l;\alpha_l]/\hat{\mathscr{D}}[\mathbf{u}_l;\alpha_l])}{\sum\hat{\mathscr{D}}[\mathbf{u}_l,\alpha_l]}$$

$$\leqslant \max_{\alpha_l}\max_{\mathbf{u}_l}\frac{\hat{\mathscr{I}}[\mathbf{u}_l]}{\hat{\mathscr{D}}[\mathbf{u}_l;\alpha_l]}\leqslant \max_\alpha \nu(\alpha)\,. \tag{20.6}$$

Hence,

$$\nu_{\mathscr{E}}\leqslant \max_\alpha \nu(\alpha)\,. \tag{20.7}$$

On the other hand, let $\mathbf{u}(x_2,x_3;\alpha)$ give $\mathscr{F}[\mathbf{u};\alpha]$ its maximum value $\nu(\alpha)$. Note that

$$\mathbf{u}e^{i\alpha x_1}+\bar{\mathbf{u}}e^{-i\alpha x_1}$$

is admissible as a competitor in $\mathbf{H}_x$ for the maximum $1/R$ of (20.4). Hence,

$$\max_\alpha \nu(\alpha)\leqslant \nu_{\mathscr{E}}\,. \tag{20.8}$$

Comparing (20.7) and (20.8) we prove (20.5).

The Fourier series reduction replaces the problem of maximizing with respect to the x-dependence of functions with the much simpler maximum problem (20.5) for the optimizing value of the number α. The relation (20.5) holds for Poiseuille flow in a straight pipe of arbitrary cross-section.

In the annulus, cylindrical polar coordinates are the preferred coordinate system. In these coordinates the azimuthal dependence of the functions

$$\mathbf{u}(x_2,x_3;\alpha)=\boldsymbol{\phi}(r,\theta;\alpha)=\textstyle\sum_{n=-\infty}^{\infty}\boldsymbol{\phi}_n(r;\alpha,n)\,e^{in\theta} \tag{20.9}$$

can again be removed by a reduction using a Fourier series. Then

$$\nu_{\mathscr{E}}=\max_{\pm n=0,1,2\ldots}\max_\alpha \nu(\alpha,n)\,. \tag{20.10}$$

It may be concluded from the reduction leading to (20.10) that the most energetic initial disturbance is always in the form of a "normal mode"

$$\mathbf{u}=\mathrm{re}\{e^{i(\alpha x+n\theta)}\boldsymbol{\phi}(r;\alpha,n)\}\,. \tag{20.11}$$

In §§ 22 and 23, we shall see that for Poiseuille flow in annular pipes, the most energetic initial disturbances has $\alpha=0$. Hagen-Poiseuille flow is an exception, but even in this case the initial disturbance with $\alpha=0$ nearly maximizes (20.2).

The normal mode (20.11) with $\alpha=0$ is a special disturbance in the class of kinematically admissible disturbances which are also independent of the axial coordinate x.

Exerise 20.1: Show that if $\nu > \nu_{\mathscr{E}}$ then

$$\mathscr{E}(t) \leqslant \mathscr{E}(0) \exp\{(\nu_{\mathscr{E}} - \nu)\hat{\Lambda} t/(b-a)^2\} \tag{20.12}$$

where

$$\frac{\hat{\Lambda}}{(b-a)^2} = \min_{\mathbf{H}_x} \hat{\mathfrak{d}}[\mathbf{u}] \quad \text{and} \quad \hat{\mathfrak{d}}[\mathbf{u}] = \langle|\nabla \mathbf{u}|^2\rangle / \langle|\mathbf{u}|^2\rangle .$$

Exercise 20.2 (Sorger, 1966): Consider the competition for the minimum of $\hat{\mathfrak{d}}[\mathbf{u}]$ for vectors of the form $\mathbf{u} = \mathbf{e}_\theta v(r)$ where $v(a) = v(b) = 0$. Show that $\hat{\mathfrak{d}}[\mathbf{e}_\theta v(r)] = \gamma_1^2$, where γ_1 is the smallest positive root of

$$J_1(\gamma a) Y_1(\gamma b) - J_1(\gamma b) Y_1(\gamma a) = 0 .$$

Prove that

$$\gamma_1^2 \geqslant \hat{\Lambda}/(b-a)^2 . \tag{20.13}$$

Exercise 20.3: Show that the average value of w on each cylinder of radius r vanishes when $\mathbf{u}$ is independent of x.

Exercise 20.4: Prove the formula which would be the analog for (20.5) when $\mathbf{u}$ is a square integrable function in an appropriate Fourier transform class.

§ 21. The Energy Eigenvalue Problem for Hagen-Poiseuille Flow

Laminar flow through a round pipe without an inner cylinder is called Hagen-Poiseuille flow (see (17.6)). In the limit $a \to 0$, $\mathscr{I} = 2U_m \langle rwu \rangle / b^2$ and we may write (20.2) as

$$\nu_{\mathscr{E}} = \frac{2U_m}{b^2} \max_{\mathbf{H}_x} \langle rwu \rangle / \langle|\nabla \mathbf{u}|^2\rangle = \nu R \max_{\mathbf{H}_x} 2\langle rwu \rangle / b^3 \langle|\nabla \mathbf{u}|^2\rangle \tag{21.1a}$$

where $R = U_m b/\nu$ and the functional $2\langle rwu \rangle / b^3 \langle|\nabla \mathbf{u}|^2\rangle$ is dimensionless. The following results can be obtained from analysis and numerical computations (Joseph and Carmi, 1969):

(i) $(2.405)^2 < \dfrac{\hat{\Lambda}}{b^2} < (3.83)^2$ (see Exercise 21.1);

where $\hat{\Lambda}$ is defined by (4.5) with l replaced by b.

(ii) $\nu_{\mathscr{E}} = U_m b / 81.49$;

(iii) The most energetic initial disturbance is a spiral mode $\alpha \neq 0$ with a first mode azimuthal periodicity ($n = 1$);

(iv) The energy of an x-independent disturbance with a first mode azimuthal periodicity ($\alpha = 0$, $n = 1$) will increase initially when $\nu > U_m b / 82.88$.

Given (i) and (ii), the inequality

$$\mathscr{E}(t) < \mathscr{E}(0) \exp\left\{\frac{-\nu}{b^2}(2.40)^2 (81.49 - U_m b/\nu) t\right\}$$

holds whenever $U_m b/\nu < 81.49$.

The proof of (i) follows directly from Exercise 21.1.

To obtain (ii), (iii) and (iv) we consider the Euler eigenvalue problem for (21.1a). This eigenvalue problem may be obtained by standard methods of the calculus of variations (see B.8 of Appendix B) in the form

$$-R\frac{r}{b}(\mathbf{e}_r u+\mathbf{e}_x w)=b^2\nabla^2\mathbf{u}-\nabla p\,, \qquad \mathbf{u}\in\mathbf{H}_x\,. \tag{21.1b}$$

Eqs. (21.1b) are next made dimensionless with b and U_m. Then, using normal modes (20.11) and eliminating the pressure and azimuthal component of velocity we obtain

$$Rn^2 f=i\alpha rDLf-(n^2+\alpha^2r^2)Lu-2\alpha^2rDu\,, \tag{21.1c}$$

and

$$R\{(n^2+\alpha^2r^2)u+i\alpha rDf\}=L^2f+4i\alpha Lu \tag{21.1d}$$

where $D=\dfrac{d}{dr}$ and $L=D^2+\dfrac{1}{r}D-\dfrac{n^2}{r^2}-\alpha^2$. In addition,

$$u=f=Df=0 \quad \text{at} \quad r=1 \quad \text{and } u \text{ and } f \text{ are bounded at} \quad r=0\,. \tag{21.1e}$$

In (21.1c, d, e) u is the dimensionless axial component of $\mathbf{u}=\boldsymbol{\phi}(r;\alpha,n)$ and $f=wr$ where w is the radial component of $\boldsymbol{\phi}(r;\alpha,n)$.

We seek $R_{\mathscr{E}}$ as the least positive eigenvalue $R(\alpha,n)$ of (21.1c, d, e) minimized with respect to the wave numbers α and n. The most energetic initial disturbance is an eigenfunction belonging to the least eigenvalue.

Boundary conditions for f, u and v can be found from kinematic conditions associated with the coordinate system (Batchelor and Gill, 1962). Suppose, without loss of generality, that at $r=0$ the velocity $\mathbf{u}=\mathbf{e}_x u_x+\mathbf{e}_t u_t$ has an axial component $\mathbf{e}_x u_x(0,\theta,x)$ and a transverse component

$$\mathbf{e}_t u_t(0,\theta,x)=\mathbf{e}_r u_r(0,\theta,x)+\mathbf{e}_\theta u_\theta(0,\theta,x)\,. \tag{21.2}$$

At $r=0$, u_t *and* u_x *are single-valued* and, therefore, independent of θ. It follows from (21.2) that when $r=0$, $u_r=u_t(x)\sin(\theta-\theta_0)$ and $u_\theta=u_t(x)\cos(\theta-\theta_0)$ where $0\leqslant\theta_0\leqslant 2\pi$ is arbitrary.

Now consider the Fourier representations

$$(u_r,u_\theta)\sim(w_n(r,x),\,v_n(r,x))\exp in\theta\,.$$

If $n\neq 1$, then the required forms for u_r, u_θ at $r=0$ imply that $w_n(0,x)=0$, $v_n(0,x)=0$. When $n=1$, it is possible to have $u_\theta\sim\cos(\theta-\theta_0)$ and $u_r\sim\sin(\theta-\theta_0)$ only if $w_1+iv_1=0$. This last condition is also implied by

$$0=\operatorname{div}\mathbf{u}=\frac{\partial w}{\partial r}+\frac{\partial u}{\partial x}+\frac{1}{r}\left(w+\frac{\partial v}{\partial\theta}\right)=0$$

at $r=0$. Of course $u_x\sim u_n(r,x)\exp in\theta$ is single-valued only when $n=0$.

In sum, at $r=0$

$$\begin{aligned} &u_n=0 \quad \text{if} \quad n\neq 0\,, \\ &w_n=v_n=f_n=Df_n=0 \quad \text{if} \quad n\neq 1\,, \\ &w_1+iv_1=f_1=0\,. \end{aligned} \tag{21.1f}$$

It follows that the only mode which does not imply that $\mathbf{u}_n=0$ at $r=0$ is the first ($n=1$). Since this is the only mode which is not constrained at $r=0$, it is always an important candidate for the most unstable mode.

Our problem now is to find the smallest positive eigenvalues $R(\alpha,n)$ of (21.1c, d, e or f). The eigenvalue $R(\alpha,n)$ is an even function of both of its variables and the lines $\alpha=0$ and $n=0$ locate local extrema of R. We seek the value

$$R_{\mathscr{E}}=\min_{\substack{\alpha>0\\ n=0,1,2,\ldots}} R(\alpha,n)=R(\tilde{\alpha},\tilde{n})\,. \tag{21.3}$$

Here we have used the fact that (α,n) symmetry implies that the search for $(\tilde{\alpha},\tilde{n})$ can proceed in the first quadrant of the (α,n) plane. An estimate like (44.24) shows that $R(\alpha,n)\to\infty$ as α^2+n^2 tends to zero or infinity. Hence, the $\min_{\alpha,n}(R(\alpha,n))$ is attained on a ray $\mathrm{Tan}^{-1}\alpha/n=\text{const}$ for $0<\alpha^2+n^2<\infty$.

On the ray $n=0$ one finds Orr's values[2]

$$180.6\cong\min_\alpha R(\alpha,0)\,,\qquad \alpha\cong 3.6\,. \tag{21.4}$$

On the ray $\alpha=0$ we may reduce (21.1c, d) to

$$L^2 f-Rn^2u=0\,, \tag{21.5a}$$

$$Lu+Rf=0 \tag{21.5b}$$

where f and u satisfy (21.1e). Elimination of u from this system leads to a sixth order system in f alone:

$$L^3 f+R^2n^2 f=0\,,\qquad f=Df=L^2 f\big|_{r=1}=0\,, \tag{21.6}$$

and f and its derivatives are bounded at $r=0$. The solution of (21.6) is

$$f(r)=A_1 J_n(\lambda_1^{1/2}r)+A_2 J_n(\lambda_2^{1/2}r)+\bar{A}_2 J_n(\lambda_3^{1/2}r)$$

where J_n is a Bessel function and

$$\lambda_1=(R^2n^2)^{1/3}\,,\qquad \lambda_2=\lambda_1 e^{-i(2/3)\pi}\,,\qquad \lambda_3=\lambda_1 e^{i(2/3)\pi}\,.$$

---

[2] The values given by (21.4) were obtained using a computer and forward integration from starting values found by the methods of Frobenius. Orr (1907) used the same method and, by hand, calculated the values 180 and $\alpha\cong 3.7$.

The overbar means complex conjugate. We normalize by choosing $A_1=1$. Then, fixing n, we form three homogeneous equations from the boundary conditions at $r=1$. These equations determine values of A_2 and R. We find that

$$R(0,1)=\min_{n=0,1,2,\ldots} R(0,n)=82.88\,, \tag{21.7}$$

and with $n=1$,

$$A_2=0.00301-i0.02455\,.$$

Since the rays $\alpha=0$ and $n=0$ are loci of stationary values of $R(\alpha,n)$, it could be conjectured that $R(0,1)=R(\tilde{\alpha},\tilde{n})$ solves (21.3). The true situation is more delicate; in fact if we allow n to take on continuous values and put $n=\varepsilon$ for small ε, it is easy to show that $R(\alpha,\varepsilon)$ is not a minimum when $\alpha=0$. In fact for very small ε, $R(\alpha,\varepsilon)$ is approximated by $R(\alpha,0)$ which, as Orr has shown (see (21.4)), is minimum when $\alpha=3.6$.

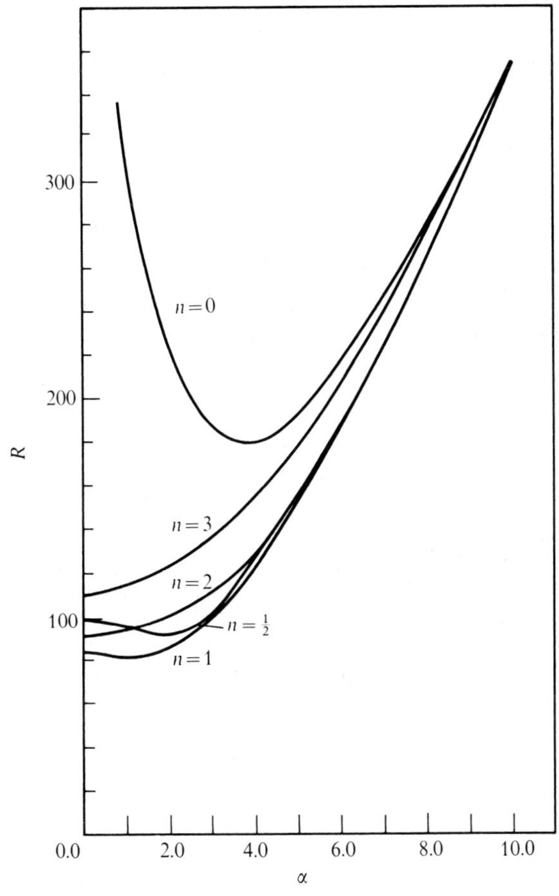

Fig. 21.1: Principal eigenvalues $R(\alpha,n)$ of Eqs. (21.1) (Joseph and Carmi, 1969)

By numerical calculation we find that among the integers n

$$R_{\mathscr{E}}=81.49, \quad (\tilde{\alpha},\tilde{n})=(1.07,1). \tag{21.8}$$

We should also note that when $n\geqslant 2$

$$\min_{\alpha\geqslant 0} R(\alpha,n)=R(0,n).$$

This means that the minimizing disturbances for $n\geqslant 2$ are all independent of the axial coordinate x (see Fig. 21.1).

The figure summarizes the results of a numerical integration of the equations. The smallest of values $R(\alpha,n)$ is 81.49. For each fixed n, the values $R(\alpha,n)\to R(\alpha,0)$ as $\alpha\to\infty$ (Orr's solution) (Joseph and Carmi, 1969).

Exercise 21.1: Show that in the region between concentric cylinders

$$\langle|\nabla\mathbf{u}|^2\rangle \geqslant \left\langle\left(\frac{\partial\mathbf{u}}{\partial r}\right)^2+\left(\frac{\partial\mathbf{u}}{\partial z}\right)^2+\frac{1}{r^2}\left(\frac{\partial\mathbf{u}}{\partial\theta}\right)^2\right\rangle$$

and that

$$\mathfrak{d}[\mathbf{u}] \geqslant \frac{\left\langle\left(\dfrac{\partial w}{\partial r}\right)^2\right\rangle}{\langle w^2\rangle} \geqslant \gamma_0^2,$$

where γ_0 is the smallest positive root of

$$J_0(\gamma a)\,Y_0(\gamma b)-J_0(\gamma b)\,Y_0(\gamma a)=0.$$

Show that when there is no inner cylinder $(a=0)$

$$(2.405)^2\leqslant\gamma_0^2\leqslant(3.83)^2. \tag{20.14}$$

The inequalities in (20.14) are actually strict (see Sorger, 1966).

§ 22. The Energy Eigenvalue Problem for Poiseuille Flow between Concentric Cylinders

(a) Parabolic Poiseuille Flow

The parabolic distribution of velocity of Hagen-Poiseuille flow is not altered by the addition of a concentric inner cylinder provided that the axial sliding speed of this cylinder is appropriately chosen:

$$U(r)=\frac{U_m}{b^2-a^2}(b^2-r^2). \tag{22.1}$$

The flow (22.1) is called *parabolic Poiseuille* flow. The energy analysis of this flow follows the one given in § 21 for the case $a=0$ except that we require that

$$\mathbf{u}\big|_{r=\tilde{a},\tilde{b}}=0\,.$$

The Euler equations for the dimensionless version of this problem (with $b-a$ as the length scale, U_m as the velocity scale and $R=U_m(b-a)/\nu$) are again (21.1c) and (21.1d). The boundary conditions

$$f=Df=u\big|_{a,b}=0 \tag{22.2}$$

are now applied at

$$\tilde{a}=\eta/(1-\eta), \qquad \tilde{b}=1/(1-\eta), \qquad \eta=a/b\,.$$

An exact solution of (21.5a, b) and (22.2) can be constructed as a linear combination of Bessel functions.

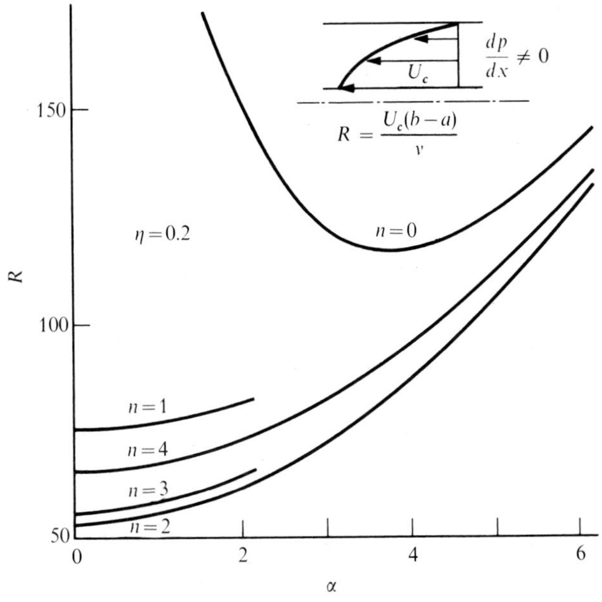

Fig. 22.1: Eigenvalues $R(\alpha,n)$ for parabolic Poiseuille flow through concentric cylinders with $\eta=0.2$ (Joseph and Munson, 1970).

Numerical analysis summarized in Fig. 22.1 indicates that the critical energy disturbance for parabolic Poiseuille flow is x-independent; this is proved in § 48 for the limiting case $\eta\to 1$. The critical energy disturbance is not x-independent for Hagen-Poiseuille flow ($\eta=0$). The relation between solutions of the stability problem for parabolic Poiseuille flow in the limit $\eta\to 0$ to solutions of the stability problem for Hagen-Poiseuille flow ($\eta=0$) is not yet known.

(b) Poiseuille Flow in an Annular Pipe

The energy stability limit for this problem is defined by (20.2) and (20.10). The Euler differential equations for the functional $R(\alpha,n)=U_m(b-a)/\nu(\alpha,n)$ are (21.1c,d) with R replaced with $-RDU/2r$ where

$$DU=-\{2r(1-\eta)^2+(1-\eta^2)/r\ln\eta\}/\{1+(1-\eta^2)/\ln(\eta)^2 \\ +(\eta^2-1)\ln[(\eta^2-1)/\ln(\eta^2)]/\ln(\eta^2)\}\,. \quad (22.3)$$ [3]

For $\alpha=0$ we are led, as before, to the system

$$L^2 f+R\frac{DU}{2r}n^2u=0,\qquad Lu-\frac{DU}{2r}Rf=0 \quad (22.4\,\text{a,b})$$

with $u=f=Df=0$ at the boundaries. Numerical analysis (Table 22.1) again indicates that the critical energy disturbance is x-independent.

Numerical integration shows that the minimum eigenvalue occurs with $\alpha=0$ and $n=2$ (Joseph and Munson, 1970).

Table 22.1: Energy stability limits (Joseph and Carmi, 1969). Critical Reynolds numbers for Poiseuille flow in an annular pipe. The eigenvalue $R(0,n;\eta)$ tends toward the pipe value 82.88 as η is decreased. For $\eta\geqslant 10^{-4}$ the smallest value of $R(\alpha,n,\eta)$ over n and α is associated with purely azimuthal solutions ($\alpha=0$). Primed values in the table mean that the search for minimum values was carried out relative to continuous n

η	R	α	n
0.0001	83.80	0	1.8′
0.001	85.10	0	1.8′
0.010	93.23	0	1.9′
0.050	98.02	0	2
0.200	100.35	0	3
0.500	99.63	0	6
0.625	99.50	0	9
0.700	99.43	0	12
0.900	99.21	0	39
1.000	99.21	0	a″

″ $\lim_{\eta\to 1}\left(\frac{n}{1-\eta}\right)=4.088$ (Busse, 1969A).

§ 23. Energy Eigenfunctions—an Application of the Theory of Oscillation Kernels

It is easy to compute the principal eigenfunctions of the problems considered in § 22 by numerical methods. It will be recalled that the principal eigenfunction of energy theory gives the form of the disturbance whose energy increases in-

[3] The Euler equations for $R(\alpha,n)$ can also be obtained from the spiral flow equations (44.12) when the pipe walls do not move ($\mathbf{V}(r)\equiv 0$).

itially at the smallest value of R. The main observation is that the principal eigenfunctions f and u are of one sign when DU is of one sign for $\tilde{a}\leqslant r\leqslant\tilde{b}$; on the other hand, for Poiseuille flow in annular pipes DU (Eq. 22.3) changes sign on $\tilde{a}\leqslant r\leqslant\tilde{b}$ and so does the energy eigenfunction u (see Fig. 23.1).

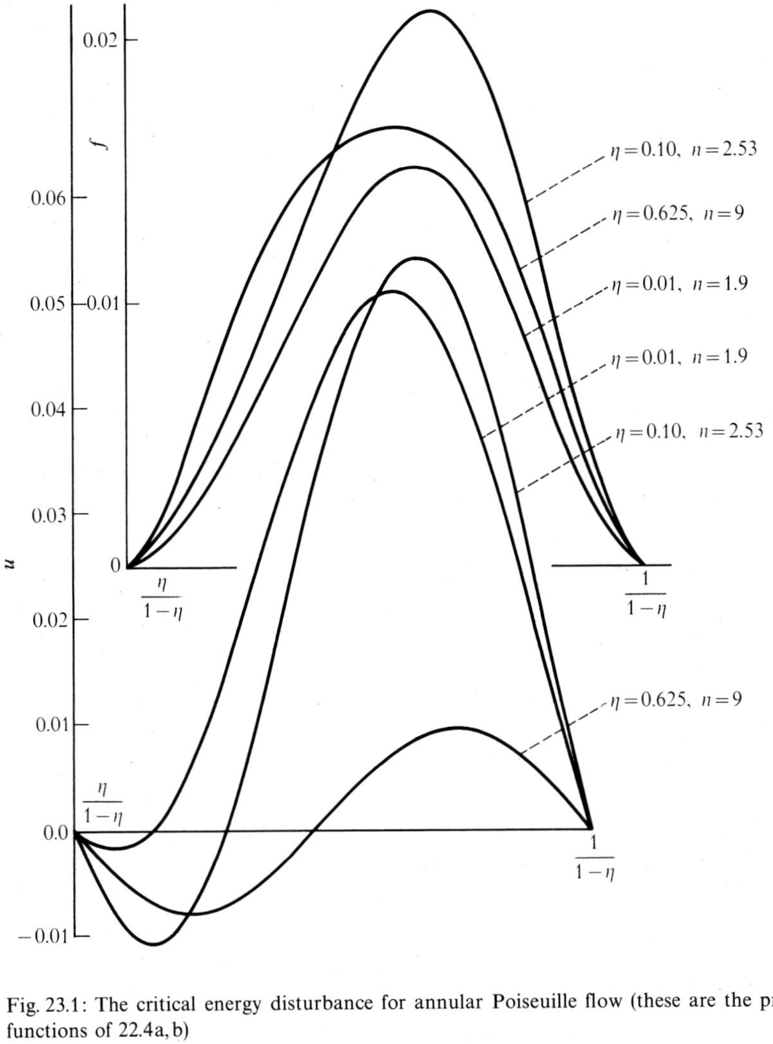

Fig. 23.1: The critical energy disturbance for annular Poiseuille flow (these are the principal eigenfunctions of 22.4a, b)

It is interesting to compare the eigenvalue problems (21.5a, b) for Hagen-Poiseuille flow (see Eq. 21.1c) and parabolic Poiseuille flow (see Eq. 22.2) with the problem (22.4a, b). The nodal properties of (21.5a, b) with boundary conditions (21.1e) or (22.2) can be completely characterized by the theory of oscillation kernels.

Consider problem (21.5a, b) relative to the boundary condition (22.2). This system can be converted into a pair of integral equations

$$u = R\int_{\tilde{a}}^{\tilde{b}} G(r,r_0;n)\, f(r_0)\, r_0 dr_0\,, \qquad f = -Rn^2\int_{\tilde{a}}^{\tilde{b}} H(r,r_0;n)\, u(r_0)\, r_0 dr_0 \tag{23.1a, b}$$

where G and H are Green functions for the problems

$$LG = \delta(r-r_0)/r \quad \text{and} \quad L^2 H = \delta(r-r_0)/r$$

where G, H, DH vanish at $r=\tilde{a}, \tilde{b}$, and $\delta(r-r_0)$ is the delta function. The pair of Eqs. (23.1a, b) can be converted into a single integral equation for u alone or for f alone. For example,

$$\begin{aligned} f &= -R^2 n^2 \int_{\tilde{a}}^{\tilde{b}} f(r_0) \left\{ \int_{\tilde{a}}^{\tilde{b}} H(r,\tau;n)\, G(\tau,r_0;n)\, \tau\, d\tau \right\} r_0 dr_0 \\ &\equiv R^2 \int_{\tilde{a}}^{\tilde{b}} f(r_0)\, K(r,r_0;n)\, r_0 dr_0\,. \end{aligned} \tag{23.2}$$

The operators L and L^2 can each be written as a repeated derivative with positive weights; for example,

$$Lu = \rho_0 D\{\rho_1 D(\rho_2 u)\}\,, \qquad \rho_0 = r^{n-1}\,, \qquad \rho_1 = r^{1-2n}\,, \qquad \rho_2 = r^n\,.$$

This fact and the boundary conditions are enough to establish that G and H are oscillatory Green functions (see Appendix D). The composition K of oscillatory Green functions is again an oscillatory kernel and this implies a positivity lemma:

There is a positive least eigenvalue R^2 *of* (23.2) *for each fixed integer* $n=0, \pm 1, \ldots$. *This eigenvalue is simple. The associated eigenfunction is one signed in* $(\tilde{a}, \tilde{b})$. *The eigenfunction* u *belonging to* $R>0$ *is of the same sign. There are no other one-signed eigensolutions for eigenvalues* $R>0$.

This same theorem applies to the energy eigenvalue problem (21.5a, b) relative to the boundary conditions (21.1e). This problem, which governs Hagen-Poiseuille flow, is a singular case in the theory of oscillation kernels[4].

It is obvious from Fig. 23.1 that the critical disturbance u of Poiseuille flow in an annular pipe is not of one sign. This disturbance satisfies an integral equation of the form (23.2) with a kernel given by a composition of oscillatory Green functions. However, the oscillation kernel theory will not apply here because $DU(r)$ given by (22.3) changes sign (see Exercise 23.1).

Exercise 23.1: Formulate problem (22.4a, b) as a single integral equation. Show that the kernel of this integral equation is not positive.

---

[4] The theory of oscillation kernels has not yet been extended to cover singular operators and does not apply directly in the present case. To apply the theory indirectly, we note that as $r\to 0$, $G(r,r_0)\to C_1 r^n$, $H(r,r_0)\to C_2 r^n + C_3 r^{n+2}$. Using these relations as boundary conditions at $r=0$, we may show (as in § 8, Chapter III, Gantmacher and Kreĭn, 1960), by direct integration of the iterated derivative form of L and L^2 that G and H are "influence functions", hence oscillation kernels.

Exercise 23.2: Show that the Green functions G and H corresponding to the problem (21.5a, b) and (21.1e) are given by

$$G = -(2n)^{-1}\{(rr_0)^n - (r_0/r)^n\} \quad \text{for} \quad r > r_0,$$

$$H = (8n(n^2-1))^{-1}\{(n-1)(rr_0)^n r^2[nr_0^2-(n+1)] + (n-1)(r_0/r)^n r_0^2$$
$$+ (n+1)(r_0 r)^n[(1-n)r_0^2+n] - (n+1)(r_0/r)^n r^2\} \quad \text{for} \quad r > r_0,$$

and

$$G(r,r_0) = G(r_0,r), \qquad H(r,r_0) = H(r_0,r).$$

§ 24. On the Absolute and Global Stability of Poiseuille Flow to Disturbances which are Independent of the Axial Coordinate

In the three previous sections we showed that the most energetic initial field frequently takes form as an x_1-independent disturbance. Now, following the plan mentioned in § 20, we wish to study the evolution of these special initial fields[5].

First we shall examine the values which $d\mathscr{E}/dt$ can take initially when the initial disturbance is x_1-independent

$$u_i(x_1, x_2, x_3, t=0) = g_i(x_2, x_3).$$

We may write the energy equations (20.1) as

$$\frac{1}{\mathscr{D}}\frac{d\mathscr{E}}{dt} = \hat{\mathscr{F}}[u_i] - \nu \tag{24.1}$$

where $\hat{\mathscr{F}}[\mathbf{u}] = \mathscr{F}[\mathbf{u}, 0]$ is defined by (20.5b). Consider the range of $\hat{\mathscr{F}}$ when its domain is restricted to x_1-independent disturbances.

Given any kinematically admissible x_1-independent disturbance $\mathbf{g} = (g_1, g_2, g_3)$, we have

$$\hat{\mathscr{F}}[g_1, g_2, g_3] = -\hat{\mathscr{F}}[-g_1, g_2, g_3] = -\hat{\mathscr{F}}[g_1, -g_2, -g_3]. \tag{24.2}$$

Moreover

$$\hat{\nu}_{\mathscr{E}} = \max_{g_i} \hat{\mathscr{F}} < \infty \tag{24.3}$$

and

$$-\hat{\nu}_{\mathscr{E}} \leq \hat{\mathscr{F}}[g_i] \leq \hat{\nu}_{\mathscr{E}}. \tag{24.4}$$

The symmetry property (24.2) follows from the fact that the indicated sign changes do not alter the conditions of admissibility (since, in any case, $\partial_\alpha g_\alpha = 0$)

---

[5] The material in §§ 24 and 25 is taken from the paper of Joseph and Hung (1971).

nor do they change the value of $\mathscr{D}$. Hence, sign$\hat{\mathscr{F}}$ = sign$\mathscr{I}$ and $\mathscr{I}$ changes sign as in (24.2). The rest of the proof is left as an exercise.

Either the energy of any given x_1-independent disturbance (g_1, g_2, g_3) or its mate $(-g_1, g_2, g_3)$ will increase initially if the viscosity is sufficiently large. To prove this choose $\mathbf{g}$ so that $\hat{\mathscr{F}} > 0$, then inspect (24.1).

The x_1-independent disturbance which makes the energy $\mathscr{E}$ increase initially in the fluid with the largest viscosity is the vector field which solves (24.3).

We shall now demonstrate that an x_1-independent disturbance is basically a transient structure which must eventually decay. But at short times the energy of the transverse motion can decay at a rate smaller than the increase in the energy of the axial motion.

To establish the above result, we consider the IBVP for x_1-independent disturbances of Poiseuille flow $U_1(x_2, x_3)$:

$$\frac{\partial u_1}{\partial t} + u_\alpha \partial_\alpha [u_1 + U_1] = \nu \partial_{\alpha\alpha}^2 u_1 , \tag{24.5 a}$$

$$\frac{\partial u_\alpha}{\partial t} + u_\beta \partial_\beta u_\alpha = -\partial_\alpha \pi + \nu \partial_{\beta\beta}^2 u_\alpha , \tag{24.5 b}$$

$$\partial_\alpha u_\alpha = 0, \quad u_i = 0 \quad \text{at} \quad r = a, b \tag{24.5 c, d}$$

and

$$u_i(0, x_2, x_3) = g_i(x_2, x_3) \quad \text{at} \quad t = 0 . \tag{24.5 e}$$

An energy equation for the transverse components of velocity can be formed using (24.5 b, c, d). In the usual way,

$$\tfrac{1}{2} \frac{d}{dt} \langle u_2^2 + u_3^2 \rangle = -\nu \langle |\partial_\beta u_2|^2 + |\partial_\beta u_3|^2 \rangle . \tag{24.6}$$

It follows from the usual energy estimates that

$$\langle u_2^2 + u_3^2 \rangle \leq \langle g_2^2 + g_3^2 \rangle \exp\left\{ \frac{-2\nu \Lambda}{(b-a)^2} t \right\} \tag{24.7}$$

where

$$\langle |\partial_\beta \theta|^2 \rangle \geq \frac{\Lambda}{(b-a)^2} \langle \theta^2 \rangle$$

for admissible θ such that $\theta = 0$ at $r = a, b$.

Eq. (24.7) proves that the *transverse velocity components of an x_1-independent disturbance must decay monotonically from the initial instant* (Joseph and Tao, 1963).

If the transverse components of the disturbances must decay independent of the size of the disturbance or the viscosity, it would be reasonable to expect an identical behavior for the axial component. There are, however, some interesting differences which we shall now delimit.

Consider the total energy of an x_1-independent disturbance. We have, from the assumption that x_1 derivatives vanish, that

$$\tfrac{1}{2}\frac{d}{dt}\langle u_1^2+u_2^2+u_3^2\rangle=-\langle u_1 u_\alpha \partial_\alpha U_1\rangle-\nu\langle|\partial_\alpha u_i|^2\rangle\,,$$

and on taking account of (24.6) and (24.7)

$$\begin{aligned}\tfrac{1}{2}\frac{d}{dt}\langle u_1^2\rangle&=-\langle u_1 u_\alpha \partial_\alpha U_1\rangle-\nu\langle|\partial_\alpha u_1|^2\rangle\\&\leqq K\langle u_1^2\rangle^{1/2}\langle u_\alpha^2\rangle^{1/2}-\frac{\nu\Lambda}{(b-a)^2}\langle u_1^2\rangle\\&\leqq K\langle u_1^2\rangle^{1/2}\langle g_2^2+g_3^2\rangle^{1/2}\exp\left\{\frac{-\nu\Lambda t}{(b-a)^2}\right\}-\frac{\nu\Lambda}{(b-a)^2}\langle u_1^2\rangle\,,\end{aligned}$$

where K is the largest of two values $(\partial_2 U_1, \partial_3 U_1)$ up to time t. Introduction of the variable $\hat{t}=\nu\Lambda t/(b-a)^2$ and $\delta=\langle u_1^2\rangle^{1/2}$ leads us to

$$\frac{d\delta}{d\hat{t}}\leqslant C_1\exp(-\hat{t})-\delta\,,$$

where C_1 is a constant. This inequality can be integrated forward from time $t=0$ to give

$$\int_0^{\hat{t}}\exp(t')\left[\frac{d\delta}{dt'}+\delta\right]dt'=\int_0^{\hat{t}}\frac{d}{dt'}(\delta\exp(t'))\,dt\leqslant C_1\hat{t}\,.$$

Hence,

$$\delta(t)\leqslant\delta(0)\,e^{-\hat{t}}+C_1\hat{t}\,e^{-\hat{t}}\,. \tag{24.8}$$

Eqs. (24.7) and (24.8) show that

$$\langle u_1^2+u_2^2+u_3^2\rangle<\tilde{C}_1^2\,\hat{t}^2 e^{-2\hat{t}}$$

at large t and this is true for all viscosities and every initial value. *Hence, Poiseuille flow is absolutely and globally stable to x_1-independent disturbances.*

Exercise 24.1: Prove that any steady parallel flow solution $\mathbf{U}(x_2,x_3)=(U(x_2,x_2),0,0)$ of (1.1) is absolutely and globally stable to x_1-independent disturbances.

§ 25. On the Growth, at Early Times, of the Energy of the Axial Component of Velocity

Now we shall restrict our considerations to the set of x_1-independent disturbances whose energy increases. We showed in § 24 that if ν is small, every x_1-independent disturbance has an initially increasing energy in the sense that if the energy of (g_1, g_2, g_3) decreases, the energy of $(-g_1, g_2, g_3)$ increases.

The point to be made about these initially increasing disturbances is that the energy of their transverse components must *always* decrease. Hence, the energy of the axial part of the disturbance must increase even more rapidly than the decrease in the transverse energy.

The evolution of $\langle u_1^2 \rangle = \delta^2$ is already fairly well represented by the estimate (24.8). The function on the right of (24.8) first increases linearly with t and then decreases.

It is more revealing to note that at $t=0$, (24.1) and (24.6) combine into

$$\tfrac{1}{2}\frac{d}{dt}\langle u_1^2 \rangle = (\hat{\mathscr{F}}[g_i] - \nu)\langle |\partial_\alpha g_1|^2 \rangle + \hat{\mathscr{F}}[g_i]\langle |\partial_\alpha g_2|^2 + |\partial_\alpha g_3|^2 \rangle . \tag{25.1}$$

When $\nu < \hat{\mathscr{F}}[g_i]$ (when the Reynolds number is large),

$$\tfrac{1}{2}\frac{d}{dt}\langle u_1^2 \rangle \geqslant \nu \langle |\partial_\alpha g_2|^2 + |\partial_\alpha g_3|^2 \rangle = -\tfrac{1}{2}\frac{d}{dt}\langle u_2^2 + u_3^2 \rangle \tag{25.2}$$

at $t=0$.

The decay of the transverse motion strongly stimulates the growth of the axial motion at early times. The largest rate of growth of the axial motion is obtained for the initial condition which maximizes $\hat{\mathscr{F}}\,(=\nu_{\mathscr{E}})$.

Stability theorem for x_1-independent disturbances:

Parallel flow with prescribed boundary values is globally and absolutely stable to disturbances of the form (24.5e). *The energy of the transverse components of such a disturbance must decay monotonically and satisfies the estimate*

$$\tfrac{1}{2}\langle u_2^2 + u_3^2 \rangle \leqslant \tfrac{1}{2}\langle g_2^2 + g_3^2 \rangle \exp\left\{ -\frac{2\nu\hat{\Lambda}}{l^2} t \right\}. \tag{25.3}$$

The axial component of this same disturbance can increase initially and at time t satisfies the estimate

$$\tfrac{1}{2}\langle u_1^2 \rangle < \tfrac{1}{2}\{\langle u_1^2 \rangle_{t=0}^{1/2} + ct\}^2 \exp\left\{ -\frac{2\nu\hat{\Lambda}}{l^2} t \right\}. \tag{25.4}$$

If $\nu_{\mathscr{E}} > \nu$, initially increasing disturbances exist. The axial energy of such increasing disturbances increases faster than the rate of decay of the energy of the transverse motion.

The first part of this theorem could be considered as a kind of nonlinear Squire's theorem (see Exercise 34.2). It eliminates the possibility that a motion of the form (24.5e) could permanently replace laminar Poiseuille flow.

The x_1-independent disturbance is an efficient motion for the transport of disturbance momentum, but in rectilinear shear flows it has no dynamic source to drive it and it decays.

In rotating flows there are forces associated with rotation which can maintain disturbances in the form of spiral vortices. These vortices do not vary along certain spiral lines and they are analogous to x_1-independent disturbances. The most reknowned of the spiral vortices are the Taylor vortices shown in Figs. 36.1. Spiral vortices associated with rotating Poiseuille flow are shown in Figs. 46.3, 4, 5.

§ 26. How Fast Does a Stable Disturbance Decay?

We shall close this chapter with a discussion of physical properties of stable disturbances whose energy increases at finite times.

The qualitative question taken up here is this: Does there exist a sharp transition, as the Reynolds number is decreased, from slow to rapid decay in the law of decay of stable disturbances? In other words, do Reynolds numbers of moderate size (say, 100 in channels) exist below which even very energetic disturbances will be immediately "puffed" out?

The question is suggested by considerations which come out of observations of the decay of large arbitrary disturbances in the entrance region of pipes and channels[6].

Let us consider Hagen-Poiseuille flow. We found, in § 21, that when $R > 81.49$, there are initial values for the disturbance such that, even if the average energy of the disturbance eventually decays, it will first increase.

On the other hand, when $R > 81.49$, the decay of some disturbances (those with large values of the vorticity, see Exercise 3.1) is arbitrarily rapid, and the energy of any disturbance decays at least as fast as

$$\tfrac{1}{2}\langle|\mathbf{u}_0|^2\rangle \exp\left\{-\frac{\nu}{b^2}(2.40)^2\left(81.49-\frac{U_m b}{\nu}\right)t\right\}.$$

What we have said seems to have some experimental justification in the work of Davies and White (1928), Naumann (1931) and R. Lindgren (1957, 1959), as well as in the earlier experiments of Grindley and Gibson (1908) and Carothers (1912). For example, Lindgren (p. 145) observes about his experiments in tubes

[6] Mention may be made here of various entrance flow studies whose aim is the prediction of the development length in pipes and channels of a profile which starts at $x=0$ as plug flow. The energy of the difference between Poiseuille and plug flow with the same mass flow is not very large, nor are the inertial nonlinearities strongly stimulated by this special initial profile. For these reasons, such studies are not central to our main concern.

Fig. 26.1: Photographs of water flow which emerges from a narrow slit (width = 6 mm) at the entrance of a round pipe (radius $b = 13.5$ mm). Panel *a* (on the left) is at the top. At the bottom of panel *a* is a support for the pipe. The seven panels are at different downstream observation stations. The flow enters from the top of panel furthest to the left. Here $R = U_m b/\nu \simeq 250$, and the flow is smooth throughout the pipe (Lindgren, 1959)

Fig. 26.2: This is as in Fig. 26.1, but $R = 1020$. The entrance flow is very agitated. Lindgren characterizes his observations at ten values of R as follows: 130—250, purely laminar flow; 510, pre-transition region; 1020, disturbed entrance flow; 1930, fading turbulent patches (spots) and flashes (slugs); 2570, transition region; 2750—2870, self-maintaining turbulent flow. At $R = 510$, the flow is like that shown here except the entrance length is shorter and somewhat less agitated (Lindgren, 1959)

that, "Depending on the disturbances level at the tube inlet the disturbance vortices occur more or less readily but ... they will not appear in the tube, however high the disturbance level might be, as long as the Reynolds number R of the flow is less than the ... (experimental) value 200". In Figs. 26.1 and 26.2, we have reproduced Lindgren's (1959) photographs of the agitated inlet at $R=250$ and $R=1020$. The theoretical value 81.49 is, of course, less than observed values (above about 200—275) at which the large disturbances are immediately puffed out.

Similar remarks are appropriate in interpreting some observations made in the (1928) experiments of Davies and White on the stability of laminar flow in channels. These observations are summarized in Fig. 26.3 in which we have (essentially) reproduced Fig. 4 of the Davies and White paper. Fig. 26.3 shows the variation of the length of pipe over which an entrance disturbance persists.

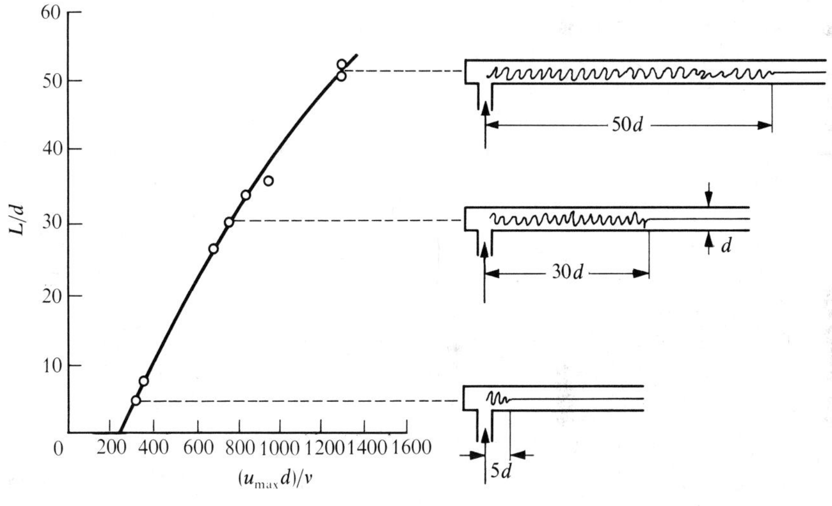

Fig. 26.3: The length of a channel over which a stable disturbance persists. The experimental apparatus is a rectangular channel of large aspect ratio. The flow is disturbed at the entrance by turning at a corner (as shown). The $R=210$ intercept is an extrapolation of the data. Theoretically, one expects the curve to pass through the origin or even to have $R=0$, $L>0$. The experiments suggests that there is a sharp change in the law of decay. The kind of disturbance which is generated at the entrance undergoes a change from slow decay to fast decay in the neighborhood of $R=200$ (Davies and White, 1928)

Davies and White make the following remark (p. 99) about the line they have drawn through the experimental points:

"The positive value of R, at which a backward extension of the curve cuts the axis of $L/2b$, suggests that, even with zero entrant length, truly viscous conditions will obtain over a certain range of velocity. In other words, it indicates that whatever may be the pre-entrant conditions, eddies

cannot be transmitted to, or exist in, the channel when the Reynolds number is less than the value given by the intersection, namely, 210. This may be a critical value in connection with viscous flow.
"In view of the importance of such a conclusion ..." [7].

The "backward extension of the curve" mentioned above is completely unacceptable from a theoretical point of view. It must be true that there is a finite development region at the entrance of the pipe for $R \geq 0$. But the experiment does show that the law of decay governing stable disturbances in the Davies-White channel does undergo a sharp change in a range of $R < 210$. These results are in agreement with later findings of Naumann (1931) and Lindgren (1957, 1959) on transition in round pipes.

Unfortunately, this comparison of entrance flow observations with the criterion $R < 81.49$ for strong stability must necessarily be tentative, since the criterion holds for the extended problem on the infinite domain and guarantees decay of the disturbance energy in time. The real problem set for $x \geq 0$ is a problem of decay in x, though time also enters. Nevertheless, there is some basis for comparing theory and experiments. In view of the restricted set of initial (entrance) conditions available in experiments and the intrinsic difference between conditions of the experiments and assumptions of the theory, the difference between theory ($R = 81.49$) and experiment ($R \cong 210$) is not large and is in good accord with one's broad understanding of the criterion. If $R > 81.49$ (in the infinite pipe), then there are initial conditions such that $\mathscr{E}(t)$ may increase initially, We think of an eddy moving downstream. Eventually it decays, but early in its history it may actually become more energetic. The presence of such eddies could certainly lead to large increases in the entrance length well beyond what would be possible with $R < 81.49$, when all possible eddies would decay right from the start.

The fact that the energy of a stable disturbance can increase seems consistent with observations of turbulent "flashes" which were already described by Reynolds (1883). R. Lindgren (1957, p. 33) describes the appearance of the flashes as follows:

"In spite of the disturbance plate at the tube entrance, visual inspections have shown purely laminar flow even close to the tube inlet, provided Reynolds numbers of the order $R \cong 200$ were not exceeded. Above this limit there appears a region of generally disturbed flow extending some distance from the inlet and in which, now and then, is observed the whiplike formation of turbulent flashes

---

[7] It is of interest that Davies and White thought this entrance flow result was their most important discovery. They stress it more than their more frequently cited transition measurement. Below the transition limit (about 2100), "... there is evidence, however of a distinct deviation from true viscous flow if initial disturbing factors are present, and the influence of such disturbing factors does not disappear entirely until a second well-defined (energy) limit is reached, which has the value of about one tenth of (2100). It would appear that below this limit eddies do not exist at any point in the pipe, and the flow is truly viscous. The suggestion is accordingly made that there may be three distinct types of flow: (a) one in which eddies cannot exist, corresponding to truly viscous flow; (b) one in which eddies may exist, due to an initial disturbance, but cannot be sustained in the pipe, the initial eddies therefore ultimately disappearing; and (c) one in which eddies once generated will be maintained without decrement throughout the pipe, corresponding to truly turbulent flow."

The divisions (a), (b) and (c) to which Davies and White were led by their experiments correspond to flows which we have called.

(a) globally and monotonically stable (in pipes $R < 81.49$),
(b) globally but not monotonically stable, ($R < R_G$) and
(c) unstable to large disturbances ($R > R_G$), where R_G is approximately 2000.

possessing turbulent structure, quite different from the rather "soft" fluctuations of the disturbed entrance flow. Increasing Reynolds numbers cause an increase of the flash frequency and extend the disturbed flow region further downstream ..."

We close this section with a remark about the frictional resistance of pipes to flow of fluid when $81.49<R<2000$. It is well known that the frictional resistance (shear stress at the wall) to such stable laminar flows should vary with the first power of the mean (over a cross section) velocity with a small correction for the effect of the entrance region. On the other hand, for larger values of R, the resistance varies with the m-th power of the mean velocity, and the range $1.7<m\leqslant 2$ is frequently cited.

The resistance in a disturbed region at the entrance could be expected to vary with the m-th power of the mean velocity with $1\leqslant m<1.7$, and such modest increases in flow resistance are observed in the experiments of Carothers (1912)[8].

Evidently no special efforts were made to have a smooth inner wall at places where pipes were jointed together, so that to a degree a new entrance is to be found at each joint. For the jointed pipes, Carothers finds a linear (laminar) law of resistance when $R<200$, and when $R>200$, there is an increase in the resistance which is roughly proportional to the mean velocity raised to the power $m=1.2$.

[8] The working fluid in these experiments was Texas oil and the experimental apparatus was on a grand scale. The largest pipe had a 10-inch diameter and was jointed in sections averaging 15 feet in length, by a collar of 6-inch length with 10.5-inch internal diameter. These "experiments" stimulate nostalgia for the age of abundance when "the oil was merely allowed to flow by gravitation from the large storage tanks (2000 gallons capacity) on the hillside to tank steamers in the harbor".

Chapter IV

Friction Factor Response Curves for Flow through Annular Ducts

In Chapter III we studied the globally stable disturbances of Poiseuille flow in the annulus whose energy could increase initially. Now we turn to the study of unstable disturbances; we seek a description in general terms of the permanent solutions which replace laminar Poiseuille flow when the Reynolds number is sufficiently large. This description is conveniently stated in terms of a response function.

§ 27. Response Functions and Response Functionals

A response function for a fluid motion can be defined as a scalar function which measures the response of the flow to the external forces which induce the motion. For example, in problems of thermal convection, which are treated in Part II, the response function can be taken as the heat transported and the external forces can be regarded as the applied temperature difference. The dimensionless response function relates the Nusselt and Rayleigh numbers. In the example which is to be considered below, flow through an annular pipe, the external force is the pressure gradient and the response function is the mass flux. The dimensionless response function relates friction factors and Reynolds numbers.

The response function is generally obtained by evaluating a response functional on a suitably defined set of solutions. In this chapter we shall study statistically stationary solutions of the Navier-Stokes equations for flow through annular ducts. These solutions are defined in § 29; their chief property is that the spatial average over cylinders of such solutions is time-independent. This is trivially true of laminar Poiseuille flow; we show in § 34 that it is true of the time-periodic motion which bifurcates from laminar Poiseuille flow. We assume that other solutions which are observed as turbulence have the property of statistical stationarity. The assumption gives a sense in which fluctuating flow in a steady environment can have steady average properties.

The purpose of this chapter is best served by drawing a distinction between laminar Poiseuille flow and all of the other statistically stationary flows, including the time-periodic bifurcating flow. We shall call all of these other flows turbulent. The analysis is conveniently framed in terms of the friction factor (f) and Reynolds number (R). The subscript l will be used to designate laminar flow. Response curves define relations between f and R.

Figs. 27.1, 2 and 3 are bifurcation diagrams in the plane of the response curve for flow through annular ducts. The essential ideas of this chapter are contained in the results summarized in these Figures.

In practice, stable turbulence appears to have the property of consistent reproducibility on the average. By this we mean that for a given pipe, there appears to be a curve, which we have called a response curve, which defines a functional relation between the Reynolds number and friction factor (between

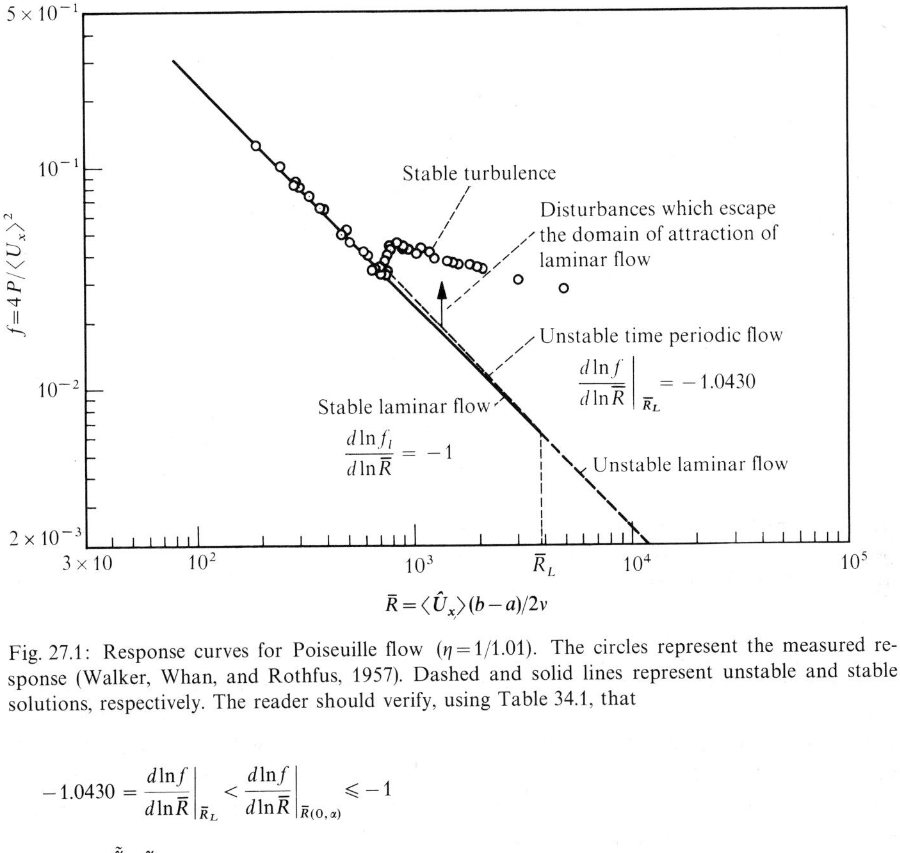

Fig. 27.1: Response curves for Poiseuille flow ($\eta = 1/1.01$). The circles represent the measured response (Walker, Whan, and Rothfus, 1957). Dashed and solid lines represent unstable and stable solutions, respectively. The reader should verify, using Table 34.1, that

$$-1.0430 = \left.\frac{d\ln f}{d\ln \bar{R}}\right|_{\bar{R}_L} < \left.\frac{d\ln f}{d\ln \bar{R}}\right|_{\bar{R}(0,\alpha)} \leqslant -1$$

for $1.021 = \frac{\tilde{\alpha}}{2} < \frac{\alpha}{2} \leqslant 1.0964$ where $\tilde{\alpha}$ is the minimizing wave number defined by

$$\bar{R}_L = \min_\alpha \bar{R}(0,\alpha) = \bar{R}(0,\tilde{\alpha})\,.$$

It follows that the slopes of successive bifurcating solutions for $\alpha/2 > 1.021$ cross each other. This suggests that the lower branch of the envelope

$$\bar{R}(\varepsilon^2) = \min_\alpha \bar{R}(\varepsilon^2,\alpha) = \bar{R}(\varepsilon^2,\alpha(\varepsilon^2))$$

of two-dimensional bifurcating solutions is taken on for wave number $\alpha(\varepsilon^2) > \tilde{\alpha}$ (Joseph and Chen, 1974)

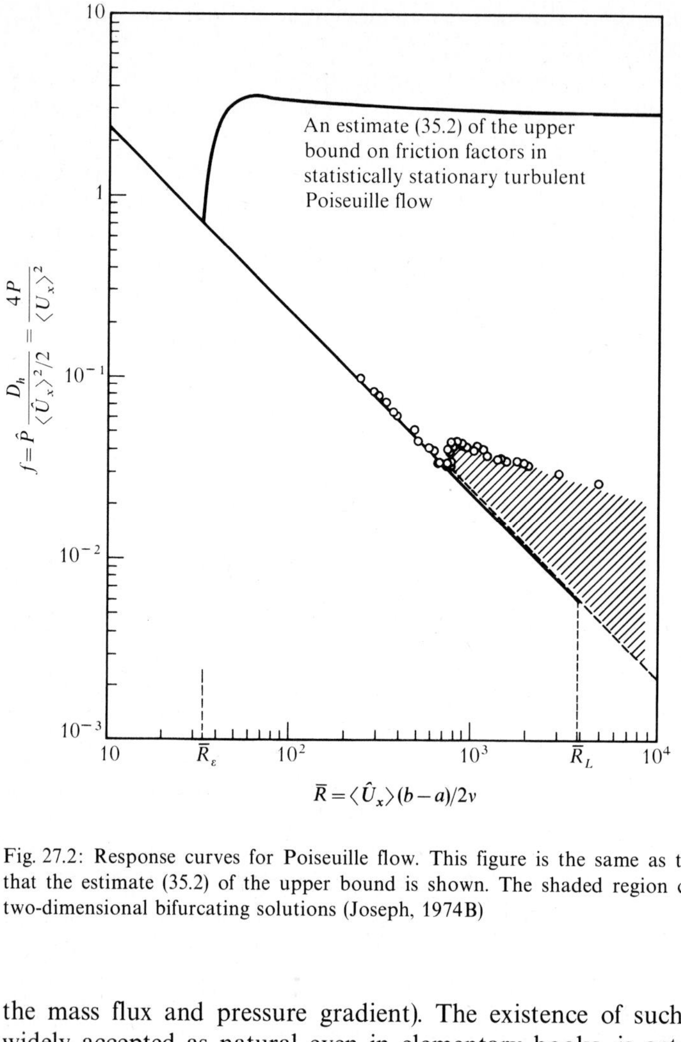

Fig. 27.2: Response curves for Poiseuille flow. This figure is the same as the previous one except that the estimate (35.2) of the upper bound is shown. The shaded region contains many unstable two-dimensional bifurcating solutions (Joseph, 1974B)

the mass flux and pressure gradient). The existence of such a curve, which is widely accepted as natural even in elementary books, is actually a remarkable event since the curve is defined over a set of fluctuating turbulent flows each of which differs from its neighbors. In this sense the response curve may be regarded as giving the steady average response of a fluctuating system subjected to steady external conditions.

§ 28. The Fluctuation Motion and the Mean Motion

Flows which take place under steady exterior conditions can be decomposed into a steady basic laminar flow plus some disturbance. We can also resolve the flow into a mean motion plus fluctuations around the mean. The equations governing the mean and fluctuating parts are called averaged equations.

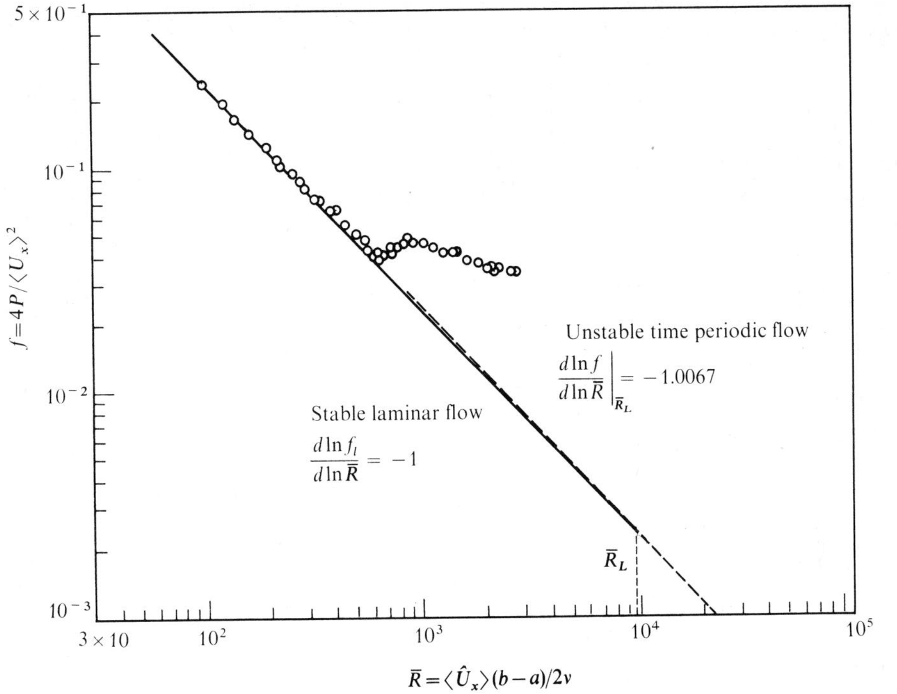

Fig. 27.3: Response curves for Poiseuille flow ($\eta = \frac{1}{2}$) (Joseph and Chen, 1974)

Consider flow which is driven through the annulus by a constant pressure gradient $\hat{P}$. When the fluctuations are zero, the flow is laminar and is given by Eq. (17.5). The equations which govern the motion of the fluid in the annulus are

$$\frac{\partial \hat{\mathbf{V}}}{\partial \hat{t}} + \hat{\mathbf{V}} \cdot \nabla \hat{\mathbf{V}} = -\nabla \hat{p} + \mathbf{e}_x \hat{P} + \nu \nabla^2 \hat{\mathbf{V}} \tag{28.1a}$$

$$\operatorname{div} \hat{\mathbf{V}} = 0 \tag{28.1b}$$

and

$$\hat{\mathbf{V}}(\hat{x}, \hat{\theta}, a, \hat{t}) = \hat{\mathbf{V}}(\hat{x}, \hat{\theta}, b, \hat{t}) = 0\,. \tag{28.1c}$$

Here $\hat{P} > 0$ is a constant whose value is determined by the applied pressure drop, and the total pressure at a point in the fluid is

$$\hat{p}(\hat{x}, \hat{\theta}, \hat{r}, \hat{t}) - \hat{P}\hat{x}$$

where $\hat{p}(\hat{x}, \hat{\theta}, \hat{r}, \hat{t})$ is uniformly bounded for $-\infty < \hat{x} < \infty$.

We shall work with a dimensionless statement of the problem 28.1. The dimensional variables

$$[\hat{x}, \hat{\theta}, \hat{r}, \hat{t}, \hat{V}_x, \hat{V}_\theta, \hat{V}_r, \hat{p}, \hat{P}] \tag{28.2a}$$

are related to dimensionless variables

$$[x, \theta, r, t, V_x, V_\theta, V_r, p, P] \tag{28.2b}$$

by multiplying the dimensionless variables by the scale factors

$$\left[l, 1, l, \frac{l^2}{\nu}, \frac{\nu}{l}, \frac{\nu}{l}, \frac{\nu}{l}, \frac{\nu^2}{l^2}, \frac{\nu^2}{l^3}\right] \tag{28.2c}$$

where

$$l = b - a\,.$$

The domain occupied by the fluid described in dimensionless variables, is

$$\mathscr{V} = [x, \theta, r : -\infty < x < \infty,\ 0 \leqslant \theta \leqslant 2\pi,\ \tilde{a} \leqslant r \leqslant \tilde{b}]$$

where $\eta = a/b$, $\tilde{a} = \eta/(1-\eta)$ and $\tilde{b} = 1/(1-\eta)$. We shall also need to designate the cylinder average

$$\bar{f}(r,t) = \lim_{L\to\infty} \frac{1}{2L} \int_{-L}^{L} \left\{ \frac{1}{2\pi} \int_0^{2\pi} f(x,\theta,r,t)\, d\theta \right\} dx$$

and the overall average

$$\langle \bar{f} \rangle = \langle f \rangle = \frac{2}{\tilde{b}^2 - \tilde{a}^2} \int_{\tilde{a}}^{\tilde{b}} \bar{f} r\, dr\,.$$

The dimensionless parameter

$$\langle \bar{V}_x \rangle = \frac{b-a}{\nu} \left[\frac{2}{b^2 - a^2} \int_a^b r \bar{\hat{V}}_x dr \right] \equiv 2\bar{R} \tag{28.3}$$

is proportional to a Reynolds number based on the average velocity and one-half the gap thickness. We shall refer to $\langle \bar{V}_x \rangle$ as a dimensionless mass-flux. Eq. (17.8) gives the ratio $V_x(r)/\langle \bar{V}_x \rangle$ of the velocity of laminar Poiseuille flow to the mean velocity.

A slightly different interpretation of the cylinder and overall average will be used in the limit $\eta \to 1$, $(x, \theta, r) \to (x, y, z)$, defined in Exercise 17.1. In this case

$$\bar{f}(z,t) = \lim_{L\to\infty} \left\{ \frac{1}{4L^2} \int\int_{-L}^{L} f(x,y,z,t)\, dx\, dy \right\}$$

and

$$\langle\bar{f}\rangle=\langle f\rangle=\int_{-1/2}^{1/2}\bar{f}(z,t)\,dz\,.$$

In the dimensionless variables the equations which govern the motion of the fluid are

$$\frac{\partial\mathbf{V}}{\partial t}+\mathbf{V}\cdot\nabla\mathbf{V}=-\nabla p+\mathbf{e}_x P+\nabla^2\mathbf{V}$$

and

$$\operatorname{div}\mathbf{V}=0 \quad \text{in} \quad \mathscr{V} \tag{28.4}$$

and

$$\mathbf{V}=0 \quad \text{at} \quad r=\tilde{a},\tilde{b}\,.$$

The motion may be decomposed into mean and fluctuating parts

$$[V_x,V_\theta,V_r,\pi]=[\bar{V}_x+u,\bar{V}_\theta+v,\bar{V}_r+w,\bar{p}+p] \tag{28.5}$$

where the overbar designates the cylinder average. Fluctuations, by definition, have a zero average.

An elementary consequence of the continuity equation is

$$\bar{V}_r=0\,.$$

To prove this, consider the cylinder average of the continuity equation:

$$\frac{1}{r}\frac{\partial\overline{rV_r}}{\partial r}+\frac{1}{r}\frac{\partial\overline{V_\theta}}{\partial\theta}+\frac{\partial\overline{V_x}}{\partial x}=\frac{1}{r}\frac{d}{dr}r\bar{V}_r=0\,.$$

The mean radial component of velocity is constant and this constant must be zero since $\bar{V}_r=0$ at $r=\tilde{a}$ and $r=\tilde{b}$.

To obtain the equations for the mean motion and the equations for the fluctuation motion, replace $(\mathbf{V},p)$ in (28.4) with (28.5). On taking account of $\bar{V}_r=0$, we find that

$$\frac{\partial\bar{\mathbf{V}}}{\partial t}+\frac{\partial\mathbf{u}}{\partial t}+\bar{\mathbf{V}}\cdot\nabla\bar{\mathbf{V}}+\bar{\mathbf{V}}\cdot\nabla\mathbf{u}+\mathbf{u}\cdot\nabla\bar{\mathbf{V}}+\mathbf{u}\cdot\nabla\mathbf{u}=-\nabla p-\mathbf{e}_r\frac{\partial\bar{p}}{\partial r}+\mathbf{e}_x P+\nabla^2(\bar{\mathbf{V}}+\mathbf{u}) \tag{28.6a}$$

where $\mathbf{u}=\mathbf{e}_x u+\mathbf{e}_\theta v+\mathbf{e}_r w$. The cylinder average of (28.6a) is

$$\frac{\partial\bar{\mathbf{V}}}{\partial t}+\mathbf{e}_r\left[-\frac{\bar{\mathbf{V}}_\theta^2}{r}+\overline{\mathbf{u}\cdot\nabla w}-\frac{\overline{v^2}}{r}\right]+\mathbf{e}_\theta\left[\overline{\mathbf{u}\cdot\nabla v}+\frac{\overline{vw}}{r}\right]+\mathbf{e}_x\overline{\mathbf{u}\cdot\nabla u}+\mathbf{e}_r\frac{\partial\bar{p}}{\partial r}-\mathbf{e}_x P$$

$$=\mathbf{e}_\theta\left\{\frac{1}{r}\frac{\partial}{\partial r}\left(r\frac{\partial\bar{V}_\theta}{\partial r}\right)-\frac{\bar{V}_\theta}{r^2}\right\}+\frac{\mathbf{e}_x}{r}\frac{\partial}{\partial r}\left(r\frac{\partial\bar{V}_x}{\partial r}\right). \tag{28.6b}$$

The difference {(28.6a)—(28.6b)} is

$$\frac{\partial \mathbf{u}}{\partial t}+\bar{\mathbf{V}}\cdot\nabla\mathbf{u}+\mathbf{u}\cdot\nabla\bar{\mathbf{V}}+\mathbf{u}\cdot\nabla\mathbf{u}-\mathbf{e}_r[\mathbf{u}\cdot\nabla w-v^2/r-\overline{\mathbf{u}\cdot\nabla w-v^2/r}]$$
$$-\mathbf{e}_\theta[\mathbf{u}\cdot\nabla v+vw/r-\overline{\mathbf{u}\cdot\nabla v+vw/r}]-\mathbf{e}_x(\mathbf{u}\cdot\nabla u-\overline{\mathbf{u}\cdot\nabla u})=-\nabla p+\nabla^2\mathbf{u} \tag{28.7}$$

where $\operatorname{div}\mathbf{u}=0$ and

$$\bar{\mathbf{V}}=\mathbf{u}=0 \quad \text{at} \quad r=\tilde{a},\tilde{b}\,. \tag{28.8}$$

Eq. (28.7) governs the fluctuations. Eq. (28.6b) governs the mean motion.

It is possible to consider an IBVP for the mean and fluctuation motions. Since the initial velocity field (V_x, V_θ, V_r) is arbitrary, we may assign $(\bar{V}_x, \bar{V}_\theta, 0)$ and (u, v, w) arbitrarily, provided only that the fluctuations have a zero cylinder average.

Associated with the decomposition into a fluctuation and mean motion are *two* energy identities, one for the mean motion and one for the fluctuation motion. These identities follow from $\langle\bar{\mathbf{V}}\cdot(28.6\text{b})\rangle$ and $\langle\mathbf{u}\cdot(28.7)\rangle$, respectively. We also use $\bar{V}_r=0$ and the following identity,

$$\overline{\mathbf{v}\cdot\nabla f}=\overline{\nabla\cdot(\mathbf{v}f)}=\frac{1}{r}\frac{\partial}{\partial r}(r\overline{fw})\,,$$

which is valid when $\nabla\cdot\mathbf{v}=0$. Thus,

$$\frac{1}{2}\frac{d}{dt}\langle|\bar{V}_\theta|^2+|\bar{V}_x|^2\rangle+\left\langle\frac{\bar{V}_\theta}{r}\frac{\partial}{\partial r}(r\overline{wv})+\frac{\bar{V}_\theta\overline{wv}}{r}\right\rangle+\left\langle\frac{\bar{V}_x}{r}\frac{\partial}{\partial r}(r\overline{wu})\right\rangle-P\langle\bar{V}_x\rangle$$
$$=-\left\langle\left|\frac{\partial\bar{V}_\theta}{\partial r}\right|^2+\left|\frac{\partial\bar{V}_x}{\partial r}\right|^2+\frac{\bar{V}_\theta^2}{r^2}\right\rangle \tag{28.9}$$

and

$$\frac{1}{2}\frac{d}{dt}\langle|\mathbf{u}^2|\rangle+\langle\mathbf{u}\cdot\nabla\bar{\mathbf{V}}\cdot\mathbf{u}\rangle=-\langle|\nabla\mathbf{u}|^2\rangle\,. \tag{28.10a}$$

Eq. (28.10a) is an energy identity of the same form as (3.1) because $\mathbf{u}\cdot\nabla\bar{V}\cdot\mathbf{u}=\mathbf{u}\cdot\bar{\mathbf{D}}\cdot\mathbf{u}$ where $\bar{\mathbf{D}}=\mathbf{D}[\bar{\mathbf{V}}]$ is the stretching tensor of the mean motion. However, (3.1) governs the evolution of motions fluctuating around a mean. The evolution of the mean motion is governed by (28.9) and is coupled to the fluctuations.

An evolution equation for the total energy may be formed by summing (28.9) and (28.10a). Noting first that after integration by parts, the second and third term on the left of (28.9) may be written as

$$-\left\langle wv\frac{\partial\bar{V}_\theta}{\partial r}+wu\frac{\partial\bar{V}_x}{\partial r}-\frac{\bar{V}_\theta wv}{r}\right\rangle=-\langle\mathbf{u}\cdot\nabla\bar{\mathbf{V}}\cdot\mathbf{u}\rangle\,, \tag{28.10b}$$

one finds that

$$\frac{1}{2}\frac{d}{dt}\langle|\bar{V}_\theta|^2+|\bar{V}_x|^2+|\mathbf{u}|^2\rangle=P\langle\bar{V}_x\rangle-\left\langle|\nabla\mathbf{u}|^2+\left|\frac{\partial\bar{V}_\theta}{\partial r}\right|^2+\left|\frac{\partial\bar{V}_x}{\partial r}\right|^2+\frac{\bar{V}_\theta^2}{r^2}\right\rangle. \quad (28.11)$$

The energy source for turbulent Poiseuille flow is the $P\langle\bar{V}_x\rangle$ work of the pressure gradient on the mean flow.

§ 29. Steady Causes and Steady Effects

We face the problem of describing in some useful sense all of the solutions of the Navier-Stokes equations which can arise when the steady external conditions are those which give rise to Poiseuille flow. Some progress with this hard problem can be made if we admit the basic assumption that steady external conditions have a steady effect even when the motion is fluctuating. This assumption is supported by the consistent reproducibility of certain average values in turbulent flow.

The steady effects of steady external conditions need not imply unique steady solutions. Only when $\bar{R}<\bar{R}_G$ do all flows tend to a unique steady flow. When $\bar{R}>\bar{R}_G$ there are at least two solutions which are possible: the unstable laminar flow and any one of the motions which replace laminar flow. The limiting flows which actually occur when $\bar{R}>\bar{R}_G$ are those which are in some sense stable. The stable solutions with $\bar{R}>\bar{R}_G$ need not be unique; indeed we envision stable *sets* of solutions. Though such solutions would lack uniqueness in the ordinary sense, it is consistent with observations to postulate the existence of stable sets of solutions sharing common properties in the average.

The basic property which we shall assume here is (1) that all cylinder averages are time independent. This assumption says that a consequence of steady exterior conditions (boundary conditions and pressure drop) is that cylinder averages are steady. The fluctuation fields themselves can be very unsteady. We also shall assume (2) that velocity components have a zero mean value unless a nonzero mean value is forced externally. Property (2) implies that $\bar{V}_\theta=0$. Following Howard (1963), we say that flows with properties (1) and (2) are *statistically stationary*[1].

It is necessary to add that considerable progress toward a satisfactory variational theory of turbulence can be made without either of the characterizing

---

[1] The observation that steady exterior conditions should be expected to lead to stationary turbulence needs qualification. Stationary turbulence evidently cannot exist in Hagen-Poiseuille flow and plane Couette flow when the fluctuations are infinitesimal (the linearized stability theory shows that *all* infinitesimal disturbances decay). The analysis given here applies to stationary turbulence when it exists.

It should also be noted that motions which are here called statistically stationary need not be turbulent. For example, steady laminar motions fit our definitions.

assumptions defining stationary turbulence. To relax these assumptions we introduce a time-cylinder average

$$\overline{\overline{\circ}} \equiv \lim_{T \to \infty} \int_0^T \overline{\circ}\, dt . \tag{29.1}$$

Then we prove that all of the equations from (28.5) through (30.9) hold when cylinder averages are replaced with time-cylinder averages (Exercise 30.1).

§ 30. Laminar and Turbulent Comparison Theorems

Assuming properties (1) and (2) of statistical stationarity, Eq. (28.6 b) may be written as

$$\frac{d}{dr}\left[r\overline{wu} - r\frac{d\bar{V}_x}{dr} - P[\bar{V}_x]\frac{r^2}{2}\right] = 0 , \tag{30.1}$$

$$\frac{d}{dr}[r^2\overline{wv}] = 0 , \tag{30.2}$$

$$\frac{1}{r}\frac{d}{dr}(r\overline{w^2}) - \frac{1}{r}\overline{v^2} + \frac{d\pi}{dr} = 0 , \tag{30.3}$$

where $P[\bar{V}_x]$ is the P associated with mass flux $\langle \bar{V}_x \rangle$, and Eq. (28.10 a) may be written as

$$\langle wu\, d\bar{V}_x/dr \rangle = -\langle |\nabla \mathbf{u}|^2 \rangle . \tag{30.4}$$

Eq. (30.2) shows that $\overline{wv}$ is a constant whose value is zero at the boundary and elsewhere (see Exercise 30.1).

A basic and important consequence of statistical stationarity is that (30.1) has a first integral:

$$r\overline{wu} + \frac{\tilde{b}^2 - \tilde{a}^2}{2\ln\eta}\left\langle \frac{wu}{r} \right\rangle - \frac{P[\bar{V}_x]}{2}\left[r^2 + \frac{\tilde{b}^2 - \tilde{a}^2}{2\ln\eta}\right] = r\frac{d\bar{V}_x}{dr} . \tag{30.5}$$

When the fluctuations are zero, $\bar{V}_x = U_x$ and

$$\frac{P_l[U_x]}{2}\left[r^2 + \frac{\tilde{b}^2 - \tilde{a}^2}{2\ln\eta}\right] = -r\frac{dU_x}{dr} \tag{30.6}$$

where $P_l[U_x]$ is the P for laminar flow with mass flux $\langle U_x \rangle$. Combining (30.4) and (30.5) we get

$$\frac{P[\bar{V}_x]}{2}\langle h(r)\, wu \rangle = \langle |\nabla \mathbf{u}|^2 \rangle + \left\langle \left[\overline{wu} + \frac{\tilde{b}^2 - \tilde{a}^2}{2r\ln\eta}\left\langle \frac{wu}{r} \right\rangle\right]^2 \right\rangle \tag{30.7}$$

where

$$h(r) = r + \frac{\tilde{b}^2 - \tilde{a}^2}{2r \ln \eta}.$$

Eq. (30.7) shows that

$$\langle h(r) wu \rangle \geq 0. \tag{30.8}$$

The relations (30.5), (30.6) and (30.8) lead to the following laminar-turbulent comparison theorem:

(1) *Statistically stationary, turbulent Poiseuille flow has a smaller mass flux* $\langle \bar{V}_x \rangle < \langle U_x \rangle$ *than the laminar Poiseuille flow with the same applied pressure gradient* $P[\bar{V}_x] = P_l[U_x]$; (2) *Statistically stationary turbulent Poiseuille flow has a larger applied pressure gradient* $P[\bar{V}_x] > P_l[U_x]$ *than the laminar flow with the same mass flux* $(\langle \bar{V}_x \rangle = \langle U_x \rangle)$.

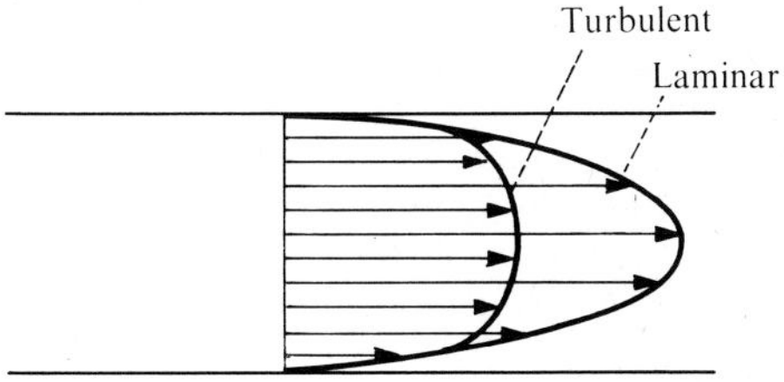

Fig. 30.1: Laminar-turbulent comparison theorem for plane-Poiseuille flow. Laminar flow and turbulent flow with the same pressure gradient have the same shear stress at the wall. The dimensionless mass efflux of the laminar flow is larger than in turbulent flow by an amount given by $\langle U_x - \bar{V}_x \rangle = \langle zwu \rangle$ where $\langle zwu \rangle$ is a positive quantity given by Eq. (30.13). A larger pressure drop is needed to drive a turbulent flow through the channel than is needed to drive a laminar flow with the same mass flux

The first of these theorems was proved by Thomas (1942); the proof given here is a modification of one by Busse (1969A). One notes that the integral of r times (30.5) plus (30.6) may be written as

$$\langle h(r) wu \rangle = 2(P - P_l) b(\eta) - 2\langle \bar{V}_x - U_x \rangle \tag{30.9}$$

where

$$b(\eta) = \tfrac{1}{8}\left[\tilde{a}^2 + \tilde{b}^2 + \frac{\tilde{b}^2 - \tilde{a}^2}{\ln \eta}\right] = \tfrac{1}{8}\left[\frac{1+\eta^2}{(1-\eta)^2} + \frac{1+\eta}{(1-\eta)\ln \eta}\right],$$

$$b(0) = \tfrac{1}{8} \geq b(\eta) \geq b(1) = \tfrac{1}{12}.$$

Eq. (30.9) together with (30.8) prove the theorem.

Exercise 30.1: Suppose that property (1) of statistical stationarity, but not property (2), holds. Show that

$$r^2 \overline{wv} - r^3 \frac{d}{dr}(\bar{V}_\theta / r) = \tilde{a}^2 \tilde{b}^2 \left\langle \frac{wv}{r^2} \right\rangle \tag{30.10}$$

and

$$\frac{P}{2}\langle h(r)\,wu\rangle = \langle|\nabla\mathbf{u}|^2\rangle + \left\langle\left[\overline{wu} + \frac{\tilde{b}^2-\tilde{a}^2}{2r\ln\eta}\left\langle\frac{wu}{r}\right\rangle\right]^2\right\rangle + \left\langle\left[\overline{wv} - \frac{\tilde{a}^2\tilde{b}^2}{r^2}\left\langle\frac{wv}{r^2}\right\rangle\right]^2\right\rangle . \tag{30.11}$$

Suppose that neither of the two properties of stationary turbulence holds. Show that all of the Eqs. (28.5) through (30.9) hold when cylinder averages are replaced with time-cylinder averages. Prove the laminar turbulent comparison theorems for time-cylinder averaged quantities.

Exercise 30.2 (Joseph, 1974B): Repeat the derivations given in this chapter when the Poiseuille flow is between parallel planes separated by a distance l. Show that all statistically stationary solutions satisfy the equations

$$\overline{wu} - \langle wu\rangle = Pz + d\overline{V}_x/dz\,,$$

$$\overline{wv} - \langle wv\rangle = d\overline{V}_y/dz\,,$$

$$\overline{w^2} - \langle w^2\rangle = -\overline{p} + \langle p\rangle\,,$$

$$\langle zwu\rangle = \langle U_x - \overline{V}_x\rangle\,, \tag{30.12}$$

$$P\langle zwu\rangle = \langle[\overline{wu} - \langle wu\rangle]^2\rangle + \langle[\overline{wv} - \langle wv\rangle]^2\rangle + \langle|\nabla\mathbf{u}|^2\rangle\,. \tag{30.13}$$

Prove the equal shear stress theorem:

Suppose that the mean velocity $\overline{V}_x(z)$ is an even function of z. Then, for a given pressure drop the wall shear stress for turbulent flow is the same as the wall shear stress for laminar flow.

A pictorial representation of the equal shear stress theorem is given in Fig. 30.1.

Exercise 30.3 (Busse, 1969A): Show that the mean wall shear stress is greater in turbulent plane Couette flow than in laminar plane Couette flow.

Exercise 30.4 (Nickerson, 1969; Busse, 1971): Compare the torque for laminar and turbulent solutions for Couette flow between rotating cylinders when the angular velocities of the outer and inner cylinders are given.

Exercise 30.5 (Thomas, 1942; Serrin, 1959A): Show that the steady laminar flow of an incompressible viscous fluid down a straight pipe of arbitrary cross-section is characterized by the property that its energy dissipation is least among all laminar (or spatially periodic) flows down the pipe which have the same total mass flux. Show that spatially periodic flow in a straight pipe with fixed pressure drop is unstable when its total exceeds that in laminar flow with the same pressure drop. Show that the pressure drop required to produce a given discharge in a pipe is greater for turbulent flow than for laminar flow.

§ 31. A Variational Problem for the Least Pressure Gradient in Statistically Stationary Turbulent Poiseuille Flow with a Given Mass Flux Discrepancy

Consider a statistically stationary flow with mass flux $\langle\overline{V}_x\rangle$ and pressure gradient $P[\overline{V}_x]$. Let $\langle U_x\rangle$ be the mass flux for laminar flow whose pressure gradient $P_l[U_x] = P[\overline{V}_x]$. A positive mass flux discrepancy μ for these flows with equal pressure gradients can be defined as

$$\langle U_x - \overline{V}_x\rangle = \tfrac{1}{2}\langle h(r)\,wu\rangle \equiv \mu \geqslant 0\,. \tag{31.1}$$

The response function for the flow $\bar{V}_x$ may be parameterized with μ; thus, setting $P[\bar{V}_x]=P(\mu)$ we may write (30.7) as

$$P(\mu) = \frac{2\langle|\nabla \mathbf{u}|^2\rangle}{\langle h(r)wu\rangle} + 4\mu \frac{\left\langle\left[\overline{wu} + \frac{\tilde{b}^2-\tilde{a}^2}{2r\ln\eta}\left\langle\frac{wu}{r}\right\rangle\right]^2\right\rangle}{\langle h(r)wu\rangle^2} \equiv \mathscr{P}[\mathbf{u};\mu] \tag{31.2}$$

where the functional $\mathscr{P}$ is originally defined for statistically stationary solutions. The functional $\mathscr{P}[\mathbf{u};\mu]$ is a homogeneous functional of degree zero in $\mathbf{u}$; $\mathscr{P}[a\mathbf{u};\mu]=\mathscr{P}[\mathbf{u};\mu]$ for any constant a.

Eq. (31.2) can be used to show that the pressure gradients $P(\mu)$ all lie above a positive lower bound $\tilde{P}(\mu)$ which increases to infinity with μ. This lower bound can be studied by variational methods. The basic idea is to consider the values which $\mathscr{P}$ takes in the class $\bar{\mathbf{H}}$ of kinematically admissible fluctuation fields

$$\bar{\mathbf{H}} = \{\mathbf{u}(x,\theta,r)\colon \operatorname{div}\mathbf{u}=0,\ \mathbf{u}|_{\tilde{a},\tilde{b}}=0,\ \bar{u}=\bar{v}=\bar{w}=0,\ \mathbf{u}\in AP(x)\cap P_{2\pi}(\theta)\}$$

where $\mathbf{u}\in AP(x)\cap P_{2\pi}(\theta)$ means that $\mathbf{u}$ is an almost periodic function of x and a periodic function of θ with period 2π. We are assuming that statistically stationary solutions exist and are elements of $\bar{\mathbf{H}}$.[2]

We shall also define

$$\mathscr{N} = \{\mathbf{u}\colon \mu=\tfrac{1}{2}\langle h(r)wu\rangle>0\},$$

the set of $\mathbf{u}$ with mass flux discrepancy $\mu>0$. Then, by (31.2),

$$\mathscr{P}[\mathbf{u};\mu]-P(\mu)=0 \tag{31.3}$$

when $\mathbf{u}$ is a statistically stationary solution. Of course $\mathbf{u}$ is also an element of the larger set of admissible fluctuations $\bar{\mathbf{H}}\cap\mathscr{N}$. Hence

$$P^{-1} \leqslant \max_{\bar{\mathbf{H}}\cap\mathscr{N}} \mathscr{P}^{-1}[\mathbf{u};\mu]. \tag{31.4}$$

Moreover,

$$\max_{\bar{\mathbf{H}}\cap\mathscr{N}} \mathscr{P}^{-1}[\mathbf{u};\mu] \leqslant \max_{\bar{\mathbf{H}}} \mathscr{P}^{-1}[\mathbf{u};\mu] \equiv \tilde{P}^{-1} \tag{31.5}$$

because all of the functions in the intersection of $\bar{\mathbf{H}}$ and $\mathscr{N}$ are in $\bar{\mathbf{H}}$.

The variational problem could be posed as

$$\tilde{P} = \min_{\bar{\mathbf{H}}} \mathscr{P}[\mathbf{u};\mu] \tag{31.6}$$

if only those $\mathbf{u}$ for which $\mu>0$ are allowed in the competition. Clearly,

---

[2] All of the required cylinder averages exist for almost periodic functions and the variational problem seems well posed in this set of functions. We do not presume to take a position on the extent to which "true turbulence" could be almost periodic. Other settings for the variational problem which lead to the same analysis could be defined; an alternate procedure which is commonly adopted is to leave the function space incompletely specified, specifying only that the fields $\mathbf{u}$ are uniformly bounded. Assumptions are then required at a later stage.

Statistically stationary turbulent Poiseuille flow with mass flux discrepancy μ *cannot exist when*

$$P<\tilde{P}. \tag{31.7}$$

Proof: Suppose $\tilde{\mathbf{u}}$ is a minimizing element for (31.6). Then, by homogeneity, there is a constant c such that $c\tilde{\mathbf{u}}\in\bar{\mathbf{H}}\cap\mathcal{N}$ also maximizes (31.4); (31.7) then follows from (31.4) and (31.5).

The minimizing functional

$$\tilde{P}(\mu)=\mathscr{P}[\tilde{\mathbf{u}};\mu]$$

has the following properties of *monotonicity and convexity*:

$$\text{(i)}\ \tilde{P}(\mu)\geqslant\tilde{P}(0)\equiv P_{\mathscr{E}} \tag{31.8}$$

where $P_{\mathscr{E}}$ is the first critical pressure gradient of energy theory,

$$\text{(ii)}\ \frac{d\tilde{P}}{d\mu}=4\frac{\left\langle\left[wu+\dfrac{\tilde{b}^2-\tilde{a}^2}{2r\ln\eta}\left\langle\dfrac{wu}{r}\right\rangle\right]^2\right\rangle}{\langle h(r)wu\rangle^2}, \tag{31.9}$$

$$\text{(iii)}\ \frac{d\tilde{P}}{d\mu}\quad\text{is a decreasing function of } \mu. \tag{31.10}$$

Proof: The first critical pressure gradient of energy theory is given by

$$P_{\mathscr{E}}^{-1}=\max_{\mathbf{H}}\frac{\langle h(r)wu\rangle}{2\langle|\nabla\mathbf{u}|^2\rangle}. \tag{31.11}$$

Noting now that $\bar{\mathbf{H}}\subset\mathbf{H}$ (since zero mean values are not required in $\mathbf{H}$) we have

$$P_{\mathscr{E}}^{-1}=\max_{\mathbf{H}}\mathscr{P}^{-1}[\mathbf{u},0]\leqslant\max_{\bar{\mathbf{H}}}\mathscr{P}^{-1}[\mathbf{u},0]=\tilde{P}^{-1}(0), \tag{31.12}$$

proving (31.8). Actually, equality holds in (31.12); to prove it, we must show that the maximizing element in $\mathbf{H}$ has a zero cylinder average (Exercise 31.1).

To prove (31.9), imagine $\mathbf{u}_1$ and $\mathbf{u}_2$ are minimizing vectors for (31.6) belonging to μ_1 and μ_2. Then

$$\begin{aligned}\mathscr{P}[\mathbf{u}_1;\mu_2]-\mathscr{P}[\mathbf{u}_1;\mu_1]&\geqslant\tilde{P}(\mu_2)-\tilde{P}(\mu_1)\\&=\mathscr{P}[\mathbf{u}_2;\mu_2]-\mathscr{P}[\mathbf{u}_1;\mu_1]\geqslant\mathscr{P}[\mathbf{u}_2;\mu_2]-\mathscr{P}[\mathbf{u}_2;\mu_1]\\&=(\mu_2-\mu_1)\frac{4\left\langle\left[\overline{wu}+\dfrac{\tilde{b}^2-\tilde{a}^2}{2r\ln\eta}\left\langle\dfrac{wu}{r}\right\rangle\right]^2\right\rangle}{\langle h(r)wu\rangle^2}.\end{aligned} \tag{31.13}$$

Dividing (31.13) by $\mu_2-\mu_1$ and passing to the limit $\mu_2\to\mu_1$ we prove (31.9). Returning to (31.13) with (31.9) we find that

$$(\mu_2-\mu_1)\frac{d\tilde{P}(\mu_1)}{d\mu}\geqslant(\mu_2-\mu_1)\frac{d\tilde{P}(\mu_2)}{d\mu}, \tag{31.14}$$

proving (31.10).

The variational problem for the values $\tilde{P}(\mu)$ have been considered by Busse (1970A, 1972A) and Howard (1972) for flow in a round pipe and by Busse (1969A, 1970A) for flow in a channel. It is well to note that the structure of the variational problem for turbulent flow has much in common with observed properties of real turbulence. We have not here emphasized this most interesting aspect of the variational theory as a model theory for turbulence. This aspect is stressed by Busse and Howard, in the notes to this chapter, and in our discussion of the turbulence problem for porous convection which is considered in Chapter XII.

Exercise 31.1: Show that the maximizing field for (31.11) has a zero horizontal mean. Hint: prove that the maximizing element in the almost periodic class has a nonzero wave number (Exercise 44.1).

Exercise 31.2 (Busse, 1969A, 1970A): Consider a homogeneous incompressible fluid between two parallel rigid plates of height d. The plates move with velocity V_0 relative to one another. Show that every statistically stationary solution of the Navier-Stokes equations has a positive fluctuation momentum transport $\langle wu\rangle>0$. Show that every statistically stationary solution with $\mu=\langle wu\rangle>0$ has

$$R=\frac{V_0 d}{\nu}>R(\mu)$$

$$R(\mu)=\min_{\bar{\mathbf{H}}}\left[\frac{\langle|\nabla\mathbf{u}|^2\rangle}{\langle wu\rangle^2}+\frac{\mu}{\langle wu\rangle^2}\langle(\overline{wu}-\langle wu\rangle)^2+(\overline{wv}-\langle wv\rangle)^2\rangle\right]. \tag{31.15}$$

Prove a monotonicity-convexity theorem for $R(\mu)$.

Exercise 31.3: Find Euler's equations for the minimum value $\tilde{P}(\mu)$ given by (31.6).

Exercise 31.4 (Hayakawa, 1970): Formulate a variational problem for the response curve relating the mass flux discrepancy to the pull of gravity in open channel flow when the axis of channel makes an angle α with the horizontal.

§ 32. Turbulent Plane Poiseuille Flow—an Lower Bound for the Response Curve

In the last section we formulated a variational problem which leads to a lower bound on the pressure gradient when the mass flux discrepancy is given. We wish to give an explicit estimate of this bound. We will restrict the analysis to the case $\eta\to 1$; that is, to plane Poiseuille flow. The main result to be obtained is first stated, then proved, below.

The response function for statistically stationary flow

$$P(\mu), \quad \mu = \langle U_x - \bar{V}_x \rangle \tag{32.1 a, b}$$

satisfies the inequality

$$P(\mu) \geqslant \tilde{P}(\mu) \geqslant \tilde{\tilde{P}}(\mu) \tag{32.2}$$

where

$$\tilde{\tilde{P}}(\mu) - 12\mu = \begin{cases} \tilde{P}(0) + 576\mu/(\tilde{P}(0)+48) & \text{for} \quad \mu \leqslant \mu^* \\ 48\sqrt{\mu} - 48 & \text{for} \quad \mu \geqslant \mu^*, \end{cases} \tag{32.3}$$

$$\mu^* = (\tilde{P}(0)+48)^2/576\,,$$

$$\tilde{P}(0) = P_{\mathscr{E}} = \min_{\mathbf{H}} \frac{\langle |\nabla \mathbf{u}|^2 \rangle}{\langle zwu \rangle} \simeq 793.6\,. \tag{32.4}$$

Before constructing the estimate (32.2) we note that $\hat{P} = 8\,U_m \nu / l^2$ (see (17.7)); hence $P = 8(U_m l/\nu)$. Therefore $P_{\mathscr{E}} = 8\,R_{\mathscr{E}}$ where $R_{\mathscr{E}}\,(\eta = 1) = 99.21$ is given in Table (22.1).

For a fixed fluid the pressure gradient P is an increasing function of the mass flux discrepancy $\langle U_x - \bar{V}_x \rangle$ on the bounding curve $\tilde{\tilde{P}}(\mu)$. The bounding curve is most easily interpreted in the coordinates of the response curve of Fig. 27.2. This interpretation will be given in § 35.

The proof of (32.2, 3) starts with the relation (30.13)

$$P = \frac{\langle |\nabla \mathbf{u}|^2 \rangle}{\langle zwu \rangle} + \mu \frac{\langle (\overline{wu} - \langle wu \rangle)^2 \rangle + \langle (\overline{wv} - \langle wv \rangle)^2 \rangle}{\langle zwu \rangle^2} \tag{32.5}$$

where

$$-\tfrac{1}{2} < z < \tfrac{1}{2}\,.$$

The inequality

$$P - 12\mu \geqslant \frac{\langle |\nabla \mathbf{u}|^2 \rangle}{\langle zwu \rangle} + \mu \frac{\langle [\overline{wu} - \langle wu \rangle - 12z\langle zwu \rangle]^2 \rangle}{\langle zwu \rangle^2} \tag{32.6}$$

follows from (32.5) in the following way: first,

$$P - 12\mu \geqslant \frac{\langle |\nabla \mathbf{u}|^2 \rangle}{\langle zwu \rangle} + \mu \frac{\langle [\overline{wu} - \langle wu \rangle]^2 \rangle}{\langle zwu \rangle^2} - 12\mu\,;$$

second, using $\langle z^2 \rangle = \frac{1}{12}$, we note that

$$\begin{aligned} \langle [\overline{wu} - \langle wu \rangle - 12z\langle zwu \rangle]^2 \rangle &= \langle \overline{wu}^2 \rangle - \langle wu \rangle^2 - 12\langle zwu \rangle^2 \\ &= \langle [\overline{wu} - \langle wu \rangle]^2 \rangle - 12\langle zwu \rangle^2\,. \end{aligned}$$

This proves (32.6). From the inequality (C.11) proved in Appendix C,

$$\frac{\langle[\overline{wu}-\langle wu\rangle-12z\langle zwu\rangle]^2\rangle}{\langle zwu\rangle^2}\geqslant\frac{576}{D+48}$$

where

$$D=\langle|\nabla\mathbf{u}|^2\rangle/\langle zwu\rangle\,.$$

It follows that

$$P-12\mu\geqslant D+(576\mu/(D+48))\geqslant\min_D[D+(576\mu/(D+48)]\,. \tag{32.7}$$

This minimum is attained when $D+48=\sqrt{576\mu}=24\mu^{1/2}$ and is equal to $2\sqrt{576\mu}-48=48\sqrt{\mu}-48$. But, by (32.4) $D>\tilde{P}(0)$; therefore, $D+48$ can never equal $24\mu^{1/2}$ if $24\mu^{1/2}<\tilde{P}(0)+48$; that is, if $\mu<(\tilde{P}(0)+48)^2/576=\mu^*$. Therefore, when $\mu<\mu^*$, $D+576\mu/(D+48)$ is an increasing function of D which must be minimum when D has its smallest value $D=\tilde{P}(0)$.

Hence, we may continue (32.7) as

$$=\begin{cases}\tilde{P}(0)+576\mu/(\tilde{P}(0)+48) & \text{for}\quad \mu\leqslant\mu^*\\ 48\sqrt{\mu}-48 & \text{for}\quad \mu\geqslant\mu^*\end{cases}$$

completing the proof of (32.2, 3).

Better estimates of the minimizing function $\tilde{P}(\mu)$ than (32.3) can be achieved through asymptotic analysis for large values of μ (see Exercise 32.2) and when certain assumptions are made about the form of the minimizing fields. Our purpose at this point is, however, adequately served by (32.3); it shows that the response curves for turbulent flow with steady averages lie in a circumscribed region of parameter space (see Fig. 27.2 and Eqs. 35.2 and 35.3).

Exercise 32.1: Prove that $\tilde{\tilde{P}}(\mu)$ is continuously differentiable at $\mu=\mu^*$.

Exercise 32.2 (Busse 1969A): Consider the variational problem for the function $P(\mu)$ when μ is large. Suspend the requirement that $\nabla\cdot\mathbf{u}=0$. Show that the solution develops boundary layers whose thickness scales with $\sqrt{\mu}$ when μ is large. Show that when μ is large

$$P(\mu)>12\mu+96(\mu/3)^{1/2}+0(\mu^{-1/2})\,.$$

Exercise 32.3 (Howard, 1972): Consider the variational problem (31.15). Show that assumption (2) of statistical stationarity implies that $\overline{wv}=0$. Show that minimizing R at a fixed value of μ is the same as maximizing μ at fixed value of R. Show that minimizing μ is equivalent to maximizing the total rate of energy dissipation. Show further that

$$\mu\leqslant 9R^2/32 \tag{32.8}$$

for all turbulent Couette flows with steady averages. Show that the estimate (32.8) is actually attained by solutions of (31.15) when the competition for the minimum is expanded to allow functions $\mathbf{u}$ which may have $\nabla\cdot\mathbf{u}\neq 0$.

Exercise 32.4: Construct an estimate like (32.3) for flow through the annulus between concentric cylinders when the radius ratio η is arbitrary $(0\leqslant\eta\leqslant 1)$.

§ 33. The Response Function Near the Point of Bifurcation

Now we turn away from energy estimates. We shall consider the linear theory of instability and the time-periodic solution which arises from this instability in the neighborhood of the point of bifurcation. We want first to show how to enrich the physical content of the perturbation theory by a proper choice of the amplitude parameter. This is accomplished by defining the amplitude as the friction factor discrepancy (see Eq. 33.13b). To obtain an expression through which the friction factor discrepancy may be related to the bifurcating solution, it is useful to compare two different resolutions of the same motion $\mathbf{V}$:

$$\mathbf{V}=\bar{V}_x(r)\mathbf{e}_x+\mathbf{u}(x,\theta,r,t)=U_x(r)\mathbf{e}_x+\mathbf{u}'(x,\theta,r,t) \tag{33.1}$$

where $U_x(r)$ is annular Poiseuille flow ((17.8) with $U=\nu U_x/l$). Eqs. (33.1) imply that

$$w=w', \quad v=v' \quad \text{and} \quad \bar{V}_x+u=U_x+u'. \tag{33.2}$$

The relation

$$\bar{V}_x(r)=U_x(r)+\bar{u}' \tag{33.3}$$

follows directly from the cylinder average of (33.1). We note that (33.3) implies that $\bar{u}'$ is time independent; in fact (33.2) implies that $\mathbf{u}'$ is statistically stationary (these requirements are verified in the remarks following (34.35)).

We shall make use of the following relation:

$$\overline{uw}=\overline{u'w'}. \tag{33.4}$$

To prove (33.4) we use (33.3) and (33.2) to write

$$\overline{uw}=\overline{uw'}=\overline{(U_x-\bar{V}_x+u')w'}=(U_x-\bar{V}_x)\overline{w'}+\overline{u'w'}=\overline{u'w'}.$$

Using (33.4) we may rewrite (30.9) as

$$\langle h(r)w'u'\rangle=2(P[\bar{V}_x]-P_l[U_x])b(\eta)-2\langle\bar{V}_x-U_x\rangle. \tag{33.5}$$

This basic relation will be used to relate the time-periodic bifurcating solutions to the response diagrams measured in the experiments.

In § 34 we shall construct statistically stationary, time-periodic flows which bifurcate from laminar Poiseuille flow at the critical point of the linear theory of stability. We shall restrict our analysis to axisymmetric solutions. For any axisymmetric motion we may introduce a stream function $\tilde{\tilde{\Psi}}$. The resolution of the motion into Poiseuille flow $\tilde{\Psi}$ plus a disturbance Ψ' may be expressed as

$$\begin{aligned}&\tilde{\tilde{\Psi}}=\tilde{\Psi}+\Psi',\\&\left(\frac{\partial\Psi'}{\partial r},\frac{\partial\Psi'}{\partial x}\right)=(ru',-rw')\end{aligned} \tag{33.6}$$

where

$$\frac{1}{r}\frac{\partial\tilde{\Psi}}{\partial r}=U_x=\langle U_x\rangle U_0(r)$$

and $U_0(r)$ is defined by (17.8). To find the problem satisfied by Ψ', make (18.2) dimensionless using the scale factors (28.2c) and set $\mathbf{u}'=\nabla\Psi'\wedge\mathbf{e}_\theta/r$. Then consider $\mathbf{e}_\theta\cdot\text{curl}$ (18.2).

$$\frac{\partial}{\partial t}\tilde{\Delta}\Psi'+\langle U_x\rangle\left\{U_0(r)\frac{\partial}{\partial x}\tilde{\Delta}\Psi'-r\frac{d}{dr}\left[\frac{1}{r}\frac{dU_0}{dr}\right]\frac{\partial\Psi'}{\partial x}\right\}$$

$$+J(\Psi',\tilde{\Delta}\Psi')-\tilde{\Delta}^2\Psi'=0 \quad \text{in} \quad \mathscr{V}, \tag{33.7a}$$

$$\Psi'|_{\tilde{a}}=\left.\frac{\partial\Psi'}{\partial r}\right|_{\tilde{b}}=\left.\frac{\partial\Psi'}{\partial r}\right|_{\tilde{a}}=0 \tag{33.7b}$$

and

$$\Psi'|_{\tilde{b}}=\tfrac{1}{2}\langle u'\rangle(\tilde{b}^2-\tilde{a}^2) \tag{33.7c}$$

where

$$J(\Psi',\tilde{\Delta}\Psi')=\frac{1}{r}\left(\frac{\partial\Psi'}{\partial r}\frac{\partial}{\partial x}\tilde{\Delta}\Psi'-\frac{\partial\Psi'}{\partial x}\frac{\partial}{\partial r}\tilde{\Delta}\Psi'\right)+\frac{2}{r^2}\frac{\partial\Psi'}{\partial x}\tilde{\Delta}\Psi'$$

and

$$\tilde{\Delta}=\frac{\partial^2}{\partial r^2}-\frac{1}{r}\frac{\partial}{\partial r}+\frac{\partial^2}{\partial x^2}.$$

The total mass flux $\langle \bar{V}_x\rangle$ of any axisymmetric statistically stationary solution of (33.7) *may be written as*

$$\langle\bar{V}_x\rangle=\langle U_x\rangle+2\bar{\Psi}'|_{\tilde{b}}/(\tilde{b}^2-\tilde{a}^2) \tag{33.8}$$

where $\langle U_x\rangle$ is the mass flux for a suitably chosen laminar flow with pressure gradient $P_l[U_x]$. The bifurcating flow and laminar flow have the same mass flux $\langle U_x\rangle=\langle\bar{V}_x\rangle$ if and only if $\bar{\Psi}'|_{\tilde{b}}=0$.

It is convenient, and completely general, to restrict one's attention to the special case $\bar{\Psi}'|_{\tilde{b}}=0$ in the construction of the bifurcating solution. To completely specify the bifurcation problem, it will be necessary to fix the spatial periodicity. Then the time-periodic solution which bifurcates from laminar flow is determined uniquely to within an arbitrary phase. This unique solution may be computed relative to a laminar flow for which $\langle U_x\rangle=\langle\bar{V}_x\rangle$. To show this we shall reduce problem (33.7) to the study of the bifurcation of laminar flow with $\langle U_x\rangle=\langle\bar{V}_x\rangle$.

The stream function for the bifurcating solution may always be written as

$$\tilde{\tilde{\Psi}}=\tilde{\Psi}(r)+\Psi', \qquad \Psi'=\Phi(r)+\Psi'' \tag{33.9a}$$

where $\Phi(r)$ is a function of r alone which can be chosen so that

$$\Phi(\tilde{a}) = \frac{d\Phi}{dr}(\tilde{a}) = \frac{d\Phi}{dr}(\tilde{b}) = 0, \qquad \Phi(\tilde{b}) = \Psi'|_{\tilde{b}} \tag{33.9b}$$

and $\Psi''(x,r,t)$ is a flow which satisfies (33.7a, b, c) and has zero mass flux:

$$\Psi''|_{\tilde{b}} = 0.$$

$\Phi(r)$ *is a Poiseuille flow. The mass flux for the Poiseuille flow* $\tilde{\Psi}(r) + \Phi(r)$ *is* $\langle \bar{V}_x \rangle$ and, as a consequence of (30.6), we have that for any laminar flow

$$\langle \bar{V}_x - U_x \rangle = \langle \bar{u}' \rangle = \left\langle \frac{1}{r}\frac{d\Phi}{dr} \right\rangle = P_l[\bar{u}']\, b(\eta). \tag{33.10}$$

Moreover,

$$\langle h(r)\, w'u' \rangle = \langle h(r)\, w''u'' \rangle = 2(P[\bar{V}_x] - P_l[\bar{V}_x])\, b(\eta). \tag{33.11}$$

The proof (33.11) is left as Exercise 33.2. The bifurcation problem for the double-primed variables may be obtained from (33.7) using (33.9a, b) and the relation

$$\Psi'' = \varepsilon \langle \bar{V}_x \rangle \Psi = 2\varepsilon \bar{R} \Psi \tag{33.12}$$

where $\bar{R}$ is defined by (28.3) and

$$\varepsilon^2 \langle \bar{V}_x \rangle^2 \langle h(r)\, wu \rangle = 2(P[\bar{V}_x] - P_l[\bar{V}_x])\, b(\eta).$$

By choosing a normalizing condition in the form

$$\langle h(r)\, wu \rangle = b(\eta)/2, \tag{33.13a}$$

we find that

$$\varepsilon^2 = f - f_l \tag{33.13b}$$

is the friction factor discrepancy where

$$f = \hat{P}[\bar{\hat{V}}_x]\, 2D_h \Big/ \left[\frac{2}{b^2 - a^2} \int_a^b r \bar{\hat{V}}_x dr\right]^2 = 4P[\bar{V}_x]/\langle \bar{V}_x \rangle^2 \tag{33.14}$$

and

$D_h = 2(b-a) = 2l$ is the hydraulic diameter (the ratio of four times the area of the cross section of the annulus to the wetted perimeter) and $P = \hat{P} l^3/\nu^2$. We rescale the time: $t \to t/\langle \bar{V}_x \rangle$.

With these definitions established we may now state the bifurcation problem in terms of the friction factor discrepancy. Thus,

$$\left[\frac{\partial}{\partial t}+U_{\mathrm{o}}\frac{\partial}{\partial x}\right]\tilde{\Delta}\Psi-r\frac{d}{dr}\left[\frac{1}{r}\frac{dU_{\mathrm{o}}}{dr}\right]\frac{\partial\Psi}{\partial x}+\varepsilon J(\Psi,\tilde{\Delta}\Psi)-\frac{1}{2\bar{R}}\tilde{\Delta}^2\Psi=0 \quad \text{in } \mathscr{V}, \tag{33.13c}$$

$$\Psi=\frac{\partial\Psi}{\partial r}=0 \quad \text{at} \quad r=\tilde{a},\tilde{b}, \tag{33.13d}$$

and

$$\Psi \text{ is } 2\pi\text{-periodic in } \alpha x \text{ and } \omega t. \tag{33.13e}$$

All solutions of (33.13) have the same mass flux $\langle U_x\rangle=\langle\bar{V}_x\rangle$. Eqs. (33.13) are a complete statement of the mathematical problem for the bifurcating solution. With α given, we seek solutions $[\Psi(x,r;\varepsilon,\alpha,\eta),\ \bar{R}(\varepsilon^2,\alpha,\eta),\ \omega(\varepsilon^2,\alpha,\eta)]$ of (33.13). The bifurcating solution is necessarily time-periodic when the solution of the spectral problem (Orr-Sommerfeld problem, see § 34(a)) is both unique and time-periodic.

There is a unique dimensionless mass flux $\langle\bar{V}_x\rangle=2\bar{R}=2\bar{R}(\varepsilon^2,\alpha,\eta)$ and associated frequencies $\omega(\varepsilon^2,\alpha,\eta)$ for which time-periodic solutions (33.13) exist. $\bar{R}(\varepsilon^2,\alpha,\eta)$ gives the response curve for Poiseuille flow near the point of bifurcation; this is the relation between the Reynolds number and the friction factor.

Exercise 33.1: Show that when $\langle U_x\rangle=\langle\bar{V}_x\rangle$,

$$R=U_m l/\nu=2U_m\bar{R}/\langle\bar{V}_x\rangle$$

and that $R=4\bar{R}$ when $\eta=0$ and $R=3\bar{R}$ when $\eta=1$.

Exercise 33.2: Prove $\Phi(r)$ is a Poiseuille flow with mass flux $\langle\overline{u}'\rangle$. Prove (33.10) and (33.11). Hint: Note that $\langle f(r)w'\rangle=0$ for any function $f(r)$ of r alone. Note also that $P_l[U_x]+P_l[\overline{u}']=P_l[U_x+\overline{u}']=P_l[\bar{V}_x]$.

§ 34. Construction of the Bifurcating Solution

The bifurcating solution may be constructed by direct application of the perturbation method described in Chapter II. It is instructive, and useful in computations, to proceed directly in the stream function formulation.

(a) The Spectral Problem

The spectral problem for the instability of annular duct flow (17.8) is obtained from (33.13c,d) by setting $\varepsilon=0$ in (33.13c). Eqs. (33.13c,d) can then be written as

$$\frac{\partial}{\partial t}\tilde{\Delta}\Psi+\mathscr{L}\Psi=0, \quad \Psi=\frac{\partial\Psi}{\partial r}=0 \quad \text{at} \quad r=\tilde{a},\tilde{b}, \frac{2\pi}{\alpha}\text{-periodicity in } x. \tag{34.1a,b,c}$$

where

$$\hat{\mathscr{L}}[U_o,\lambda] = U_o \frac{\partial}{\partial x}\tilde{\Delta} - r\frac{d}{dr}\left(\frac{1}{r}\frac{dU_o}{dr}\right)\frac{\partial}{\partial x} - \frac{\lambda}{2}\tilde{\Delta}^2 \tag{34.2}$$

and

$$\lambda = 1/\bar{R}\,.$$

To obtain the spectral problem, the solutions of (34.1) are sought in the form

$$\Psi(x,r,t;\lambda) = e^{-\sigma(\lambda)t}\psi(x,r;\lambda)\,. \tag{34.3}$$

This leads to

$$-\sigma\tilde{\Delta}\psi + \hat{\mathscr{L}}\psi = 0; \quad \psi = \frac{\partial\psi}{\partial r} = 0 \quad \text{at} \quad r = \tilde{a},\tilde{b}, \frac{2\pi}{\alpha}\text{-periodicity in } x. \tag{34.4a, b, c}$$

The values $\sigma(\lambda) = \hat{\xi}(\lambda) + i\hat{\eta}(\lambda)$ are eigenvalues of the spectral problem (34.4a, b, c). If $\hat{\xi}(\lambda) < 0$, then the flow $U_o(r)$ is unstable. For large values of λ (small $\bar{R}$), $\hat{\xi}(\lambda) > 0$ for all eigenvalues $\sigma(\lambda)$. The border between stability and instability to small disturbances is defined by the critical values $\lambda = \lambda_0 = 1/\bar{R}$ where $\hat{\xi}(\lambda_0) = 0$ (see Eq. 34.7). At criticality $\sigma(\lambda_0) = i\hat{\eta}(\lambda_0) = i\omega_0$ and $\mathscr{L}[U_o,\lambda_0] = \mathscr{L}_0$. If $\sigma(\lambda)$ is a simple eigenvalue of (34.4), then $\bar{\sigma}(\lambda) = \hat{\xi}(\lambda) - i\hat{\eta}(\lambda)$, the complex conjugate of $\sigma(\lambda)$, is also an eigenvalue. The functions $\psi(x,r;\lambda)$ and $\bar{\psi}(x,r;\lambda)$ are eigenfunctions of (34.4) belonging, respectively, to the eigenvalues σ and $\bar{\sigma}$ of (34.4).

All solutions of (34.4) are expressible as a superposition of solutions in the form

$$\psi(x,r;\lambda(\alpha)) = e^{i\alpha x}\phi(r;\lambda(\alpha))\,. \tag{34.5}$$

Substitution of (34.5) in (34.4) leads to the Orr-Sommerfeld problem for annular duct flow

$$i\alpha 2\bar{R}\left\{(U_o - \sigma/i\alpha)\mathscr{L}\phi - r\frac{d}{dr}\left(\frac{1}{r}\frac{dU_o}{dr}\right)\phi\right\} = \mathscr{L}^2\phi \tag{34.6a}$$

where

$$\phi = \frac{d\phi}{dr} = 0 \quad \text{at} \quad r = \tilde{a},\tilde{b} \tag{34.6b}$$

and

$$\mathscr{L} = \frac{d^2}{dr^2} - \frac{1}{r}\frac{d}{dr} - \alpha^2.$$

We note that eigenfunctions proportional to $\exp\{i[\alpha x - \text{im}(\sigma)t]\}$ are constant to observers translating down the axis of the pipe at a speed $dx/dt = \text{im}(\sigma)/\alpha$.

The values

$$\bar{R}(0,\alpha,\eta) \quad \text{for which} \quad \mathrm{re}[\sigma(\bar{R})]=0 \tag{34.7}$$

are the critical Reynolds numbers for disturbances which are $2\pi/\alpha$-periodic in x. The locus of values of $\bar{R}(0,\alpha,\eta)$ as α changes when η is fixed is called a neutral curve. The neutral curve $\bar{R}(0,\alpha,1)$ shown as $\hat{\xi}=0$ in Fig. 34.1 is for the plane Orr-Sommerfeld problem $(D\equiv d/dz)$

$$L(\alpha R,\alpha^2,\sigma/\alpha)\phi \equiv \frac{-i}{2\alpha\bar{R}}(D^2-\alpha^2)^2\phi-\left(U_0-\frac{\sigma}{i\alpha}\right)(D^2-\alpha^2)\phi+D^2U_0\phi=0,$$
$$\phi=D\phi=0 \quad \text{at} \quad z=\pm\tfrac{1}{2} \tag{34.8a,b}$$

governing the linearized stability of the plane Poiseuille flow $U_0=\frac{3}{2}(1-4z^2)$. The minimum critical value

$$\bar{R}_L(\eta)=\min_{\alpha<0}\bar{R}(0,\alpha,\eta)=\bar{R}(0,\tilde{\alpha},\eta) \tag{34.9}$$

determines the length $2\pi/\tilde{\alpha}$ of the most unstable axisymmetric wave. The values $\bar{R}_L(\eta)$ are shown in the top curve of Fig. 35.1.

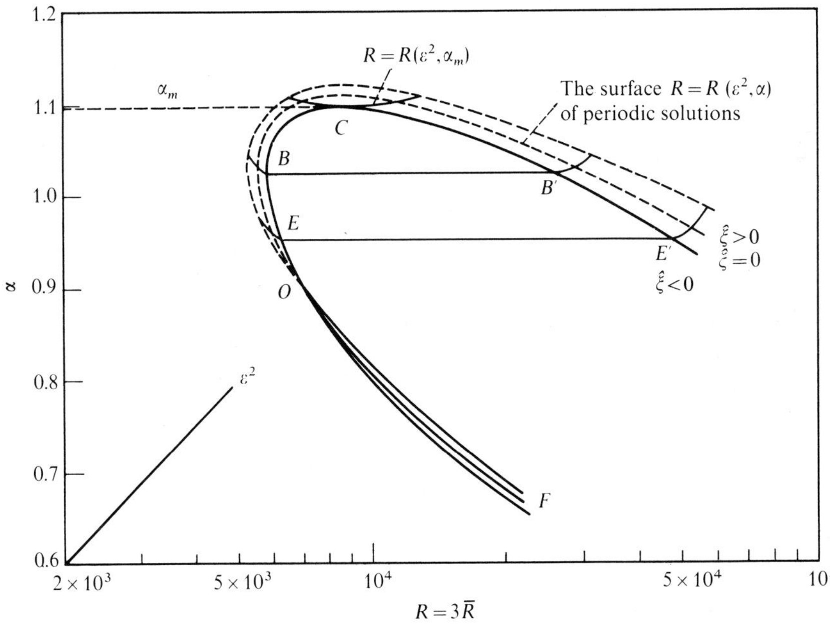

Fig. 34.1: Schematic sketch of the surface of periodic solutions which bifurcate from plane Poiseuille flow. The dotted lines indicate subcritical bifurcation (Chen and Joseph, 1973)

(b) The Perturbation Series

As in the general theory of Chapter II, the construction of the bifurcating solution requires that solvability conditions for the perturbation problem be expressed as a condition of orthogonality to adjoint eigenfunctions. To define these eigenfunctions we first introduce the scalar product

$$(f,g)=\langle f,\bar{g}\rangle \tag{34.10}$$

which is defined for f and g which are $2\pi/\alpha$-periodic in x and $g=f=\partial g/\partial r=\partial f/\partial r=0$ at $r=\tilde{a},\tilde{b}$. The operator $-\bar{\sigma}\tilde{\Delta}^*+\hat{\mathscr{L}}^*$, the adjoint of $-\sigma\tilde{\Delta}+\hat{\mathscr{L}}$, is defined by the requirement that

$$(f,[-\sigma\tilde{\Delta}+\hat{\mathscr{L}}]g)=([-\bar{\sigma}\tilde{\Delta}^*+\hat{\mathscr{L}}^*]f,g)\,. \tag{34.11}$$

With the aid of (34.2) and (34.11) one finds that

$$\hat{\mathscr{L}}^* f=-\tilde{\Delta}^*\left(U_0\frac{\partial f}{\partial x}\right)+r\frac{d}{dr}\left(\frac{1}{r}\frac{dU_0}{dr}\right)\frac{\partial f}{\partial x}-\frac{\lambda}{2}\tilde{\Delta}^{*2} f$$

where

$$\tilde{\Delta}^*=\frac{\partial^2}{\partial r^2}+\frac{3}{r}\frac{\partial}{\partial r}+\frac{\partial^2}{\partial x^2}\,.$$

The adjoint eigenvalue problem is therefore defined by

$$\begin{gathered}-\bar{\sigma}\tilde{\Delta}^*\psi^*+\hat{\mathscr{L}}^*\psi^*=0,\qquad \psi^*=\frac{\partial\psi^*}{\partial r}=0\quad\text{at}\quad r=\tilde{a},\tilde{b}\,,\\ \psi^*\ \text{is}\ \frac{2\pi}{\alpha}\text{-periodic in } x.\end{gathered} \tag{34.12}$$

Bifurcation analysis gives the solutions of (33.13a, b, c) of permanent form which are $2\pi/\alpha$-periodic in x and continuous in ε at the point $\varepsilon=0$. Such solutions are time-periodic when the frequency at criticality is not zero, $\hat{\eta}(\bar{R}(0,\alpha,\eta))=\omega_0(\alpha,\eta)\neq 0$. We seek time-periodic solutions with frequencies $\omega(\varepsilon^2,\alpha,\eta)$ which reduce continuously to the time-periodic solution of the linear theory with frequency ω_0. Such a solution of permanent form is possible only for certain "critical values" $\bar{R}(\varepsilon^2,\alpha,\eta)$. When $\varepsilon=0$, such a solution of permanent form is a neutral solution; it can exist only when $\bar{R}=\bar{R}(0,\alpha,\eta)$. The neutral solution loses its stability when $\bar{R}$ is increased if, at criticality,

$$\hat{\xi}'(\alpha,\eta)=\frac{d\hat{\xi}}{d\lambda}=\frac{-1}{\bar{R}^2}\frac{d\hat{\xi}}{d\bar{R}}>0.$$

To obtain a formula for the derivative σ' of $\sigma(\lambda)$ we differentiate (34.4) with respect to $\bar{R}$ and apply the solvability condition. This leads to

$$\sigma'=\hat{\xi}'+i\hat{\eta}'=-\tfrac{1}{2}(\tilde{\Delta}^2\psi,\psi^*)/(\tilde{\Delta}\psi,\psi^*)\,. \tag{34.13}$$

As in Chapter II we set $\omega t = s$ and reformulate the problem (33.13) as

$$\mathscr{J}\Psi + \varepsilon J(\Psi, \tilde{\Delta}\Psi) = 0, \quad \Psi = \frac{\partial \Psi}{\partial r} = 0 \quad \text{at} \quad r = \tilde{a}, \tilde{b}, \tag{34.13a, b}$$

$\Psi(x,r,s)$ is 2π-periodic in αx and s,

$$\frac{b(\eta)}{2} = -\left\langle \frac{h(r)}{r^2} \frac{\partial \Psi}{\partial r} \frac{\partial \Psi}{\partial x} \right\rangle \tag{34.13c}$$

where

$$\mathscr{J}[\omega, \lambda] = \omega \frac{\partial}{\partial s} \tilde{\Delta} + \hat{\mathscr{L}}[U_0, \lambda]. \tag{34.14}$$

The solution of (34.13) can be constructed as a Taylor series

$$\begin{bmatrix} \Psi(x,r,s;\varepsilon,\alpha,\eta) \\ \omega(\varepsilon^2,\alpha,\eta) \\ \lambda(\varepsilon^2,\alpha,\eta) \end{bmatrix} = \sum_{l=0} \varepsilon^l \begin{bmatrix} \Psi_l(x,r,s;\alpha,\eta) \\ \omega_l(\alpha,\eta) \\ \lambda_l(\alpha,\eta) \end{bmatrix}. \tag{34.15}$$

Substitution of the series (34.15) into (34.13) gives a sequence of problems for the Taylor coefficients, as in Chapter II. We list the first three problems:

$$\mathscr{J}_0 \Psi_0 = 0, \quad \Psi_0 = \frac{\partial \Psi_0}{\partial r} = 0 \quad \text{at} \quad r = \tilde{a}, \tilde{b}; \quad \frac{b(\eta)}{2} = -\left\langle \frac{h(r)}{r^2} \frac{\partial \Psi_0}{\partial r} \frac{\partial \Psi_0}{\partial x} \right\rangle; \tag{34.16}$$

$$\begin{gathered} \mathscr{J}_0 \Psi_1 + \mathscr{J}_1 \Psi_0 + J_0 = 0, \quad \Psi_1 = \frac{\partial \Psi_1}{\partial r} = 0 \quad \text{at} \quad r = \tilde{a}, \tilde{b}, \\ 0 = \left\langle \frac{h(r)}{r^2} \left(\frac{\partial \Psi_0}{\partial r} \frac{\partial \Psi_1}{\partial x} + \frac{\partial \Psi_1}{\partial r} \frac{\partial \Psi_0}{\partial x} \right) \right\rangle; \end{gathered} \tag{34.17}$$

$$\begin{gathered} \mathscr{J}_0 \Psi_2 + \mathscr{J}_1 \Psi_1 + \mathscr{J}_2 \Psi_0 + J_1 = 0, \quad \Psi_2 = \frac{\partial \Psi_2}{\partial r} = 0 \quad \text{at} \quad r = \tilde{a}, \tilde{b}, \\ 0 = \left\langle \frac{h(r)}{r^2} \left(\frac{\partial \Psi_0}{\partial r} \frac{\partial \Psi_2}{\partial r} + \frac{\partial \Psi_1}{\partial r} \frac{\partial \Psi_1}{\partial x} + \frac{\partial \Psi_2}{\partial r} \frac{\partial \Psi_0}{\partial x} \right) \right\rangle \end{gathered} \tag{34.18a}$$

where

$$\begin{gathered} \mathscr{J}_0 = \mathscr{J}[\omega_0, \lambda_0], \quad \mathscr{J}_l = \omega_l \frac{\partial}{\partial s} \tilde{\Delta} - \frac{\lambda_l}{2} \tilde{\Delta}^2, \\ J_0 = J(\Psi, \tilde{\Delta}\Psi_0), \quad J_1 = J(\Psi_1, \tilde{\Delta}\Psi_0) + J(\Psi_0, \tilde{\Delta}\Psi_1). \end{gathered} \tag{34.18b}$$

Only two solutions of (34.16) are possible when $\sigma = \sigma(\lambda_0)$ $(\lambda_0 = \bar{R}^{-1}(\alpha))$ is the only eigenvalue and is a simple eigenvalue of $\mathscr{L}_0$. These are Z_1 and $Z_2 = \bar{Z}_1$ where

$$Z_1 = e^{-is} \psi_0(x, r; \lambda_0(\alpha, \eta), \eta).$$

The real-valued solution

$$\Psi_0 = 2\,\mathrm{re}(Z_1) \tag{34.18c}$$

is unique up to arbitrary shifts in the origin of s. The amplitude of the solution is fixed by the last of the conditions (34.16).

The problem which is adjoint to (34.16) relative to the scalar product,

$$[\cdot,\cdot] = \frac{1}{2\pi}\int_0^{2\pi}(\cdot,\cdot)\,ds\,,$$

is

$$\mathscr{J}_0^*\Psi_0^* = \left(-\omega_0\frac{\partial}{\partial s}\tilde{\Delta} + \hat{\mathscr{L}}_0^*\right)\Psi_0^* = 0\,, \qquad \Psi_0^* = \frac{\partial \Psi_0^*}{\partial r} = 0 \quad \text{at} \quad r = \tilde{a}, \tilde{b}\,. \tag{34.19}$$

Suppose that $\pm i\omega_0$ are simple eigenvalues of $\mathscr{L}_0$ and consider the equations $(n = 1, 2, \ldots)$

$$\mathscr{J}_0\Psi_n = f_n\,, \qquad \Psi_n = \frac{\partial \Psi_n}{\partial r} = 0 \quad \text{at} \quad r = \tilde{a}, \tilde{b} \tag{34.20}$$

where the f_n are 2π-periodic in αx and s and Ψ_n has this periodicity. As in Chapter II, the inhomogeneous problems (34.20) are solvable if and only if

$$[f_n, Z_1^*] = [f_n, Z_2^*] = 0\,. \tag{34.21}$$

where $Z_1^* = e^{-is}\Psi_0^*$, $Z_2^* = \bar{Z}_1^*$ and Ψ_0^* is an eigenfunction of (34.12) at criticality. The solutions $\Psi_n(x,r,s)$ are not unique; $\Psi_n + A_n\Psi_0$ is also a solution. The values A_n are however uniquely determined by the normalizing condition

$$\left\langle \frac{h(r)}{r^2}\left(\frac{\partial \Psi}{\partial r}\,\frac{\partial \Psi}{\partial x}\right)_n \right\rangle = 0\,. \tag{34.22}$$

Application of the first of the solvability conditions (34.21) (with $[f_n, z_1^*] \equiv [f_n]$) to (34.17) leads one to

$$[\mathscr{J}_1\Psi_0] + [J_0] = 0\,, \qquad [\mathscr{J}_1\Psi_1] + [\mathscr{J}_2\Psi_0] + [J_1] = 0\,. \tag{34.23}$$

Since J_0 contains no terms proportional to the factor e^{-is} of Z_1^* we find, as in § 12, that $[J_0] = 0$. This implies that $\omega_1 = \lambda_1 = \mathscr{J}_1 = 0$; in fact, by (12.6) and (12.7), $\omega_{2l+1} = \lambda_{2l+1} = 0$ and

$$\begin{bmatrix} \omega(\varepsilon^2, \alpha, \eta) \\ \lambda(\varepsilon^2, \alpha, \eta) \end{bmatrix} = \sum_{l=0} \varepsilon^{2l} \begin{bmatrix} \omega_{2l} \\ \lambda_{2l} \end{bmatrix}. \tag{34.24}$$

The first nonlinear correction is given by (34.23),

$$-i\omega_2+\lambda_2\sigma'+[J_1]/(\tilde{\Delta}\psi_0,\psi_0^*)=0\,, \tag{34.25}$$

and can be found from the solutions of (34.16) and (34.17).

To compute (34.25) we must compute solutions at orders zero and one. At criticality $\sigma=i\omega_0$ and

$$\begin{bmatrix}\psi_0(x,r)\\ \psi_0^*(x,r)\end{bmatrix}=\begin{bmatrix}\phi_0(r)\\ \phi_0^*(r)\end{bmatrix}e^{i\alpha x}. \tag{34.26}$$

The Orr-Sommerfeld system (34.27b) and its adjoint (34.27a) are:

$$\left(U_0-\frac{\omega_0}{\alpha}\right)\left(\phi_0^{*\prime\prime}+\frac{3}{r}\phi_0^{*\prime}-\alpha^2\phi_0^*\right)+2U_0'\left(\phi_0^{*\prime}+\frac{2}{r}\phi_0^*\right)$$
$$=-\frac{\lambda_0}{2i\alpha}\left[\phi_0^{*iv}+\frac{6}{r}\phi_0^{*\prime\prime\prime}+\left(\frac{3}{r^2}-2\alpha^2\right)\phi_0^{*\prime\prime}-\frac{1}{r}\left(\frac{3}{r^2}-6\alpha^2\right)\phi_0^{*\prime}+a^4\phi_0^*\right], \tag{34.27a}$$

$$\left(U_0-\frac{\omega_0}{\alpha}\right)\left(\phi_0''-\frac{1}{r}\phi_0'-\alpha^2\phi_0\right)-\left(U_0''-\frac{1}{r}U_0'\right)\phi_0$$
$$=\frac{\lambda_0}{2i\alpha}\left[\phi_0^{iv}-\frac{2}{r}\phi_0'''+\left(\frac{3}{r^2}-2\alpha^2\right)\left(\phi_0''-\frac{1}{r}\phi_0'\right)+\alpha^4\phi_0\right]. \tag{34.27b}$$

$$\phi_0=\phi_0'=\phi_0^*=\phi_0^{*\prime}=0\quad\text{at}\quad r=\tilde{a},\tilde{b}\,. \tag{34.27c}$$

Here the primes denote differentiation and ϕ_0 is normalized in accord with (34.16),

$$\left[\frac{4\alpha}{\tilde{b}^2-\tilde{a}^2}\int_{\tilde{a}}^{\tilde{b}}\text{im}(\phi_0\bar{\phi}_0')\,dr+\frac{2\alpha}{\ln\eta}\int_{\tilde{a}}^{\tilde{b}}\text{im}(\phi_0\bar{\phi}_0')\frac{dr}{r^2}\right]=b(\eta)/2\,. \tag{34.27d}$$

Next, attention is directed to the first-order system. With $\mathscr{J}_1=0$, one can write (34.17) as

$$\mathscr{J}_0\Psi_1+J_0=0\,,\qquad \Psi_1=\frac{\partial\Psi_1}{\partial r}=0\quad\text{at}\quad r=\tilde{a},\tilde{b} \tag{34.28a, b}$$

$$0=\left\langle\frac{h(r)}{r^2}\left(\frac{\partial\Psi_0}{\partial r}\frac{\partial\Psi_1}{\partial x}+\frac{\partial\Psi_1}{\partial r}\frac{\partial\Psi_0}{\partial x}\right)\right\rangle. \tag{34.28c}$$

Now, from (34.26) one finds

$$J_0=J(\Psi_0,\tilde{\Delta}\Psi_0)=Ae^{2i(\alpha x-s)}+\bar{A}e^{-2i(\alpha x-s)}+B+\bar{B} \tag{34.29}$$

where

$$A=\frac{i\alpha}{r}\left[\left(\phi_0'+\frac{2}{r}\phi_0\right)\left(\phi_0''-\frac{1}{r}\phi_0'-\alpha^2\phi_0\right)-\phi_0\left(\phi_0'''-\frac{1}{r}\phi_0''+\frac{1}{r^2}\phi_0'-\alpha^2\phi_0'\right)\right]$$

and

$$B=\frac{i\alpha}{r}\left[\left(\bar{\phi}_0'-\frac{2}{r}\bar{\phi}_0\right)\left(\phi_0''-\frac{1}{r}\phi_0'-\alpha^2\phi_0\right)+\bar{\phi}_0\left(\phi_0'''-\frac{1}{r}\phi_0''+\frac{1}{r^2}\phi_0'-\alpha^2\phi_0'\right)\right].$$

By linearity, one may write

$$\Psi_1=\phi_{11}(r)e^{2i(\alpha x-s)}+\bar{\phi}_{11}(r)e^{-2i(\alpha x-s)}+\phi_{12}+\bar{\phi}_{12}+a_0\Psi_0$$

where $a_0\Psi_0$ is a multiple of the solution of the system (34.27). Application of (34.28c) leads to $a_0=0$. This results in

$$\Psi_1=\phi_{11}(r)e^{2i(\alpha x-s)}+\bar{\phi}_{11}(r)e^{-2i(\alpha x-s)}+\phi_{12}+\bar{\phi}_{12}\,. \tag{34.30}$$

Substitution of (34.30) and (34.29) into (34.28a, b) leads to

$$\mathscr{I}_0(\phi_{11}e^{2i(\alpha x-s)})+Ae^{2i(\alpha x-s)}=0\,,\qquad \phi_{11}=\phi'_{11}=0\quad\text{at}\quad r=\tilde{a},\tilde{b} \tag{34.31a, b}$$

and

$$\mathscr{I}_0\phi_{12}+B=0\,,\qquad \phi_{12}=\phi'_{12}=0\quad\text{at}\quad r=\tilde{a},\tilde{b}\,. \tag{34.32a, b}$$

Working out (34.31a) and (34.32a), one obtains

$$\left(U_0-\frac{\omega_0}{\alpha}\right)\left(\phi''_{11}-\frac{1}{r}\phi'_{11}-4\alpha^2\phi_{11}\right)-\left(U''_0-\frac{1}{r}U'_0\right)\phi_{11}$$
$$-\frac{\lambda_0}{4i\alpha}\left[\phi^{iv}_{11}-\frac{2}{r}\phi'''_{11}+\left(\frac{3}{r^2}-8\alpha^2\right)\left(\phi''_{11}-\frac{1}{r}\phi'_{11}\right)+16\alpha^4\phi_{11}\right]=-\frac{A}{2i\alpha}$$

and

$$\frac{\lambda_0}{2}\left(\phi^{iv}_{12}-\frac{2}{r}\phi'''_{12}+\frac{3}{r^2}\phi''_{12}-\frac{3}{r^3}\phi'_{12}\right)=B\,.$$

To calculate λ_2 and ω_2 from (34.25), one begins with the evaluation of $[J_1]$. By employing (34.18c), (34.26) and (34.30) one finds

$$[J_1]=\frac{2}{\tilde{b}^2-\tilde{a}^2}\int_{\tilde{a}}^{\tilde{b}}H\bar{\phi}^*_0\,r\,dr \tag{34.33}$$

where

$$H=\frac{i\alpha}{r}\left\{(\phi'_{12}+\bar{\phi}'_{12})\left(\phi''_0-\frac{1}{r}\phi'_0-\alpha^2\phi_0\right)-\phi'_{11}\left(\bar{\phi}''_0-\frac{1}{r}\bar{\phi}'_0-\alpha^2\bar{\phi}_0\right)\right.$$
$$-2\phi_{11}\left[\bar{\phi}'''_0-\frac{1}{r}\bar{\phi}''_0+\left(\frac{1}{r^2}-\alpha^2\right)\bar{\phi}'_0\right]+\frac{4}{r}\phi_{11}\left(\bar{\phi}''_0-\frac{1}{r}\bar{\phi}'_0-\alpha^2\bar{\phi}_0\right)$$
$$+2\bar{\phi}'_0\left(\phi''_{11}-\frac{1}{r}\phi'_{11}-4\alpha^2\phi_{11}\right)-\phi_0\left[\phi'''_{12}+\bar{\phi}'''_{12}-\frac{1}{r}(\phi''_{12}+\bar{\phi}''_{12})+\frac{1}{r^2}(\phi'_{12}+\bar{\phi}'_{12})\right]$$
$$+\bar{\phi}_0\left[\phi'''_{11}-\frac{1}{r}\phi''_{11}+\left(\frac{1}{r^2}-4\alpha^2\right)\phi'_{11}\right]+\frac{2}{r}\phi_0\left[\phi''_{12}+\bar{\phi}''_{12}-\frac{1}{r}(\phi'_{12}+\bar{\phi}'_{12})\right]$$
$$\left.-\frac{2}{r}\bar{\phi}_0\left(\phi''_{11}-\frac{1}{r}\phi'_{11}-4\alpha^2\phi_{11}\right)\right\}.$$

Eq. (34.33) can be further simplified by integration by parts

$$[J_1]=\frac{2\alpha i}{\tilde{b}^2-\tilde{a}^2}\left\{\int_{\tilde{a}}^{\tilde{b}}\bar{\phi}^{*\prime}_0[-\bar{\phi}_0\phi''_{11}-\bar{\phi}'_0\phi'_{11}+2\bar{\phi}''_0\phi_{11}+\phi_0(\phi''_{12}+\bar{\phi}''_{12})-\phi'_0(\phi'_{12}+\bar{\phi}'_{12})]\,dr\right.$$
$$+\int_{\tilde{a}}^{\tilde{b}}\bar{\phi}^*_0\left[-(\phi'_{12}+\bar{\phi}'_{12})\left(\frac{1}{r}\phi'_0+\frac{3}{r^2}\phi_0-\alpha^2\phi_0\right)+\frac{3}{r}\phi_0(\phi''_{12}+\bar{\phi}''_{12})\right.$$
$$+\phi_{11}\left(\frac{6}{r}\bar{\phi}''_0-\frac{6}{r^2}\bar{\phi}'_0-6\alpha^2\bar{\phi}'_0+\frac{4}{r}\alpha^2\bar{\phi}_0\right)$$
$$\left.\left.-\phi_{11}\left(\frac{1}{r}\bar{\phi}'_0-\frac{3}{r^2}\bar{\phi}_0+3\alpha^2\bar{\phi}_0\right)-\frac{3}{r}\bar{\phi}_0\phi''_{11}\right]dr\right\}. \tag{34.34}$$

Using (34.34), the values of λ_2 and ω_2 can be evaluated from (34.25). By separating (34.25) into real and imaginary parts, we find that

$$\lambda_2 = -\operatorname{re}([J_1]/(\tilde{\Delta}\psi_0,\psi_0^*))/\operatorname{re}\sigma',$$
$$\omega_2 = \lambda_2 \operatorname{im}\sigma' + \operatorname{im}([J_1]/(\tilde{\Delta}\psi_0,\psi_0^*))$$

where $\sigma' = -\frac{1}{2}(\tilde{\Delta}\psi_0,\psi_0^*)/(\tilde{\Delta}\psi_0,\psi_0^*)$ at criticality.

We use (34.10) and (34.26) to find that

$$(\tilde{\Delta}\psi_0,\psi_0^*) = -\frac{2}{\tilde{b}^2-\tilde{a}^2}\int_{\tilde{a}}^{\tilde{b}}\left(\phi_0'\bar{\phi}_0^{*\prime} + \frac{2}{r}\phi_0'\bar{\phi}_0^* + \alpha^2\phi_0\bar{\phi}_0^*\right)r\,dr$$

and

$$(\tilde{\Delta}^2\psi_0,\psi_0^*) = \frac{2}{\tilde{b}^2-\tilde{a}^2}\int_{\tilde{a}}^{\tilde{b}}\left[\phi_0''\bar{\phi}_0^{*\prime\prime} + \frac{4}{r}\phi_0''\bar{\phi}_0^{*\prime} - \left(\frac{3}{r^2}-2\alpha^2\right)\phi_0'\bar{\phi}_0^{*\prime} + \frac{4}{r}\alpha^2\phi_0'\bar{\phi}_0^* + \alpha^4\phi_0\bar{\phi}_0^*\right]r\,dr\,.$$

The reader will readily formulate the problems which govern the higher order ε derivatives of the time-periodic bifurcating solution. It is clear from the constructions given in this section that the stream function series (34.15) involve x and s only in the combination $\alpha x - s = \theta$. It follows from this that

$$\Psi(x,r,s;\varepsilon) = \hat{\hat{\Psi}}(r,\theta;\varepsilon)\,. \tag{34.35}$$

Functions of x and s which are in the form given by (34.35) are statistically stationary.

Table 34.1: Parameter values at criticality for $\eta = 1/1.01$ (Joseph and Chen, 1974)

$\frac{\alpha}{2}$	$\bar{R}\left(0,\frac{\alpha}{2},\eta\right)$	$c_r = \frac{\omega_0}{\alpha}$	$\frac{\xi'}{2}$	λ_2	$\frac{\omega_2}{2}$	$\operatorname{re}\frac{2[J_1]}{(\tilde{\Delta}\psi_0,\psi_0^*)}$
0.650	14923	0.2483	200.63	−0.3821	339.89	76.66
0.700	10889	0.2733	162.89	−0.3726	286.26	60.70
0.750	8299	0.2974	135.55	−0.3448	257.55	46.73
0.800	6583	0.3203	114.91	−0.2886	248.50	33.16
0.850	5424	0.3417	98.72	−0.1911	257.50	18.86
0.900	4641	0.3612	85.46	−0.0314	287.17	2.68
0.950	4136	0.3782	73.94	0.2290	347.27	−16.93
1.000	3875	0.3918	62.92	0.6815	467.35	−42.88
1.020[a]	3846[a]	0.3959[a]	58.22[a]	0.9696[a]	551.11[a]	−56.44[a]
1.025	3848	0.3967	56.97	1.0580	577.77	−60.28
1.050	3924	0.3996	49.96	1.6718	773.16	−83.53
1.075	4206	0.3987	40.03	3.0007	1238.91	−120.13
1.090	4676	0.3935	29.08	5.6075	2234.54	−163.08
1.095	5064	0.3886	21.46	9.0251	3602.77	−193.69
1.09707	6000	0.3771	4.88	53.4886	22073.0	−261.13
1.09094	7333	0.3626	−16.78	−20.8865	−9156.86	−350.42
1.07654	9333	0.3450	−47.18	−10.1464	−4835.26	−478.69
1.05163	12667	0.3229	−94.31	−7.2881	−3877.26	−687.36
1.02528	16667	0.3036	−147.24	−6.3231	−3696.57	−931.01

[a] Parameters evaluated at this minimum critical point are designated with tilda overbar.

The values of parameters through second order have been computed by Joseph and Chen (1974) for $\eta = 1/1.01$, $\frac{1}{2}$ and $\frac{1}{3}$. Table 34.1 summarizes the results for $\eta=1/1.01$. The scale factors used by Joseph and Chen are slightly different from the ones used here; their $[t, x, \hat{\Psi}, \sigma, \alpha, \varepsilon, \lambda]$ correspond to our $[2t, 2x, 4\Psi, \sigma/2, \alpha/2, \varepsilon, \lambda]$.

(c) Some Properties of the Bifurcating Solution

Some of the more important properties of bifurcating solutions can be understood by studying Fig. 34.1. In the figure we have sketched the surface $\bar{R}(\varepsilon^2, \alpha, 1)=\bar{R}(\varepsilon^2, \alpha)$ of periodic solutions. The trace of this surface in the plane $\varepsilon=0$ is the neutral curve. Laminar Poiseuille flow is unstable for points $(\bar{R}, \alpha)$ which lie inside the neutral curve; points outside the neutral curve are deemed stable by the linear theory of stability. A glance at Fig. 34.1 shows that neighboring the arc OCE' there are $(\bar{R}, \alpha)$ points which lie on the surface of periodic solutions whose projections on the plane $\varepsilon=0$ are outside the neutral curve. Such points are in the $(\bar{R}, \alpha)$ region $\hat{\xi}>0$ of linearized stability for laminar flow. Motions existing in this $(\bar{R}, \alpha)$ region are called subcritical. On the other hand, the points on the surface of periodic solutions neighboring OF are in the unstable region for laminar flow and are supercritical.

The properties of the sketch may be verified by inspecting the values in Table 34.1 noting that $\partial\bar{R}(0, \alpha)/\partial\varepsilon=0$ and

$$\frac{\partial\bar{R}(0, \alpha)}{\partial(\varepsilon^2)}=\frac{1}{2}\,\frac{\partial^2\bar{R}(0, \alpha)}{\partial\varepsilon^2}=-\tfrac{1}{2}\,[\bar{R}(0, \alpha)]^2\,\lambda_2\,.$$

At the point C, $\hat{\xi}'=0$, and it is preferable to represent the surface of periodic solutions as $f(\alpha, \bar{R}, \varepsilon^2)=0$. This may be solved for $\alpha(\bar{R}, \varepsilon^2)$; $\partial\alpha/\partial\bar{R}=0$ and $\partial\alpha/\partial\varepsilon^2>0$ at C. It follows that nonlinear periodic solutions exist with larger wave numbers $(\alpha(R, \varepsilon^2)>\alpha_m)$ than the largest value $\alpha=\alpha_m$ found on the neutral curve.

The bifurcation theory assumes that the eigenvalues σ are simple at criticality. The simplicity assumption has not yet been justified theoretically[3]. Computations for axisymmetric bifurcations are consistent with the assumption of simplicity and it has always been possible to find one and only one normalized solution of the Orr-Sommerfeld problem for each pair $(\bar{R}, \alpha)$. Given simplicity, theory guarantees that, apart from phase, only one axisymmetric periodic solution (with α fixed) bifurcates.

The Floquet analysis of the stability of bifurcating solutions which was developed in § 14 applies here. According to the factorization theorem, subcritical

[3] The restriction of the stability and bifurcation analysis to axisymmetric disturbances is artificial and undesirable. Three-dimensional disturbances, even when spatially periodic, will introduce another wave-number parameter and σ, at criticality, may lose simplicity for certain values of the parameters. The restriction to two-dimensional solutions for the plane Poiseuille flow problem is less restrictive than might at first be supposed. This follows from Squire's theorem (Exercise 34.2) which guarantees that the values $\bar{R}(\alpha)$ lying on the lower branch of the neutral curve (arc CF) are the smallest of the critical values $\bar{R}$ for three-dimensional disturbances. It follows that the two-dimensional disturbances will bifurcate first at the smallest values of $\bar{R}(\alpha)$; in particular, the solution which bifurcates at the minimum critical value $\bar{R}_L$ (at B) is two dimensional.

solutions are unstable so long as the friction factor discrepancy (ε^2) is a decreasing function of the Reynolds number R. For larger values of ε^2 the friction factor bifurcation curve is expected to turn back around (see Figs. 27.1 and 27.3) and to regain stability. Stability is in the sense of linear theory; moreover, here, stability is only with respect to small two-dimensional disturbances, 2π-periodic in αx. Therefore, the recovery of stability of supercritical solutions with larger amplitudes (see Fig. 15.1) may hold only in a restricted class of disturbances.

The numerical results of Table 34.1 show that for most values of α on the neutral curve of Fig. 34.1, the bifurcating solution is subcritical, unstable, and will not be seen in experiments. It is possible that the unstable time-periodic bifurcating solution can be observed as a transient when the parameters lie on or near the surface of periodic solutions.

Exercise 34.1: Formulate the linearized stability problem for Poiseuille flow through annular ducts for three-dimensional disturbances. Show that the assumption that the velocity components of the disturbance are independent of θ implies that the azimuthal (θ) component of the disturbance must vanish.

Exercise 34.2 (Squire, 1933): Consider the linearized stability problem for plane Poiseuille flow. Show that for disturbances which are proportional to $\exp[i(\alpha x+\beta y)]$,

$$L(\alpha\bar{R},\alpha^2+\beta^2,\sigma/\alpha)w=0, \qquad w=Dw=0 \quad \text{at} \quad z=\pm\tfrac{1}{2}.$$

Prove that $\bar{R}(0,\alpha)$ on the lower branch of the neutral curve for two-dimensional ($\beta=0$) disturbances (CF in Fig. 34.1) is also the locus of the smallest critical values for three-dimensional disturbances ($\beta\neq 0$). Proofs of this theorem are in Lin (1955), Stuart (1963) and Betchov and Criminale (1967).

Exercise 34.3 (Yih, 1973): Prove that

$$\min[U_x]\leqslant \operatorname{im}(\sigma/\alpha)\leqslant \max[U_x]$$

for all solutions of (34.8 a, b) with $\hat{\xi}\leqslant 0$.

Exercise 34.4: Assuming that σ is a simple eigenvalue at criticality, prove that subcritical bifurcating solutions are unstable.

Exercise 34.5: Show that bifurcating time-periodic Poiseuille flow has time-independent horizontal averages. Is this flow statistically stationary?

§ 35. Comparison of Theory and Experiment

This section is divided into two parts: in part (a) we compare the results of stability theories with experiments. The main conclusion is that though laminar flow is conditionally stable at subcritical values of the Reynolds number, the attracting radius δ of the laminar solution is very small; most natural disturbances are attracted to other solutions. In part (b) we discuss these other solutions. The bifurcating, time-periodic solutions are unstable and therefore cannot attract disturbances. The experimental evidence is that large disturbances are attracted to a stable set of solutions which we have called stable turbulence. Solutions in the attracting set seem to share a common response curve.

(a) Instability of Laminar Poiseuille Flow

The points in the shaded region of Fig. 35.1 are experimentally determined stability limits (collected by R. Hanks, 1963). Since the disturbances in these experiments were not controlled, we shall assume that the experimental limits in the shaded region approximate global limits of stability. This assumption presupposes that all disturbances which could destabilize a flow are actually present in the experiments.

The curves $R_{\mathscr{E}}(\eta)$ and $R_L(\eta)$ give numerically computed critical values of the energy and linear theory of stability. The values $R_{\mathscr{E}}(\eta)$ were obtained by the methods of §§ 21 and 22. When $R > R_{\mathscr{E}}(\eta)$ the energy of some disturbances will increase initially. The azimuthal periodicity of the most energetic initial disturbance is given by integers n corresponding to disturbances proportional to $\cos n\theta$. The critical values of n are shown on the curve $R_{\mathscr{E}}(\eta)$.

The critical values $R_L(\eta)$ are minimum critical values of $U_m(b-a)/\nu$ defined by (34.9) and in Exercise 33.1. The computation of these values (Mott and Joseph, 1968) assumes axial symmetry. This assumption can be justified when $\eta=1$ (Exercise 34.2). When $\eta=0$ (no inner cylinder) the numerical analyses of Davey and Drazin (1969) and Salwen and Grosch (1972) indicate that Hagen-Poiseuille flow is stable to all disturbances. Assuming that this same absolute stability holds in the limiting case, $R_L(\eta)\to\infty$ as $\eta\to 0$, we see that the curve $R_L(\eta)$ of Fig. (35.1) is correctly given at its endpoints independent of symmetry assumptions[4].

The experimentally determined values of the stability limit for Hagen-Poiseuille flow are approximately $U_m(b-a)/\nu = R \simeq 2000$. Older references to experiments are found in Bateman (1932). The form of the unstable disturbance which persists at the smallest R is unknown. On the one hand, in the experiments of Leite (1959) disturbances which were initially introduced with no symmetry became more axially symmetric as they moved downstream. On the other hand, Fox, Lessen and Bhat (1968) suggest that the most unstable disturbance is dominated by the $\cos\theta$ term of its Fourier decomposition. The critical disturbance of energy theory (see § 21) is also a spiral mode with an $n=1$ azimuthal periodicity. The axial wave length of the critical disturbance of energy theory is five times smaller than the one observed by Fox, Lessen and Bhat.

The only extensive experimental studies bearing on the stability of flow in annular ducts with $0<\eta<1$ are those of Walker, Whan and Rothfus (1957). Many of the experimental points on Fig. 35.1 are taken from these studies.

The stability limit $R\simeq 2100$ for Poiseuille flow in a channel $(\eta\to 1)$ was observed by Davies and White (1928) and is widely cited. Beavers, Sparrow and

[4] The critical values $R_L(\eta)$ given in Fig. 35.1 may not be the true critical values. To obtain the true values it is necessary to study the spectral problem generally, without restricting the analysis to axisymmetric disturbances.

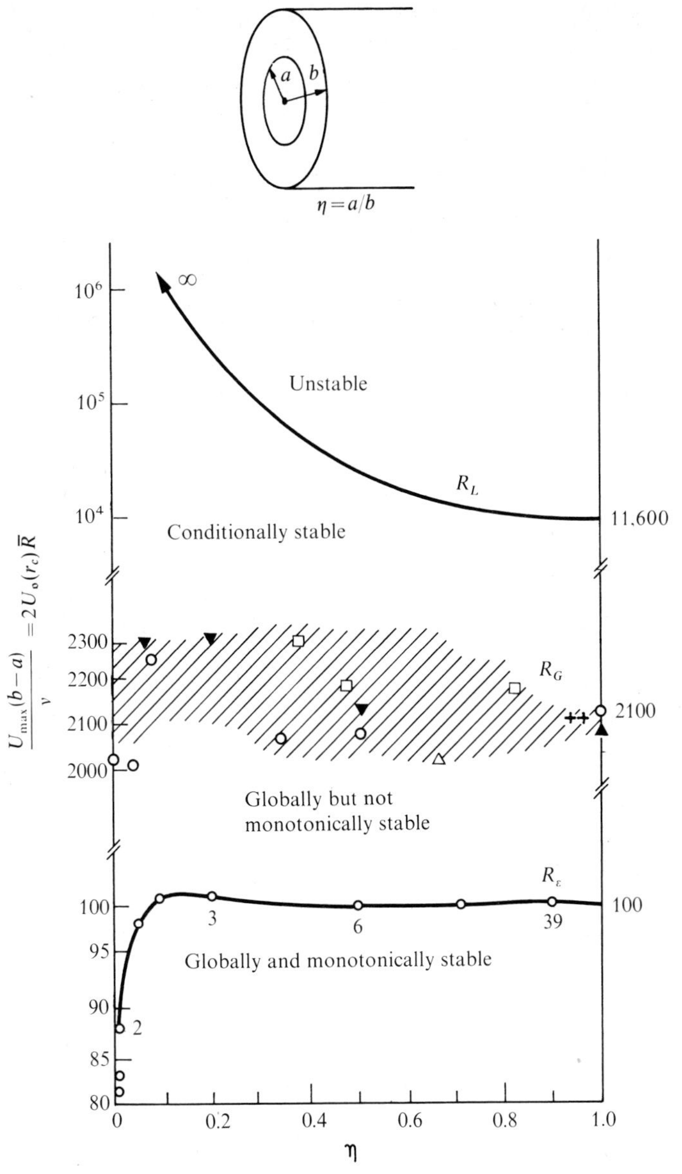

Fig. 35.1: Stability limits for Poiseuille flow; $\bar{R} = \langle U_x \rangle / 2\nu$ and $\langle U_x \rangle = \langle U \rangle$ is defined under (17.8)

Magnuson (1970) have observed that the limit which is observed depends on how the entrance is disturbed. Their limits range from $1650 < U_m(b-a)/\nu < 2500$. The lower value is achieved when the entrance is smooth; the higher value is achieved when there are sharp bumps inside the channel entrance. The Davies and White inlet was very disturbed because the flow was required to turn 90° before entering the channel (Fig. 26.3). Kao and Park (1970) claim some agreement between their observations and spatially damped eigenfunctions of linearized theory. A damped disturbance must eventually become small enough for the linear theory to apply.

A glance at Fig. 35.1 shows that instability is observed at subcritical values of R which would be judged stable by linear theory. This does not mean that the linearized theory is out of agreement with experiments; the linearized theory predicts instability when $R>R_L$; the instability is observed. Linear theory guarantees stability for laminar flow with $R<R_L$ only in a conditional sense, for small disturbances. The conditional stability of laminar Poiseuille flow with $R<R_L$ can be observed in experiments in which great care is taken to suppress the size of disturbances. In round pipes laminar flow has been observed even when $R>40{,}000$ (see Schlichting, 1960, Chapter XVI.1 for a detailed report and references). Experiments indicate that natural disturbances are not attracted by laminar flow when $R<R_L$. This is why it is so hard to obtain laminar flow in experiments with R larger than, say, 2400.

(b) Description of the Diagrams

Figs. 27.1, 2, 3, may be regarded as bifurcation diagrams in the plane $(\bar{R}, f)$ of the response curve. The experimentally observed values are taken from the paper of Walker, Whan and Rothfus (1957). The coordinates Re_2 and f_2 used by these authors are related to $\bar{R}$ and f by

$$\bar{R}=Re_2/4F_1(\eta), \quad f=4f_2/F_1(\eta)$$

where

$$F_1(\eta)=\frac{1+(1-\eta^2)/2\ln\eta}{1-\eta}.$$

The response curve for laminar flow $f_l=2/\bar{R}b(\eta)$ appears as a straight line in a log-log plot:

$$\frac{d\ln f_l}{d\ln\bar{R}}=-1. \tag{35.1}$$

The 45° laminar line is a lower bound for the friction factors in all possible motions with steady average values.

In Fig. 27.2 we have also graphed an upper bound for friction factors in turbulent flow with steady averages. To convert the bound (32.3) into (R, f) coordinates we note that (32.3) holds for flows with equal pressure gradients:

$$P[\bar{V}_x]=P_l[U_x]=12\langle U_x\rangle.$$

Then, using (33.14) and $\langle\bar{V}_x\rangle=2\bar{R}$ we note that

$$f=P/\bar{R}^2=12\langle U_x\rangle/\bar{R}^2$$

and

$$\mu = \langle U_x \rangle - \langle \bar{V}_x \rangle = f\bar{R}^2/12 - 2\bar{R} > 0 .$$

These two relations are used to change variables $(\mu, P) \to (\bar{R}, f)$ in (32.3); we find

$$f \leqslant \tilde{f} \equiv \begin{cases} \dfrac{24}{\bar{R}} + \dfrac{12}{\bar{R}^2}(\bar{R}_{\mathscr{E}}+2)(\bar{R}-\bar{R}_{\mathscr{E}}) & \text{for} \quad \bar{R}_{\mathscr{E}} \leqslant \bar{R} < \gamma \\ 3 + \dfrac{36}{\bar{R}} + \dfrac{12}{\bar{R}^2} & \text{for} \quad \bar{R} \geqslant \gamma \end{cases} \tag{35.2}$$

where

$$\gamma = 2(\bar{R}_{\mathscr{E}} + 1)$$

and $\bar{R}_{\mathscr{E}} \simeq 793.6/24$ is the energy stability limit.

Summarizing; statistically stationary flow through a channel can exist only if

$$f_t = \frac{24}{\bar{R}} \leqslant f(\bar{R}) \leqslant \tilde{f}(\bar{R}) . \tag{35.3}$$

The inequality (35.3) is shown as a shaded region in Fig. 27.2.

The slope of the response function for the time-periodic bifurcating solution at the minimum critical Reynolds number is given in Table 35.1 and is drawn as a dashed line (subcritical, unstable bifurcation) on Figs. 27.1 and 27.3.

Table 35.1 (Joseph and Chen, 1974): Values of the slope of the response curve at the minimum critical Reynolds number

η	1/1.01	$\frac{1}{2}$	$\frac{1}{3}$
$b(\eta)/2\lambda_2(\tilde{\alpha}, \eta)$	0.0430	0.0067	0.0035
$\left.\dfrac{d \ln f}{d \ln \bar{R}}\right\|_{\bar{R}=\bar{R}_L}$	−1.0430	−1.0067	−1.0035

The slopes given in Table (35.1) are computed as follows: One notes that

$$\varepsilon^2 = f - f_l = (\lambda - \lambda_0)/\lambda_2 + O(\varepsilon^4)$$

where $f_l = 2/\bar{R}(\varepsilon^2, \alpha, \eta)\, b(\eta)$ and $\lambda = 1/\bar{R}$. Hence, to within terms of $O(\varepsilon^4)$,

$$\ln f = \ln f_l + \ln\left[1 + \frac{b(\eta)}{2\lambda_2(\alpha, \eta)}\left(1 - \frac{\bar{R}(\varepsilon^2, \alpha, \eta)}{\bar{R}(0, \alpha, \eta)}\right)\right].$$

Differentiation with respect to $\ln\bar{R}(\varepsilon^2,\alpha,\eta)$, when α and $\bar{R}(0,\alpha,\eta)$ are fixed gives

$$\left.\frac{d\ln f}{d\ln\bar{R}}\right|_{\varepsilon=0}=-1-\frac{b(\eta)}{2\lambda_2(\alpha,\eta)}. \tag{35.4}$$

Bifurcation always takes place above the 45° line shown in Figs. 27.1, 2, 3. There are a continuum of solutions (depending on α) which bifurcate for each value $\bar{R}(0,\alpha,\eta)\geq\min_\alpha\bar{R}(0,\alpha,\eta)\equiv\bar{R}_L$. In fact there are two values of $\bar{R}$ for each α (see Fig. 34.1) and two different solutions, each with a different slope (35.4), bifurcate at each $\bar{R}>\bar{R}_L$. It is instructive to use Table 34.1 to locate the points and slopes (35.4) of bifurcating solutions at criticality (see Exercise 35.1).

Further study of bifurcating solutions in two dimensions requires the computation of the envelope $\lambda(\varepsilon^2,\alpha(\varepsilon^2))$ of bifurcation curves $\lambda(\varepsilon^2,\alpha)$ depending on the parameter α. The envelope condition

$$\frac{\partial\lambda}{\partial\alpha}=0 \tag{35.5}$$

is to be an identity in ε^2. Points on the envelope give the smallest values $\bar{R}(\varepsilon^2,\alpha(\varepsilon^2))=\lambda^{-1}(\varepsilon^2,\alpha(\varepsilon^2))$ for which two-dimensional bifurcating solutions can be found. At second order in powers of ε^2 we have

$$\lambda(\varepsilon^2,\alpha(\varepsilon^2))=\lambda(0,\tilde{\alpha})+\lambda_2(\tilde{\alpha})\varepsilon^2+(\lambda_4(\tilde{\alpha})+\lambda_{2,\alpha}(\tilde{\alpha})\alpha_2)\varepsilon^4+O(\varepsilon^6) \tag{35.6}$$

where the derivatives

$$\begin{aligned}&\lambda_2=\partial\lambda/\partial(\varepsilon^2), \qquad \lambda_4=\tfrac{1}{2}\partial^2\lambda/\partial(\varepsilon^2)^2,\\ &\lambda_{2,\alpha}=\tfrac{1}{2}\partial^2\lambda/\partial\alpha\partial(\varepsilon^2), \qquad \alpha_2=d\alpha/d(\varepsilon^2)\end{aligned} \tag{35.7}$$

are all evaluated at $\varepsilon=0$, $\alpha=\tilde{\alpha}$, the minimum point on the neutral curve (see Table 34.1). These derivatives can all be computed by analytic perturbation theory; in particular the value of α_2 arises from differentiation of the identity (35.5) with respect to ε^2 at $\varepsilon^2=0$:

$$\lambda_{,\alpha\alpha}\alpha_2+\lambda_{2,\alpha}=0$$

where $\lambda_{,\alpha\alpha}=\partial^2\lambda/\partial\alpha^2$.

(c) Inferences and Conjectures

We shall now summarize what can be inferred from analysis and experiments about flow through annular ducts. The major points are most graphically explained as an interpretation of results shown in Figs. 27.1 and 27.2.

First, when $\bar{R}<\bar{R}_{\mathscr{E}}\simeq 33.3$ (or $f>f_{\mathscr{E}}\simeq 24/\bar{R}_{\mathscr{E}}$), all disturbances of plane Poiseuille flow decay monotonically from the initial instant; if $\bar{R}_{\mathscr{E}}<\bar{R}<\bar{R}_G$, then all disturbances decay eventually but the decay need not be monotonic. Plane Poiseuille flow can be unstable when $\bar{R}>\bar{R}_G$. Experiments suggests that $\bar{R}_G\simeq 650$. For $\bar{R}_G<\bar{R}<\bar{R}_L$ laminar plane Poiseuille flow is stable to small disturbances; the experiments indicate that the stable disturbances must be very small. If natural disturbances are suppressed, however, one can achieve laminar flow with $\bar{R}_G<\bar{R}<\bar{R}_L$. For $\bar{R}>\bar{R}_L$, laminar flow is unstable. At $\bar{R}=\bar{R}_L$ a time-periodic solution bifurcates from laminar flow. The bifurcating solution is subcritical and unstable. The unstable bifurcating solutions cannot attract disturbances and solutions which are not attracted to laminar Poiseuille flow snap through the bifurcating solutions and are attracted to a stable set of solutions with much larger values of the friction factor discrepancy[5].

The experiments suggests that when $\bar{R}>\bar{R}_G$, there is a stable set of solutions, called stable turbulence, which appear to share a common response curve. The surprising and noteworthy observation is that at a given $\bar{R}>\bar{R}_G$, it is possible to reproduce the same value (as far as "sameness" can be ascertained from experiments) of the friction factor discrepancy. The surprise stems from the fact that the solutions which are observed are turbulent and all different; despite this, each of these infinitely many turbulent solutions leads to an apparently common value of the friction factor.

Assuming that the stable turbulence is statistically stationary, the response function is bounded from above by a curve (35.4) which may be computed from the variational theory of turbulence. The response of statistically stationary solutions is bounded from below by the response of laminar flow (the 45° line in Figs. 27.1, 2, 3). There are surely very many statistically stationary solutions in the region between the 45° laminar response line and the upper bound. The bifurcating solution shown as a heavy dashed line in Figs. 27.1, 3 is but one example; there are also at least a continuum of solutions depending on α with $\bar{R}(0,\alpha)>\bar{R}_L$ which bifurcate subcritically. Many of these solutions are demonstrably unstable and all of them may be unstable.

The significance of the solution bifurcating from Poiseuille flow at the lowest critical value $\bar{R}_L$ is that it is the first solution to bifurcate and it is the only one of the bifurcating solutions with $\varepsilon\to 0$ which could be stable to disturbances with different wave numbers (Exercise 35.2). The heavy dashed line shown in Figs. 27.1 and 27.3 gives the slope of the response curve of the two-dimensional bifurcating solution at the point of bifurcation.

The physical significance of the bifurcation analysis appears to be limited in the case of subcritical bifurcations by the fact that subcritical bifurcating solutions are unstable when ε is small. This limitation may be more apparent than real; it is almost certain that higher order corrections (say (35.6), up to terms

[5] It may be possible to observe the time-periodic bifurcating solution as a transient of the snap-through instability. Small disturbances of laminar flow with a fixed value of $\bar{R}(R_G<\bar{R}<\bar{R}_L)$ which are marginally attracted from Poiseuille flow may take on the properties of the two-dimensional bifurcating solution; given $\bar{R}$ the solution could be expected to oscillate with a frequency $\omega(\varepsilon^2(\bar{R}))$. The measured friction factor for this flow would be given by $f(\bar{R})=f_l(\bar{R})+\varepsilon^2(\bar{R})=24/\bar{R}+\varepsilon^2(\bar{R})$. This transient time-periodic solution might exist for a time before being destroyed by instabilities.

of $O(\varepsilon^4)$) of the bifurcation curves in Figs. 27.1 and 27.6 would cause the dashed lines representing bifurcation to turn around. Then, assuming the conditions which are necessary for the factorization theorem of § 14 to control stability, we would get a stable two-dimensional time-periodic bifurcating flow on the upper branch of the bifurcation curve (see Fig. 15.1). The numerical computations of Zahn, Toomre, Spiegel and Gough (1974) for bifurcating Poiseuille flow give results consistent with the picture implied by the factorization theorem[6]. They find that the bifurcation curve turns around, that the lower branch is stable and the upper branch is unstable.

The conclusion for stability on the upper branch of the bifurcation curve has a limited scope; it is not known how generally the conclusion applies. The conclusion rests on restricting disturbances along the lines specified at the end § 34.

If we suppose that the upper branch of subcritical time-periodic bifurcation is stable we may describe the associated physics as a *snap-through instability:* Conditionally stable laminar flow is stable but not every stable; it is hard to get disturbances which are attracted to laminar flow when the Reynolds number is in the conditionally stable region of Fig. 35.1. These disturbances cannot be attracted to the time-periodic solution on the lower branch of the bifurcation curve; instead disturbances are attracted to stable solutions on the upper branch. This picture would be perfect if two-dimensional time-periodic solutions with large amplitudes were observed in the subcritical case. Though such time-periodic solutions are not observed the true physics can be interpreted from the hypothesis that there are, in fact, something like two stable solutions at a given $\bar{R} < \bar{R}_L$. The hypothesis, of course, applies to spatially periodic disturbances in infinitely long pipes and comparisons with experiments in pipes of finite length are at best suggestive. In finite pipes, when $R_G < R < R_L$, there also seem to be two "stable" solutions, one of which is laminar (Wygnanski and Champagne, 1973; Wygnanski, Sokolov, Friedman, 1975). The flow is spatially segregated into distinct patches of traveling packets of laminar and turbulent flow (turbulent "puffs" when R is near R_G, and "slugs" at higher values of R). The transition from laminar to turbulent flow at a fixed place occurs suddenly as a puff or slug sweeps over the place, and the reverse transition occurs just as suddenly when it leaves the place. These observations suggest a sort of cycling in "phase space" between two distinct relatively stable but weakly attracting solutions.

To be more definite about transition to turbulence under subcritical conditions it is necessary to know more about the envelope of solutions on the upper branch of the bifurcation curve and about possible branching of subcritical time-periodic solutions. Without such results we are forced nearly all the way back to the *principle of equi-conjecture.* If you don't know anything, all conjectures are equally probable.

---

[6] They expand the solutions into a Fourier series in the downstream coordinate. This leads to an infinite system of coupled nonlinear equations which are then drastically truncated. The truncated equations are solved numerically. The stability of the time-periodic solutions are tested by numerical integration of the initial value problem. They find that solutions on the lower branch of the bifurcation curve are unstable and the solutions on the upper branch are stable. It is necessary to add that though the computations of Zahn, et al., proceed from truncated equations, the factorization theorem applies equally to the full equations and to the truncated equations.

Exercise 35.1: Use (35.4) and Table 34.1 to show that two solutions bifurcate above the 45° line in Fig. 27.1 for each value $\bar{R}(0,\alpha)>\bar{R}_L$. Identify subcritical and supercritical bifurcating solutions. Locate the points on the neutral curve of Fig. 34.1 for which bifurcation is supercritical. Show that the slope of the response curve (35.4)

$$\left.\frac{d\ln f}{d\ln \bar{R}}\right|_{\bar{R}(0,\alpha)}$$

of solutions bifurcating from values $\bar{R}(0,\alpha)>\bar{R}_L$ on the arc BC of the neutral curve of Fig. 34.1 are closer to -1 than the value given in Fig. 27.1.

Exercise 35.2: Show that all the supercritical bifurcating solutions with wave length $2\pi/\alpha$ are unstable to disturbances with wave lengths $2\pi/\beta$ where β is such that $\bar{R}(0,\beta)<\bar{R}(0,\alpha)$.

Hint: see Chapter XI.

Notes for Chapter IV

§ 28. The theory leading to equations for the mean and fluctuating motions is one of the many contributions of Reynolds (1895).

The value of the nonlinear Reynolds equations was recognized and applied by Heisenberg (1924) in an important paper in which most of the essential details of the asymptotic solutions of the Orr-Sommerfeld equation are laid out. The first nonlinear theory of stability based on Reynolds' nonlinear equations follows out of approximations employed by Meksyn and Stuart (1951).

§ 30. Reynolds (1895) studies the stability of Poiseuille flow. He seeks the smallest value of ν (the largest R) for which it is possible to find a kinematically admissible $\mathbf{u}$ such that the "discriminating equation"

$$\nu\langle|\nabla\mathbf{u}'|^2\rangle = -\langle\overline{u_i'u_j'}D_{ij}[\bar{V}_x]\rangle \tag{*}$$

holds. In fact, Reynolds does not treat this problem but finally finds the least ν from

$$\nu\langle|\nabla\mathbf{u}'|^2\rangle = -\langle\overline{u_i'u_j'}D_{ij}[U_x]\rangle \tag{**}$$

where $\mathbf{D}[U_x]$ is the stretching tensor of the basic parabolic Poiseuille flow and not the mean motion.

To obtain (**) from (*) one first notes that if the disturbance is infinitesimal, the fluctuation may be dropped relative to the quadratic terms; then $D_{ij}[\bar{V}_x]=D_{ij}[U_x]$. For infinitesimal fluctuations the mean motion coincides with the basic motion. Reynolds reasons that the fluctuation dissipation integrals (for example, the last term of (30.7)) which have been neglected in going from (*) to (**) are of one sign and always contribute to a decrease in $\langle|\mathbf{u}|^2\rangle$ (see (28.10)). Therefore, he reasons, the criterion which is obtained from (**) should suffice for stability of the nonlinear problem.

Reynolds did not possess a mathematically sound procedure for finding *the criterion* from the *discriminating equation.* Rather, he guesses at the correct form for the fluctuations. The correct (variational) procedure was discovered by Orr (1907) who remarks (p. 123) that, "In the applications of the method by Reynolds, Sharpe and H.A. Lorentz, the character of the disturbance is to a certain extent assumed, and apparently somewhat arbitrarily; and I proceed ... to conduct similar in-investigations, while endeavouring to avoid any such arbitrary choice."

Though the general form of the discriminating equation in a bounded domain was known to Reynolds, Orr considers only parallel flows (rectilinear motions). Later authors Sharpe (1905), H.A. Lorentz (1907), Kármán (1924), Harrison (1921) and Havelock (1921) for the most part follow Orr's lead and consider parallel flow. An exception is Tamaki (1920) who considers the stability of Couette flow between rotating cylinders. Tamaki gives (incorrectly) the general partial differential (Euler) equations for the energy problem. The correct equations are given by Harrison (1921) (see Bateman, 1932, for a review of early work).

Orr was fully aware that his treatment of Reynolds "linearized" equations were supposed to apply to the full problem associated with (*). The basis of his belief that his numbers "are true least values" is that they guarantee the decrease of $\langle|\mathbf{u}|^2\rangle$ even when the "stabilizing" fluctuation dissipation terms are neglected. Though he has not, in fact, calculated true least values in any case tried by him, this is a computational matter. It does not detract from the admirable understanding which he had of the broad meaning of his criterion or the variational procedures used by him to obtain this criterion.

§ 31, 35. Howard's analysis was stimulated by earlier ideas of Malkus (1954). Malkus, like Reynolds, formulated the turbulence problem in terms of a spatially averaged mean-motion and a fluctuation from this mean. He also noted that it is possible to obtain a first integral of the equations governing the mean temperature. Malkus' formulation of the turbulence problem is in the spirit of Reynolds' earlier formulation of the turbulence problem in shear flow. Neither author shows that the integrals which they form are bounded functionals on a suitably defined space of kinematically admissible functions. Malkus conjectured that the turbulent solutions actually realized in convection when the imposed temperature differences are fixed are the ones which maximize the heat transported. This strict selection principle for turbulent solutions is now generally regarded as false; the solutions which are realized are those which are stable and possess some finite attracting radius. Solutions which maximize the heat transported may not be stable, or if stable, they need not be the only stable solutions. For example, in problems of generalized convection in closed containers heated from below (see Chapter X) there can be two stable solutions (corresponding to up-flow and down-flow in the container); in the usual case the two solutions will transfer a different amount of heat when subjected to the same prescription of temperature on the boundary. The up and down solutions in a circular cylinder of finite height heated from below which was studied Liang, Vidal and Acrivos (1969) provides an excellent example of stability of solutions with different transport capabilities (see Fig. 77.3).

Despite the fact that the selection principle conjectured by Malkus is not strict it has been, at least, historically important; it leads directly, through Howard, to the variational theory of turbulence. The selection principle may yet find strict realization in circumscribed situations which generalize Malkus' idea to other response functionals. Howard (1972), for example, sees the possibility of a sort of asymptotic selection principle at large Reynolds numbers—at least one might investigate the truth of such a simplified selection principle with a mathematical theory that allows for definite and possibly negative conclusions.

One such circumscribed possibility is the one discussed in this chapter. The selection hypothesis here is that there are no statistically stationary solutions, stable or unstable, which lead to higher values of the friction factor at a given mass flux than the solutions which appear on the experimental circles of Figs. 27.1, 27.2 and 27.3. The hypothesis is precise and totally non-trivial, for at least a continuum of statistically stationary solutions (the bifurcating ones) exist which do not maximize the friction factor. These solutions are unstable and are not selected. To establish the hypothesis just framed we should need to demonstrate that there are no statistically stationary solutions with friction factors larger than the ones in the experiments; this demonstration is quite beyond us.

To an extent, however, the investigation of response functionals by Howard's methods and by Busse's (1969A, 1970A) extension of these methods to shear flows is one way of investigating the hypothesis that realized motions do tend to select states on which suitably selected response functionals are maximized. Such investigations at the very least lead to rigorous upper bounds which are not without physical interest. It hardly needs saying that turbulence is such a difficult and important subject that mathematically rigorous formulations of even small parts of the whole are especially precious.

In fact, it seems true that the variational theory, even without a selection principle, has rather more physical content than one might have first believed possible. Much of this content stems from the view that the solutions of the variational equations already exhibit features which model turbulent mechanisms of a subtle kind. This property of the variational theory as a "model theory" for turbulence has been repeatedly stressed by Busse and stems largely from the structure of the "multi-α" solutions which were discovered by him. It is premature to discuss these solutions here; a rather full account is given in Chapter XII. In addition to the work which is treated in Chapter XII, the interested reader will profit from the study of Busse's (1970A) paper and from Howard's (1972) review of the subject founded by him. A most striking results of Busse's (1970A) application of his multi-α

solutions to turbulent Couette flow is that the mean velocity profile for the maximizing solution is not flat in the interior as popularly supposed but has a slope which is just one-quarter of laminar Couette flow. As Howard notes, "While this is at variance with popular ideas about what turbulent flow "ought" to be like, Reichardt's experiments seem to agree remarkably well with it".

§§ 33, 34. The basic results for these sections are taken from the papers of Joseph and Sattinger (1972), Joseph and Chen (1974) and Joseph (1974B). The formulation of bifurcation theory in the parameters of the response curve is useful, perhaps crucial, in probing the significance of perturbation theory in studying flow. Certain of the qualitative results about bifurcation of plane Poiseuille flow expressed in terms of an arbitrary amplitude A, were obtained by Reynolds and Potter (1967) using their own formal extension of the Stuart-Watson method of amplitude expansions. More complete numerical results for this problem have been obtained by McIntire and Lin (1972) who also consider the stability of Poiseuille flow of a non-Newtonian (second-order) fluid. Pekeris and Shkoller (1967, 1969) have also studied the time-periodic solutions which bifurcate from plane Poiseuille flow. Related to these studies are the interesting but still tentative perturbation analyses of plane Couette flow by Ellingsen, Gjevik and Palm (1970) and of Hagen-Poiseuille flow by Davey and Nguyen (1971). Both of these flows are absolutely stable to infinitesimal disturbances and are therefore hard to treat by perturbation methods. Andreichikov and Yudovich (1972) have applied the rigorous version of the Liapounov-Schmidt theory which applies to time-periodic bifurcation problems (see notes for § 9 of Chapter II) to the problem of bifurcating Poiseuille flow.

Chapter V

Global Stability of Couette Flow between Rotating Cylinders

In this chapter we shall consider the stability of flow between concentric rotating cylinders. In the idealized problem the cylinders are infinitely long and rotate about their common axis with constant angular velocities Ω_1 at $r=a$ and Ω_2 at $r=b$ $(b>a)$. The motion of the fluid satisfies the Navier-Stokes equations between the cylinder and the adherence condition at the cylinder walls.

§ 36. Couette Flow, Taylor Vortices, Wavy Vortices and Other Motions which Exist between the Cylinders

There is a unique, steady solution of the Navier-Stokes equations which depends on r alone and takes on prescribed values on the cylinders:

$$\mathbf{U}=\mathbf{e}_\theta(Ar+B/r)=\mathbf{e}_\theta V(r) \tag{36.1}$$

where

$$A=\frac{b^2\Omega_2-a^2\Omega_1}{b^2-a^2}, \qquad B=\frac{-a^2b^2(\Omega_2-\Omega_1)}{b^2-a^2}$$

where (r,θ,x) are polar cylindrical coordinates. The flow (36.1) is called Couette flow between rotating cylinders. This idealized flow is actually a good representation of the flow which is observed between cylinders of finite length, away from the ends, when

$$R_B=\frac{|B|}{\nu}=\frac{a^2b^2|\Omega_1-\Omega_2|}{\nu(b^2-a^2)} \quad \text{is small}. \tag{36.2}$$

Couette flow is unstable when R_B is large; then there can be many other stable flows which need not be uniquely determined by the given data. Two other stable flows are shown in Fig. 36.1 a, b. In Fig. 36.1 a you see Taylor vortices; this beauti-

a

b

c

Fig. 36.1: (a) Taylor vortices (R. Block, 1973). Laminar Couette flow between rotating cylinders has lost its stability and is replaced by the secondary motion shown in the photograph. The inner cylinder is rotating and the outer cylinder is stationary. (b) Wavy vortices (R. Block, 1973). The secondary flow in Taylor vortices has lost its stability and is replaced by a time periodic motion in which undulations of the vortices propagate around the cylinders. The inner cylinder is rotating and the outer cylinder is stationary. (c) A transient motion which arises as an instability following start-stop motion of the outer cylinder (D. Coles, 1965)

fully regular, steady axisymmetric motion takes form as a stack of tori of approximately square cross-section. In Fig. 36.1b you see wavy vortices; these vortices bifurcate from Taylor vortices when, under certain conditions, the Taylor vortices lose stability. The Taylor vortices arise as a steady bifurcation of Couette flow; the wavy vortices arise as a time-periodic bifurcation of Taylor vortices.

Couette flow will not bifurcate into Taylor vortices under all circumstances. If the absolute value of

$$\mu \equiv \Omega_2/\Omega_1 \tag{36.3}$$

is large, the loss of stability of Couette flow can be more complicated and various types of unsteady and nonaxisymmetric motions may arise (Taylor, 1923; Coles, 1965; Krueger, Gross and DiPrima, 1966; DiPrima and Grannick, 1971; Snyder, 1968A, 1970). Besides the various stable steady and unsteady motions which can be seen when Couette flow loses stability, there are still more motions which satisfy the equations of motion but are unstable and are not observed.

§ 37. Global Stability of Nearly Rigid Couette Flows

Energy stability analysis (Chapter I) shows that Couette flow is monotonically and globally stable whenever

$$\nu > \nu_{\mathscr{E}} = \max \frac{-\langle \mathbf{u}\cdot \mathbf{D}[\mathbf{U}]\cdot \mathbf{u}\rangle}{\langle |\nabla \mathbf{u}|^2\rangle} \tag{37.1}$$

where $\mathbf{U}$ is given by (36.1),

$$\mathbf{D}[\mathbf{U}] = -(\mathbf{e}_\theta \mathbf{e}_r + \mathbf{e}_r \mathbf{e}_\theta) B/r^2 \tag{37.2}$$

and the maximum is taken over solenoidal vectors $\mathbf{u}(r, \theta, x) = (w, v, u)$ which vanish at $r = a, b$, are periodic in θ and almost periodic in x. The angle brackets designate volume-averaged integrals (defined by (19.1) with $L_x \to \infty$).

When $\Omega_1 = \Omega_2$, $B = 0$ and the fluid between the cylinders rotates as a rigid body. For rigid motions $\nu_{\mathscr{E}} = 0$. Rigid rotation of the fluid between the cylinders is absolutely, monotonically and globally stable.

When B is small the Couette flow is nearly a rigid rotation. This nearly rigid rotation is also monotonically and globally stable. To prove this assertion we note that

$$\left|\frac{wv}{r^2}\right| \leq \frac{1}{2}\left(\frac{w^2}{r^2} + \frac{v^2}{r^2}\right),$$

and

$$\langle|\nabla \mathbf{u}|^2\rangle = \left\langle |\nabla w|^2 + |\nabla v|^2 + |\nabla u|^2 + \frac{v^2}{r^2} + \frac{w^2}{r^2} - \frac{2v}{r^2}\frac{\partial w}{\partial \theta} + \frac{2w}{r^2}\frac{\partial v}{\partial \theta}\right\rangle$$

$$\geqslant \left\langle \left(\frac{\partial w}{\partial r}\right)^2 + \left(\frac{\partial v}{\partial r}\right)^2 + \frac{1}{r^2}\left(\frac{\partial w}{\partial \theta} - v\right)^2 + \frac{1}{r^2}\left(\frac{\partial v}{\partial \theta} + w\right)^2 \right\rangle$$

$$\geqslant \left\langle \left(\frac{\partial w}{\partial r}\right)^2 + \left(\frac{\partial v}{\partial r}\right)^2 \right\rangle .$$

It follows that

$$\frac{-\langle \mathbf{u}\cdot\mathbf{D}\cdot\mathbf{u}\rangle}{\langle|\nabla \mathbf{u}|^2\rangle} = \frac{2B\langle wv/r^2\rangle}{\langle|\nabla \mathbf{u}|^2\rangle} \leqslant |B| \frac{\langle w^2/r^2\rangle + \langle v^2/r^2\rangle}{\left\langle \left(\frac{\partial w}{\partial r}\right)^2 + \left(\frac{\partial v}{\partial r}\right)^2 \right\rangle} .$$

Since $\sum a_n / \sum b_n \leqslant \max_n a_n/b_n$ when $b_n > 0$ we may continue the inequality:

$$\leqslant |B| \max_\phi \lambda^{-2}[\phi] \tag{37.3}$$

where $\phi = 0$ at $r = a, b$,

$$\lambda^{-2}[\phi] = \frac{\langle \phi^2/r^2\rangle}{\langle |D\phi|^2\rangle}$$

and $D\phi = d\phi/dr$.

The Euler eigenvalue problem for (37.3) is

$$\frac{1}{r}D(rD\phi) + \lambda^2\phi/r^2 = 0, \qquad \phi = 0 \quad \text{at} \quad r = a, b. \tag{37.4}$$

The eigenfunctions of (37.4) are

$$\sin\{\lambda \ln(r/b)\}$$

with eigenvalues

$$\lambda^2 = n^2\pi^2 \Big/ \left(\ln\frac{a}{b}\right)^2 .$$

The smallest eigenvalue is

$$\lambda^2 = \pi^2 \Big/ \left(\ln\frac{a}{b}\right)^2 . \tag{37.5}$$

It now follows from (37.1), (37.3) and (37.5) that Couette flow is monotonically and globally stable whenever[1]

$$\frac{|B|}{\nu}=\frac{a^2b^2|\Omega_2-\Omega_1|}{\nu(b^2-a^2)}<\frac{\pi^2}{(\log b/a)^2}\,. \tag{37.6}$$

Nearly rigid Couette flows are those for which $|\Omega_2-\Omega_1|$ is small; such flows are globally stable.

The variational problem (37.1) for $\nu_{\mathscr{E}}$ was first considered by Tamaki and Harrison (1920) and by Harrison (1921). Harrison (1921) corrected basic errors in the equations given by Tamaki and Harrison. However, Harrison, like Orr (see Notes to IV, § 30), solves the variational problem for the wrong disturbances. He considers roll disturbances whose axes are parallel to cylinder generators. These rolls are not the form of the critical energy disturbance when three dimensional disturbances are assumed and they do not give the smallest critical viscosity of energy theory.

Serrin (1959 B) was the first to consider the energy stability of Couette flow to axisymmetric disturbances; $\mathbf{u}(r,\theta,x)=(w(r,x),\ v(r,x),\ u(r,x))$. Axisymmetric competitors for the maximum of (37.1) satisfy the following equations:

$$\operatorname{div}\mathbf{u}=\frac{1}{r}\frac{\partial rw}{\partial r}+\frac{\partial u}{\partial x}=0\,,$$

$$-(v\mathbf{e}_r+w\mathbf{e}_\theta)B/r^2=-\mathbf{e}_r\frac{\partial p}{\partial r}-\mathbf{e}_x\frac{\partial p}{\partial x}+\nu_{\mathscr{E}}\nabla^2\mathbf{u}\,,$$

$$\mathbf{u}(a,\theta,x)=\mathbf{u}(b,\theta,x)=0\,,$$

$\mathbf{u}$ is a periodic function of x.[2]

Eliminating p and u we find that

$$\begin{aligned}\nu_{\mathscr{E}}\mathscr{L}^2w&=-B\frac{\partial^2 v}{\partial x^2}\Big/r^2\\ \nu_{\mathscr{E}}\mathscr{L}v&=-Bw/r^2\end{aligned} \tag{37.8}$$

where

$$\mathscr{L}=\frac{1}{r}\frac{\partial}{\partial r}\left(r\frac{\partial}{\partial r}\right)-\frac{1}{r^2}+\frac{\partial^2}{\partial x^2}$$

and

$$w=\partial_r w=v=0\quad\text{at}\quad r=a,b\,.$$

[1] The estimate (37.6) was derived by Serrin (1959B) by a different method. A slightly sharper, but more complicated, estimate is given by (44.25).

[2] Serrin's analysis is framed for functions which are periodic in x. The same analysis can be justified for almost periodic functions (see the derivation leading to (20.7) and Exercise 44.1).

The Eq. (37.8) are linear and the coefficients are independent of x. Hence, Fourier components of the solution must separately satisfy (37.8). Each of these components is in the form

$$(w, v)=(\hat{w}(r), \hat{v}(r)) \cos(\alpha x+\beta) \tag{37.9}$$

where β is arbitrary. The functions $\hat{w}(r)$ and $\hat{v}(r)$ are governed by ordinary differential equations.

Serrin solves the ordinary differential equations when the cylinders are close together, $a/b \to 1$. Then,

$$\frac{B}{r^2} \to \frac{4B}{(a+b)^2} \approx \frac{B}{ab},$$

$$\mathscr{L} \to L = \frac{d^2}{dr^2} - \alpha^2$$

and $(\hat{w}, \hat{v})$ are governed by

$$\begin{aligned} \nu_{\mathscr{E}} L^2 \hat{w} &= \alpha^2 B \hat{v}/ab \\ \nu_{\mathscr{E}} L \hat{w} &= -B \hat{w}/ab \end{aligned} \tag{37.10}$$

where $\hat{w}$, $D\hat{w}$ and $\hat{v}$ vanish at $r=a, b$. The eigenvalue problem (37.10) arises in the study of Bénard convection (see § 62); its properties are well-known and it can be solved by separating variables (see Chandrasekhar, 1961) or by numerical integration (see Table 48.1). The minimum eigenvalue for (37.10) is

$$\left(\frac{B}{\nu_{\mathscr{E}}}\right)^2 = \frac{a^2 b^2 (1708)}{(b-a)^4}$$

where the minimum is taken first over all eigenvalues when α is fixed and then over all $\alpha \in \mathbb{R}$. The minimizing $\alpha = \alpha_{\mathscr{E}}$ is given by

$$\alpha_{\mathscr{E}}^2 = (3.117/(b-a))^2 .$$

Recalling that $b+a \approx 2a$ Serrin finds that

$$\frac{|\Omega_2 - \Omega_1|}{\nu_{\mathscr{E}}} = \frac{2\sqrt{1708}}{a(b-a)}$$

is the limiting value $(a \to b)$ for global stability with monotonic decay.

Serrin notes that the solution of the general problem (37.1) cannot be *assumed* to be axisymmetric. This general problem is formidable, and the best completely rigorous result known for it is (44.25), which follows on the earlier and simpler estimate (37.6). But the problem is amenable to numerical analysis (Hung, 1968). Hung finds that among all the possible solutions, the Taylor vortices give the lowest energy limit.

In Table 37.1 we have listed the parameter values associated with the principal eigenvalue $R_{\mathscr{E}} = \dfrac{B}{\nu_{\mathscr{E}}}$ of (37.8). All disturbances of Couette flow decay when $R < R_{\mathscr{E}}$.

Table 37.1: Critical parameters of energy theory (Hung, 1968)

η	$\alpha_{\mathscr{E}}(b-a)$	$R_{\mathscr{E}}$
0.10	3.66	187.84
0.30	3.28	105.28
0.50	3.22	89.72
0.70	3.15	84.48
0.90	3.12	82.80

The stability limits of Table 37.1 appear as a band in the (Ω_2, Ω_1) plane. These energy bands are shown in Figs. 37.1.

By integrating the shear stresses of a fluid in Couette flow at the cylinder walls we find a linear relation between torque and viscosity. This simple relation leads directly to the Couette viscometer. But the linear relation holds only so long as the flow is stable. In the early days it seems to have been believed, following an erroneous observation of Mallock (1888), that Couette flow is always turbulent when the inner cylinder rotates and the outer one is at rest. For this reason the inner cylinder is put to rest and the outer one is rotated in the Couette viscometer. It is not possible to keep laminar flow when the outer cylinder is rotating too fast; Couette (1890), for example, finds turbulence when $\Omega_2 b(b-a)/\nu > 1925$ in water and 1907 in air. But it certainly is possible to keep laminar flow at higher speeds by rotating only the outer cylinder than by rotating only the inner cylinder. Thus, unlike the criteria of classical energy theory, the stability of Couette flow is not symmetric to interchanges of Ω_1 and Ω_2. The energy theory shows, however, that no matter what the disturbance, nearly rigid Couette flows (with B/ν small enough) are stable.

In the present formulation of the variational problem for the torque in statistically stationary turbulent Couette flow (Nickerson, 1969; Busse, 1971) the torque depends only on the magnitude of the angular velocity $|\Omega_1 - \Omega_2|$. Like the energy theory to which it reduces under linearization, the present formulation of the variational theory does not distinguish between rotating the inner cylinder with outer cylinder fixed and rotating the outer cylinder with inner cylinder fixed.

§ 38. Topography of the Response Function, Rayleigh's Discriminant

It is instructive to view the result about the global stability of nearly rigid Couette flow in the context of the response function shown in Fig. 38.1. The response function gives the ratio of the actual torque required to rotate the cylinders to the

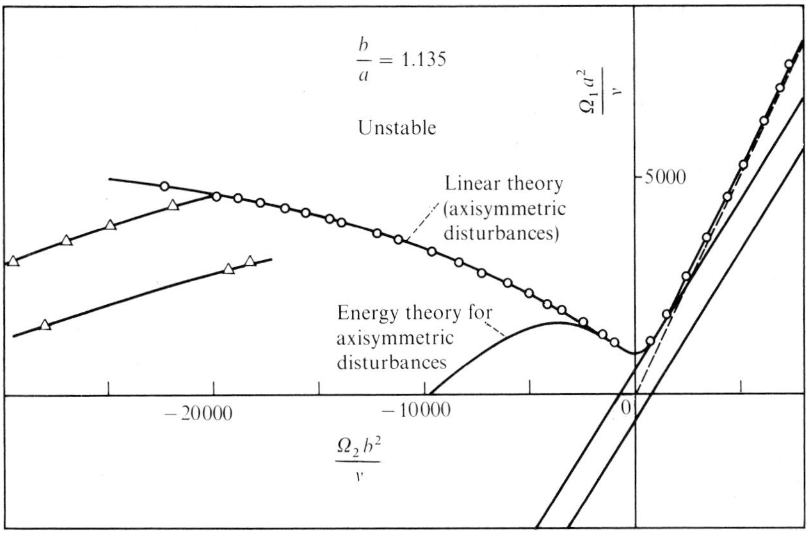

Fig. 37.1.a: Stability regions for Couette flow between rotating cylinders. The circles and triangles are observed points of instability in the experiments of D. Coles (1965) (Joseph and Hung, 1971)

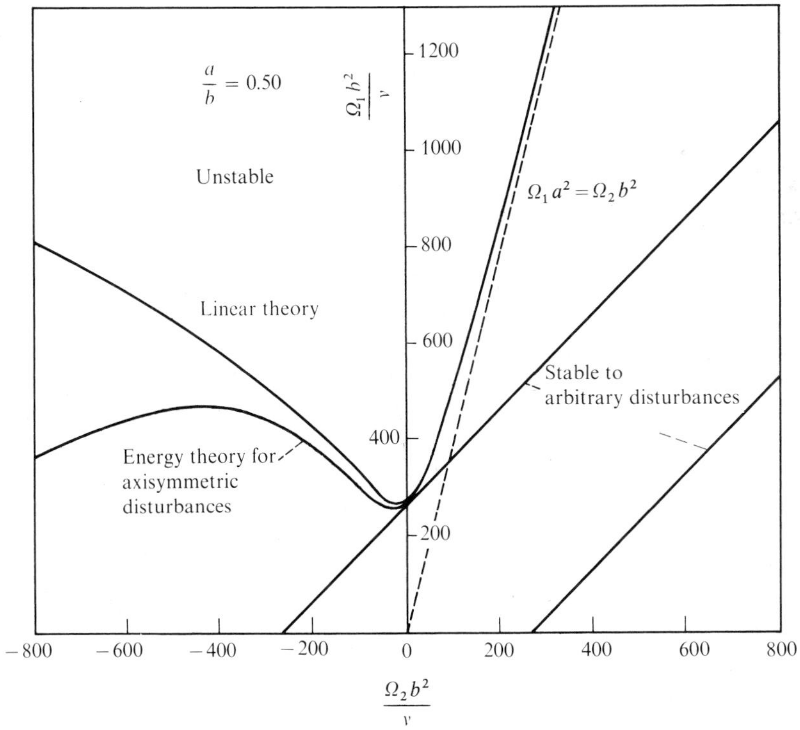

Fig. 37.1.b: Stability regions for Couette flow between rotating cylinders (Joseph and Hung, 1971)

torque which would be required to rotate them at the same speed if the flow inside were Couette flow. In Fig. 38.1, there is a flat valley where Couette flow is stable and there are bifurcating mountains where other solutions with higher torques are stable. In § 37 we proved that if you stay near to the path of solid body rotation you will avoid the mountains. It is clear, however, that the topography is not symmetric about the path of solid body rotation and it is to this asymmetry that we now turn our attention.

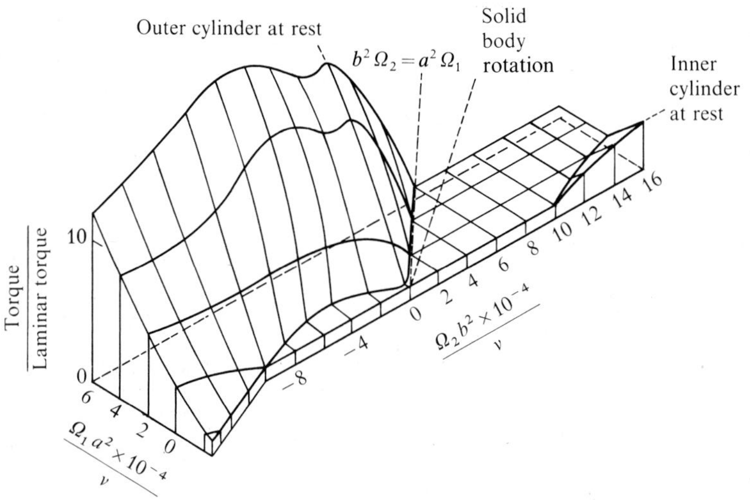

Fig. 38.1: Stability of laminar Couette flow between rotating cylinders. This is a pictorial representation of the results of a large number of experiments (Wendt, 1933). The figure is a slight modification of one drawn by Coles (1965). The picture should be symmetric to reflection through the origin since the flow is unchanged by changing the signs of Ω_1 and Ω_2. The energy band would then appear as nearly the widest band along solid rotation which would carry one through the mountain pass. This figure should be compared with Figs. 37.1a, b. Far to the left and right of the line $\Omega_2=0$, the instability appears to be strongly subcritical

You will notice that there is a second line through the valley of Couette flow, the line $A=0$ or $b^2\Omega_2=a^2\Omega_1$. This line is called Rayleigh's line and it borders the foothills to the bifurcating mountains. When $A=0$, $\mathbf{U}=\mathbf{e}_\theta B/r$ is a potential flow and the angular momentum of this flow is a constant, equal to B, throughout the flow. The line $A=0$ is called Rayleigh's line because of a physical stability argument which was first advanced by Rayleigh. Rayleigh's argument requires axisymmetry; symmetry is essential for the argument and for the physics.

The first part of the argument omits the effect of viscosity. Following Rayleigh (1916) and Kármán (1934) we consider an inviscid disturbance of an axisymmetric flow having circular streamlines. The rotating flow is in equilibrium when the centrifugal force $\rho v^2(r)/r$ is balanced by a radial pressure gradient; an inviscid axisymmetric disturbance of this equilibrium must conserve angular momentum (see Exercise 38.1) so that a disturbance which moves out from r_1 to r_2 will acquire a new velocity $\tilde{v}=r_1 v(r_1)/r_2$. The centrifugal force $\rho\tilde{v}^2/r_2$ associated with the disturbance will exceed the equilibrium pressure gradient $\rho v^2(r_2)/r_2$ and the disturbance will continue to accelerate outward if $\rho r_1^2 v^2(r_1)/r_2^3 > \rho v^2(r_2)/r_2$; that is, if $r_1^2 v^2(r_1) > r_2^2 v^2(r_2)$, or if

$$\frac{d}{dr}(r^2 v^2) = 2r^3 \Omega\zeta < 0 \tag{38.1}$$

where $\Omega = v/r$ is the angular velocity and $\zeta = \frac{1}{r}\frac{d}{dr}(rv)$ is the local vorticity.[3] Rayleigh's criterion (38.1) asserts that axisymmetric rotating flows are unstable when the angular momentum is a decreasing function of the distance from the axis of rotation. Stated in another way this criterion judges rotating flows stable when the rotation Ω and the vorticity ζ have the same sign.

Applied to the cylinder problem, the criterion states that the potential flow is neutrally stable, the flow with the inner cylinder at rest is stable, and that the class of unstable flows includes all flows for which the two cylinders rotate in opposite directions. This proves too much since the motion can be stable when the cylinders counter-rotate. Even in this counter-rotating case, however, the criterion (38.1) is not silent. In the words of D. Coles (1965):

"In a brilliant contribution to the literature of fluid mechanics, Taylor (1923) improved this stability criterion for Couette flow to take account of viscosity, verified his calculations experimentally, and described the secondary flow which appears after the first onset of instability. This secondary flow consists of a regular cellular vortex structure in which closed ring vortices alternating in sign are wrapped around the axis of rotation. To a good approximation, the secondary vorticity is confined to the part of the fluid where there is instability according to the inviscid criterion. If the cylinders are rotating in the same direction this is the whole of the fluid, but if they are rotating in opposite directions, it is only the region between the inner cylinder and the surface of vanishing tangential velocity."

The restoration of the effect of viscosity on the description of the physical mechanism of instability starts with the observations that (a) if there is an imbalance of radial forces, there will be a radial motion and this motion will be opposed by viscous forces, and (b) if the radial displacement takes a long time, angular momentum will not be conserved; instead, viscosity will restore the equilibrium distribution of angular momentum.

Suppose that a ring of fluid of thickness d is displaced outward a distance d in a time t which is much smaller than the diffusion time d^2/ν which is required to bring the angular velocity of the spinning ring to its equilibrium value at radius $r+d$. Then, the ring centered at $r+d$ will have the velocity

$$\tilde{v} = v(r)r/(r+d) \tag{38.2}$$

---

[3] The form $\Omega\zeta < 0$ of Rayleigh's criterion was given first in the important big work of D. Coles (1965) on non-uniqueness and hysteresis and transition in the flow between rotating cylinders. He gives additional physical arguments, explains the generation of Taylor vortices, and shows (Coles, 1967) how these arguments may be used to derive scaling laws to collapse Taylor boundaries for different values of a/b onto a single curve.

consistent with conservation of angular momentum and, in addition, the inequality $t \ll d^2/\nu$ holds. The viscous resistance may be estimated as $\rho\nu w/d^2$ where $w \sim d/t$ is the average radial velocity. The centrifugal force excess on the ring at $r+d$ is estimated, in the same spirit, as

$$-\frac{dp}{dr}\bigg|_{r+d} + \rho\frac{\tilde{v}^2}{r} = \frac{\rho}{r}[\tilde{v}^2 - v^2(r+d)] = \frac{\rho}{r(r+d)^2}[r^4\Omega^2(r) - (r+d)^4\Omega^2(r+d)] \quad (38.3)$$

where $\rho\tilde{v}^2/r$ is the centrifugal force implied by conservation of angular momentum and dp/dr is evaluated for the undisturbed swirling flow. Given that $t < d^2/\nu$, the viscous resistance will overcome the centrifugal force excess for fields $\Omega(r)$ such that

$$\frac{\rho}{r(r+d)^2}[r^4\Omega^2(r) - (r+d)^4\Omega^2(r+d)] < \frac{\rho\nu^2}{d^3} < \frac{\rho\nu}{td} \sim \frac{\rho\nu w}{d^2}. \quad (38.4)$$

If the equilibrium distribution of angular momentum increases outward $(\Omega\zeta > 0)$ the left side of (38.4) is negative and the inequality (38.4) must hold. This means that equilibrium fields with $\Omega\zeta > 0$ should be stable to axisymmetric disturbances. On the other hand, the argument also shows how flows with an adverse distribution of angular momentum $(\Omega\zeta < 0)$ can be stable when the "Taylor number"

$$\frac{d^3}{r(r+d)^2\nu^2}[r^4\Omega^2(r) - (r+d)^4\Omega^2(r+d)]$$

is small enough.

The physical description of instability which was just given has a very important but limited application[4]. It does not explain the complete topography

[4] A precise statement of the limits of validity of the criterion (38.1), extending a result of Synge (1938A, B), is given in § 40. The substantial advantage of casual physical arguments like the one just given is that they lead to hypotheses about mechanisms of instability in situations in which exact analysis is difficult and in new situations not yet treated by analysis. In Chapter VI we shall show how the Rayleigh criterion, written in the form, $\Omega\zeta < 0$, as suggested by Coles (1965), can be generalized to spiral flows. This generalization requires that we think of a spiral ribbon which is generated by a moving radius perpendicular to the cylinder axis which rotates and translates down this axis at constant speed. One can introduce a spiral family of orthogonal surfaces in the cylinder in this way. Given any cylinder the radius vector will cut every generator of the cylinder at a constant angle provided that the radius vector rotates with angular velocity $\mathring{\theta}$ and translates with velocity $\mathring{x}$ where the ratio $\mathring{x}/\mathring{\theta}$ is constant. The orthogonal family of cylinders is generated by the motion of the radius vector which makes an angle Ψ with cylinder generators on the cylinders and the two ribbons are mutually orthogonal at all r. Geometry shows that $\operatorname{Tan}\Psi = r\mathring{\theta}/\mathring{x}$ is linear in r.

In the description of the physical mechanism of instability of spiral flow, we suppose that the angle Ψ is the spiral angle giving the direction along which the disturbance pressure does not vary. The test for instability of the inviscid fluid is now carried out relative to the ribbon on which the disturbance does not vary (see Exercise 38.2). In § 52 we show that for the narrow-gap problem (rotating plane Couette flow) instability can be expected when $\Omega\zeta < 0$ $(-2\Omega\zeta = \mathbb{F} > 0)$ where Ω and ζ are components of angular velocity and vorticity normal to the spiral ribbon.

A striking example of the mechanism of instability which was just described is exhibited in Figs. 46.3, 4, 5. There we have pictures of the spiral vortices which replace Poiseuille flow in a rotating

of the bifurcating mountains shown in Fig. 38.1. Despite the fact that the criterion (38.1) for the stability of Couette flow, for which

$$\Omega\zeta = 2A(A + B/r^2),$$

is always satisfied when $\Omega_1 = 0$, Couette flow is unstable when $|\Omega_2|$ is sufficiently large (see Fig. 38.1). In these cases, and all those in which μ is large and negative, the Rayleigh mechanism does not operate strongly, and more complicated time-dependent and subcritical motions are observed. Moreover, Rayleigh's argument does not explain why nonaxisymmetric disturbances are suppressed when $\mu > 0$ but lead to time-periodic bifurcation when μ is less than a certain negative value (see § 39 for a fuller discussion of this point).

Exercise 38.1: Show that the angular momentum of an incompressible fluid particle in inviscid, axisymmetric flow is conserved.

Exercise 38.2: Suppose that the pressure is constant on spiral lines in the intersection of cylinders and spiral ribbons with constant $\dot{\theta}/\dot{x} = \operatorname{Tan}\Psi/r$. Show that $u_\theta \operatorname{Tan}\Psi + u_x$ is conserved on incompressible particles in inviscid flow.

§ 39. Remarks about Bifurcation and Stability

The spectral problem for periodic disturbances of Couette flow is:

$$-\sigma\tilde{\mathbf{u}} + \frac{V(r)}{r}\left[\frac{\partial\tilde{\mathbf{u}}}{\partial\theta} - \mathbf{e}_r\tilde{v}\right] + \mathbf{e}_\theta\tilde{w}DV = -\nabla\tilde{p} + \nu\nabla^2\tilde{\mathbf{u}} \tag{39.1}$$

$$\operatorname{div}\tilde{\mathbf{u}} = 0, \quad \tilde{\mathbf{u}}\big|_{a,b} = 0,$$

$\tilde{\mathbf{u}}$ is $2\pi/\alpha$-periodic in x and 2π-periodic in θ.

In general, the eigenvalues $\sigma = \hat{\xi} + i\hat{\omega}$ are complex and for stability $\hat{\xi}(\nu, A, B) > 0$ for all eigenvalues σ. At criticality, $\hat{\xi}(\nu_L, A, B) = 0$, and the eigenvalue problem (39.1) may be reduced to ordinary differential equations by Fourier decomposition into normal modes proportional to $e^{i(\alpha x + n\theta + \omega t)}$. Elimination of the pressure and the axial disturbance u, using

$$-i\alpha r u = Df + inv \quad \text{where} \quad f = wr,$$

annulus as an instability. For the Poiseuille flow, the discriminant $\Omega\zeta$ can have the correct (negative) sign for instability only if the spiral disturbance angle is of opposite sign on the two walls. These two different spirals, one on each wall, are observed in Nagib's (1972) experiment and can be seen in Figs. (46.4, 5).

Another striking example of Rayleigh's mechanism is exhibited in the photographs (Figs. 53.12 and 53.13) of Taylor instability near the equator of rotating spheres. The Taylor vortices appear near the equator of the sphere where $\Omega\zeta < 0$; near the poles $\Omega\zeta > 0$ and a vertical secondary motion does not develop (see Exercise 53.3).

leads to the following:

$$\begin{aligned}
&\nu\left\{L^2 f-4L\left(\frac{Df+inv}{r}\right)\right\}=i\kappa Lf-\frac{2inBDf}{r^3}-\frac{2i\kappa}{r}(Df+inv)\\
&\qquad -2in\frac{ADf}{r}+\frac{2(n^2+\alpha^2r^2)}{r^2}\left(Ar+\frac{B}{r}\right)v\,,\\
&\nu\left\{L\left[(n^2+\alpha^2r^2)\frac{v}{r}\right]-inL\left(\frac{Df}{r}\right)+\frac{2in\alpha^2 f}{r^2}-\frac{2\alpha^2}{r}D(rv)\right\}\\
&\qquad =\frac{\kappa}{n}(n^2+\alpha^2r^2)\frac{Df+inv}{r}-\alpha^2 r\frac{\kappa}{n}Df+2\alpha^2 Af
\end{aligned}\tag{39.2}$$

where $\kappa=n\left(A+\frac{B}{r^2}\right)-\hat{\omega}$, $L=\frac{1}{r}D(rD)-\frac{n^2}{r^2}-\alpha^2$ and $f=Df=v=0$ at $r=a,b$.

Eqs. (39.2) determine eigenvalues

$$\nu(\alpha,n)=\nu(-\alpha,n)\,. \tag{39.3}$$

Assuming that $\nu(\alpha,n)$ is a simple eigenvalue of (39.2) we have real eigenfunctions of (39.1) in the form, say,

$$\begin{aligned}
r\tilde{w}(r,\theta,x)&=f(r)\exp i(\alpha x+n\theta+i\hat{\omega}t)+\text{c.c.}\\
&\quad+cf(r)\exp i(-\alpha x+n\theta+i\hat{\omega}t)+\text{c.c.}
\end{aligned}\tag{39.4}$$

where c is an arbitrary complex constant and "c.c." stands for complex conjugate. In general, $\nu(\alpha,n)$ is a multiple (double) eigenvalue corresponding to the superposition of two waves.

Results for (39.2) may, without loss of generality, be stated under the restriction $\Omega_1>0$; the stability boundary for $\Omega_1<0$ is obtained by reflection through the origin in the (Ω_2,Ω_1) plane. In § 37 we showed that there are no solutions of (39.2) when $|\Omega_2-\Omega_1|<\nu\tilde{c}(a,b)$ where $\tilde{c}(a,b)$ is a constant depending on a and b alone which is defined by the variational problem (37.1,2). In § 40 we will show that there are no neutral or amplified ($\hat{\xi}\leq 0$), axisymmetric ($n=0$) solutions of (39.2) when $A>0$, $\mu>0$ (this is the region between the lines defining solid rotation and potential flow). Taylor found axisymmetric, steady solutions ($n=\hat{\omega}=0$) for $A<0$, $\mu>0$ and for $A<0$, $\mu<0$ including large values of $(-\mu)$. Reid (1960) considered inviscid ($\nu=0$) Couette flow in the limit $\eta\to 1$ and showed that there is a single set of steady Taylor vortices ($\hat{\omega}=0$) when $0<\mu<1$. But when the cylinders counter-rotate, $\mu<0$, there is a double set of eigenvalues; one for which $\hat{\omega}=0$ and another for which $\hat{\omega}\neq 0$. The occurence of nonsteady modes in this problem led Reid to suggest that modes which are not Taylor vortices might be found for the full viscous problem when $\mu<0$. Yih (1972B) has given a demonstration which he says shows that when $A<0$ and $\mu>0$ then all axisymmetric eigenfunctions of (39.2) have $\hat{\omega}=0$; if $A<0=n<\mu$ then $\hat{\omega}=0$.

Krueger, Gross and DiPrima (1966) showed that the critical eigenfunctions of (39.2) are not axisymmetric and are not steady when $\mu < -0.73$. This result is in good agreement with the experiments of Snyder (1968 A, B) which show that the bifurcating solution for negative values of μ are not Taylor vortices and are not steady. The experimental circles of D. Coles (1965), which lie on the Taylor boundary of Fig. 37.1 a, are not all Taylor vortices. Coles, in a footnote to the paper of Krueger, Gross and DiPrima, says that there is changeover from closed rings to a weak helical structure, presumable unsteady, for $-\mu$ in the range 0.75—0.80. This shows how dots on lines from Taylor's theory can deceive people into thinking that there is better than actual agreement between the linear axisymmetric theory and experiments. For very large values of $-\mu$ the bifurcating solutions (triangles in Fig. 37.1 a) are unsteady and appear to be strongly subcritical.

Bifurcating Taylor vortices are a realization of the theory of bifurcation in the case of a simple eigenvalue passing through zero at criticality. This problem was discussed in § 15. For the idealized Taylor problem, $\nu_1=0$, $\nu_2 \neq 0$ and the solution bifurcates supercritically. Most bifurcation results are restricted to the case $\mu>0$, $Ar+B/r>0$, $A<0$.

Bifurcation theory for Couette flow is not well developed when μ is negative; the problem is complicated by multiple eigenvalues. In the study of bifurcation from multiple eigenvalues it is necessary to determine the number of solutions which bifurcate (see Chapter X) and to compute the different bifurcating branches. The constants c in (39.4) are undetermined in the linear theory and are selected so as to satisfy requirements for bifurcation. It is not possible to state the stability properties of the bifurcating solutions (the spiral vortices) in advance; stable subcritical solutions and unstable supercritical solutions can possibly bifurcate from an eigenvalue of higher multiplicity (McLeod and Sattinger, 1973). Some bifurcation results for the multiple eigenvalue problem with $\mu<0$ have been given by DiPrima and Grannick (1971). One interesting result is that Taylor vortices bifurcate *subcritically* when $-0.73<\mu<-0.70$ and $\eta=0.95$. The bifurcation picture for $\mu<-0.73$ is very complicated.

Bifurcation theory for Couette flow when $\mu>0$ is well developed and in good agreement with experiments. This bifurcation can be called a nonlinear Taylor problem since it leads to steady, supercritical Taylor vortices. The nonlinear Taylor problem was studied first by Stuart (1958). He considers the Reynolds energy equation for fluctuations. He imagines that near critically the disturbances are close to the eigenfunctions of spectral problems but have an unknown amplitude $A(t)$. (This kind of approximation is called a shape assumption.) The approximation leads to an ordinary differential equation for the amplitude A. The results are in good agreement with experiments. A. Davey (1962) worked with an on expansion of the Stuart-Watson type and carried the analysis through terms of order A^3 in the amplitude equation (see Eq. (***) in Notes for Chapter II). Reynolds and Potter (1967 B) computed the amplitude expansions through order A^7. The results are consistent with each other and in good agreement with experiments (see Stuart, 1971 for a thorough discussion).

Mathematically rigorous theories for bifurcating Taylor flow have been given by Velte (1966), Yudovich (1966 A), Ivanilov and Iakolev (1966) and Kirchgässner and Sorger (1968, 1969). The analysis of the authors named last is the most complete and includes comparisons of computations with experiments as well as theorems. These analyses showed that bifurcation into Taylor vortices is supercritical: $\frac{B}{\nu}=R(\varepsilon,\alpha)>R_L=B/\nu_L$ for each preassigned cell size of height π/α. Here ε can be taken as a measure of the amplitude of the bifurcating solution and $R_L(\alpha)=R(0,\alpha)$.

To understand the motion which ensues when the basic laminar flow loses stability it is necessary to study the stability of the bifurcating solution. Solutions which bifurcate subcritically from a simple eigenvalue are always unstable. Though supercritical solutions which bifurcate from a simple eigenvalue are stable to small disturbance of a restricted kind, other unstable disturbances may destabilize the supercritical solution. This kind of destabilization is possible when simplicity is achieved by an artificial mathematical restriction which need not be satisfied by the flow. For example, simplicity can be obtained by fixing the spatial periodicity of the bifurcating flow. This flow might be stable to disturbances with the same periodicity but unstable to other physical disturbances with a different periodicity; this process can be called wave number selection through stability.

Kirchgässner and Sorger (1968) and Kogleman and DiPrima (1970), following earlier work of Eckhaus (1965), have studied the problem of wave number selection through stability. Solutions with cell size π/α bifurcate supercritically for each and every $\alpha\in\mathbb{R}^+$. But in this continuum of possible cells the only one which can be stable when $\varepsilon\to 0$ is the cell of height π/α_L where α_L is the critical size and is given as a root of the equation

$$R_L(\alpha_L)=\min_\alpha R_L(\alpha)\,.$$

The mathematical analysis which leads to this result can be extended and the extension is of general applicability (see Chapter XI). For now it will suffice to say that when $\varepsilon\neq 0$ there is a curve $R(\varepsilon,\alpha(\varepsilon))$ on the surface $R(\varepsilon,\alpha)$ of bifurcating solutions, containing the point $R_L(\alpha_L)$ which separates the stable from the unstable bifurcating solutions. Stability does select allowed cell sizes but the selection does not generally lead to a uniquely determined single cell size. The same wave number selection through instability of supercritical bifurcating solutions is true for problems of thermal convection. Thermal convection problems can be complicated because the disturbances which lead to bifurcation need not be two-dimensional. The same complication may be important in wave number selection of Taylor vortices.

The problem of bifurcation of Taylor vortices into wavy vortices, a secondary bifurcation which is described by the photos in Fig. 36.1, has been considered by Davey, DiPrima and Stuart (1968) and by Eagles (1974). The results are in good agreement with experiments but there are many assumptions in the analysis which need justification.

Different aspects of secondary and repeated bifurcation of flow between rotating cylinders have been studied in the experiments of Coles (1965) and of

Gollub and Swinney (1975). Coles' study draws attention to the marked degree on non-uniqueness and hysteresis which characterize the spatial structure of these flows. For supercritical speeds of a rotating inner cylinder up to about ten times critical, Coles finds that in one and the same apparatus the number of vortices and the number of waves traveling around these vortices are not uniquely determined by the speed. The number of Taylor cells in his apparatus range from 18 to 32 and the number of waves which travel along the axis of the cells range from 3 to 7. Moreover, "as many as 20 or 25 different states (each state being defined by the number of Taylor cells and the number of tangential waves) have been observed at a given speed".

Gollub and Swinney do not report observations of the spatial structure of the flow. Instead, they monitor the radial component of velocity at a fixed point using an optical heterodyne technique. Gollub and Swinney report five distinct supercritical transitions (see their Fig. 2).

(1) Bifurcation of Couette flow into Taylor vortices (bifurcation of steady solutions into steady solutions).

(2) Bifurcation of Taylor vortices into wavy vortices (bifurcation of steady solutions into time-periodic solutions).

(3) Bifurcation of wavy vortices into wavy vortices with two rationally independent frequencies (bifurcation of a periodic solution into a quasi-periodic solution). As the angular velocity is increased the second frequency disappears. After this they find

(4) Bifurcation of wavy vortices into wavy vortices with two rationally related frequencies (bifurcation of a periodic solution into a periodic solution).

(5) Bifurcation of the periodic solution with two rationally related frequencies into a non-periodic attractor with phase mixing (see Notes to § 16).

Gollub and Swinney say that these transitions do not exhibit hysteresis though the first four bifurcations do depend on the height of the fluid between the cylinders. The fifth transition is reported to be "sharp, reversible and non-hysteretic" and independent of the height of the fluid between the cylinders.

Gollub and Swinney interpret their observations as giving qualitative support to the ideas of Ruelle and Takens about repeated supercritical branching. The nature of the fifth transition does support certain of these ideas but not others. However the second transition from torus T^1 to T^2 seems to contradict the notion "genericity" based on residual sets. Without going into this, it will suffice to note that generically quasi-periodic attractors are not expected on the torus T^2. The disappearance of the second frequency in the experiments is another interesting feature which may be hard to interpret. After the disappearance of this frequency, we again have, after transition (3), a periodic solution (T^1). Of course it is not possible to tell from experiments whether two frequencies are rationally independent but in either case the bifurcation (5) from T^1 or T^2 into a non-periodic attractor is not consistent with Ruelle-Takens. Many of the most interesting properties of stability of flow between rotating cylinders still elude analysis.

In the remaining three sections of this chapter we will confine our attention to the study of one of the simplest of the stability problems that can be defined for the flow between cylinders: the stability of Couette flow to axisymmetric disturbances. The generalized energy theory to be developed now, unlike the one given in § 37, is not symmetric to interchanges of Ω_1 and Ω_2, and the nonlinear stability criteria which follow from it agree with the observations of instability of nearly potential Couette flow and with Taylor's theory.

§ 40. Generalized Energy Analysis of Couette Flow; Nonlinear extension of Synge's Theorem

In this section and § 41 and § 42, the consequences of assuming from the outset that the disturbance motion which replaces Couette flow has a Taylor vortex form are explored. There is an important range of the parameters in which the axisymmetric form of the disturbance flow is just the one which is observed in experiments; in this parameter range the Taylor vortex form for the disturbance is theoretically "right" in the sense that, as far as the results of energy and linear analysis are known, both the stability (energy) limit and instability (linear) limit are taken on for axisymmetric disturbances.

The present energy analysis differs from the one in § 37 in that, at the outset, a disturbance is assumed in axisymmetric form. In the analysis of § 37 no assumption about the form of the disturbance is made at the outset, but the *solution* vector field for the energy problem is an axisymmetric field.

In the restricted class of axisymmetric disturbances, one can achieve big improvements in the energy criteria for stability to nonlinear disturbances.

The basic equations for axisymmetric (Taylor vortex) disturbances of Couette flow are

$$\frac{\partial w}{\partial t}+(\mathbf{u}\cdot\nabla_2)w-\frac{v^2}{r}-2v\left[A+\frac{B}{r^2}\right]=-\partial_r p+\nu\left(\nabla_2^2-\frac{1}{r^2}\right)w\,, \tag{40.1 a}$$

$$\frac{\partial v}{\partial t}+(\mathbf{u}\cdot\nabla_2)v+\frac{wv}{r}+2Aw=\nu\left(\nabla_2^2-\frac{1}{r^2}\right)v\,, \tag{40.1 b}$$

and

$$\frac{\partial u}{\partial t}+(\mathbf{u}\cdot\nabla_2)u=\partial_x p+\nu\nabla_2^2 u\,, \tag{40.1 c}$$

where

$$(\mathbf{u},v)\in H_{\text{II}}[\mathbf{u},v:\mathbf{u}=\mathbf{e}_r w+\mathbf{e}_x u,\ \nabla_2\cdot r\mathbf{u}=0,\ \mathbf{u}|_{\tilde{a},\tilde{b}}=v|_{\tilde{a},\tilde{b}}=0],$$

$$\mathbf{u},\ v \text{ are periodic in } x \text{ with period } 2\pi/\alpha]$$

$$\nabla_2=\mathbf{e}_r\partial_r+\mathbf{e}_x\partial_x\,,$$

$$\nabla_2^2=\frac{1}{r}\partial_r(r\partial_r)+\partial_{xx}^2\,.$$

The feature of these equations which forms the basis of our analysis is the absence of a pressure gradient term in (40.1b). A consequence of the absence of a pressure gradient in the circumferential direction is the existence of *two* energy equations

$$\frac{1}{2}\frac{d}{dt}\langle w^2+u^2\rangle-\left\langle\frac{wv^2}{r}\right\rangle-2\left\langle\left(A+\frac{B}{r^2}\right)wv\right\rangle=-v\left\langle|\nabla_2 w|^2+|\nabla_2 u|^2+\left|\frac{w}{r}\right|^2\right\rangle \tag{40.2}$$

and

$$\frac{1}{2}\frac{d}{dt}\langle v^2\rangle+\left\langle\frac{wv^2}{r}\right\rangle+2A\langle wv\rangle=-v\left\langle|\nabla_2 v|^2+\left|\frac{v}{r}\right|^2\right\rangle. \tag{40.3}$$

Here, the angle bracket is the volume integral over a period cell.

It is convenient to work with the linear combination

$$(40.2)+\lambda(40.3),$$

where λ is a positive coupling parameter. Then setting

$$\phi=\sqrt{\lambda}v,$$

we come to the single equation

$$\frac{d\mathscr{E}}{dt}+\left(1-\frac{1}{\lambda}\right)\left\langle\frac{w\phi^2}{r}\right\rangle=-\mathscr{H}[w,\phi,\lambda]-v\mathscr{D}_{\mathrm{II}}, \tag{40.4}$$

where

$$\mathscr{E}=\tfrac{1}{2}\langle w^2+u^2+\phi^2\rangle,$$

$$\mathscr{D}_{\mathrm{II}}=\left\langle|\nabla_2 w|^2+|\nabla_2 u|^2+\left|\frac{w}{r}\right|^2+|\nabla_2\phi|^2+\left|\frac{\phi}{r}\right|^2\right\rangle,$$

and

$$-\mathscr{H}=\frac{2(1-\lambda)}{\sqrt{\lambda}}A\langle w\phi\rangle+\frac{2B}{\sqrt{\lambda}}\left\langle\frac{w\phi}{r^2}\right\rangle.$$

The conditional energy stability theorem (I) for Couette flow which we shall prove from (40.4) is best stated in two parts. First we state the part about the *stability criterion.*

Let $\lambda>0$ *be preassigned and suppose that* $v>v_{\mathscr{E}}(\lambda)$ *where*

$$v_{\mathscr{E}}=\max_{\mathbf{H}_{\mathrm{II}}}-\mathscr{H}/\mathscr{D}_{\mathrm{II}} \tag{40.5}$$

and $\mathbf{H}_{\mathrm{II}}$ *is the set of kinematically admissible axisymmetric vectors. Then, every axisymmetric disturbance of Couette flow satisfies the inequality*

$$\frac{d\mathscr{E}}{dt}+\frac{(\lambda-1)}{\lambda}\left\langle\frac{w\phi^2}{r}\right\rangle\leqslant-(v-v_{\mathscr{E}})\mathscr{D}_{\mathrm{II}}. \tag{40.6}$$

Since axisymmetric solutions of the IBVP for (40.4) are in H_{II}, we must have that $-\mathscr{H}\leqslant v_{\mathscr{E}}\mathscr{D}_{\mathrm{II}}$ where $\mathscr{H}$ and $\mathscr{D}_{\mathrm{II}}$ are evaluated for solutions. Combining this with (40.4) we arrive at (40.6).

The last part of the stability theorem introduces the "critical amplitude". We get nonlinear stability when the axially symmetric disturbances are initially below a certain finite size.

The decay constant $\hat{\Lambda}$ which will appear in the statement of the last part of the theorem is defined by

$$\frac{(b-a)^2}{\hat{\Lambda}}=\max_{\mathbf{H}_{\mathrm{II}}}\mathscr{E}/\mathscr{D}_{\mathrm{II}}. \tag{40.7}$$

The constant $\hat{\lambda}$ is defined by

$$\frac{(b-a)^2}{\hat{\lambda}}=\max\langle\phi^2\rangle/\langle|\nabla_2\phi|^2\rangle,$$

where the maximum is taken over functions ϕ which are $2\pi/\alpha$-periodic in x and vanish at $r=a$ and $r=b$. The constants $\hat{\Lambda}$ and $\hat{\lambda}$ are the positive zeros of certain Bessel functions.

Suppose that

$$v>v_{\mathscr{E}}(\lambda) \tag{40.8}$$

and

$$\mathscr{E}(0)<(F/G)^2, \tag{40.9}$$

where

$$F=(v-v_{\mathscr{E}})\hat{\Lambda}^{1/2}/(b-a)$$

and

$$G=\frac{|\lambda-1|}{\lambda\sqrt{2a^3}}\sqrt{\frac{\alpha(b-a)}{\pi\hat{\lambda}^{1/2}}+1}.$$

Then $\mathscr{E}(t)$ *decays to zero monotonically according to the law*

$$\frac{\mathscr{E}(t)}{[F-G\mathscr{E}^{1/2}(t)]^2}\leqslant\frac{\mathscr{E}(0)}{[F-G\mathscr{E}^{1/2}(0)]^2}\exp\left\{\frac{-F\hat{\Lambda}^{1/2}t}{(b-a)}\right\}. \tag{40.10}$$

To prove the theorem, we need first to prove an imbedding inequality:

Let $\hat{\phi}(r, x)$ be a smooth function which vanishes at $r=a$ and $r=b$ and which is periodic in x with period $2\pi/\alpha$. Then

$$\langle\hat{\phi}^4\rangle \equiv \int_a^b \int_0^{2\pi/\alpha} \hat{\phi}^4 r\,dr\,dx \leqslant \frac{\alpha\langle\hat{\phi}^2\rangle^{3/2}}{2\pi a}\left\langle\left|\frac{\partial\hat{\phi}}{\partial r}\right|^2\right\rangle^{1/2} + \frac{\langle\hat{\phi}^2\rangle}{a}\left\langle\left|\frac{\partial\hat{\phi}}{\partial r}\right|^2\right\rangle^{1/2}\left\langle\left|\frac{\partial\hat{\phi}}{\partial x}\right|^2\right\rangle^{1/2}$$

$$\leqslant \frac{\langle\hat{\phi}^2\rangle\langle|\nabla_2\hat{\phi}|^2\rangle}{2a}\left\{\frac{\alpha(b-a)}{\pi\hat{\lambda}^{1/2}}+1\right\}. \tag{40.11}$$

The proof of (40.11) is given in Appendix C.

Using (40.11), we form the estimate

$$\frac{|\lambda-1|}{\lambda}\left\langle\frac{w\phi^2}{r}\right\rangle \leqslant \frac{|1-\lambda|}{\lambda a}\sqrt{\langle w^2\rangle\langle\phi^4\rangle} \leqslant G\sqrt{\langle w^2\rangle\langle\phi^2\rangle\langle|\nabla_2\phi|^2\rangle} \leqslant G\mathscr{E}\sqrt{\mathscr{D}_{\mathrm{II}}}. \tag{40.12}$$

The energy inequality can then be written as

$$\frac{d\mathscr{E}}{dt} \leqslant [-(v-v_{\mathscr{E}})\sqrt{\mathscr{D}_{\mathrm{II}}}+G\mathscr{E}]\sqrt{\mathscr{D}_{\mathrm{II}}}.$$

When (40.8) holds, this inequality may be written as

$$\frac{d\mathscr{E}}{dt} \leqslant [-F\sqrt{\mathscr{E}}+G\mathscr{E}]\sqrt{\mathscr{D}_{\mathrm{II}}},$$

where we have used (40.7). Now, if in addition to (40.8),

$$\mathscr{E}(0) < \left(\frac{F}{G}\right)^2 = \frac{2a^3\hat{\Lambda}(v-v_{\mathscr{E}})^2\lambda^2}{|1-\lambda|^2(b-a)^2}\Bigg/\left[1+\frac{\alpha}{\pi}\frac{(b-a)}{\hat{\lambda}^{1/2}}\right], \tag{40.13}$$

we may again continue the inequality as

$$\frac{d\mathscr{E}}{dt} \leqslant [-F\mathscr{E}+G\mathscr{E}^{3/2}]\hat{\Lambda}^{1/2}/(b-a). \tag{40.14}$$

Eq. (40.10) now follows from (40.14) by integration.

The energy stability theorem for axisymmetric disturbances gives rise to a stability criterion $v>v_{\mathscr{E}}$ and a side condition (40.13) on initial conditions for which stability can be guaranteed. We note that large initial values $\mathscr{E}(0)$ are allowed by (40.13) if $v>v_{\mathscr{E}}$ is fixed and either $a/b\to1$ or $\lambda\to1$. When $\lambda=1$, the evolution equation (40.4) is just the one which was considered by Serrin (1959 B). When $\lambda=1$, axial symmetry is not required and the resulting stability criterion is global; it holds for all $\mathscr{E}(0)$ and for all possible disturbances. On the other hand, when $\lambda\neq1$, the theorem gives certain stability to axisymmetric disturbances with initial energies which tend to zero as $v-v_{\mathscr{E}}\to0$.

In § 41 we shall compute, using the "best" λ, a conditional energy stability boundary (see Figs. 37.1, a, b) which is virtually indistinguishable from the linear stability boundary in the region

$$A<0\,, \quad (a^2\Omega_1 > b^2\Omega_2 \geqslant 0)\,.$$

When $A=0$, we have

$$-\mathscr{H} = \frac{2B}{\sqrt{\lambda}}\left\langle \frac{w\phi}{r^2}\right\rangle,$$

and since $-\sqrt{\lambda}(\mathscr{H}/\mathscr{D}_{\mathrm{II}})$ is bounded above, we may verify that

$$\nu_{\mathscr{E}}(\sqrt{\lambda}\to\infty)\to 0\,.$$

Hence, for $A=0$ there is absolute conditional stability to axisymmetric disturbances.

Recall Rayleigh's inviscid stability criterion

$$A>0\,, \quad (b^2\Omega_2 > a^2\Omega_1,\ \Omega_1 > \Omega_2)$$

for axisymmetric disturbances. The following *conditional energy stability theorem (II) for Couette flow* holds when $A>0$ and $\Omega_1>\Omega_2$.

Let

$$\mathscr{E}_V(t) = \tfrac{1}{2}\left\langle A w^2 + A u^2 + \left(A + \frac{B}{r^2}\right) v^2 \right\rangle,$$

$$\mathscr{D}_V = \left\langle A|\nabla_2 w|^2 + A|\nabla_2 u|^2 + A\left|\frac{w}{r}\right|^2 + (Ar^2+B)\left|\nabla_2 \frac{v}{r}\right|^2 \right\rangle,$$

$$G_V = \frac{B}{\sqrt{2}\,a^{5/2}A^{1/2}\Omega_1}\left\{\frac{\alpha}{\pi}\frac{b-a}{\hat{\lambda}^{1/2}} + 1\right\}^{1/2}$$

and

$$\frac{(b-a)^2}{\Lambda_V} = \max_{\mathbf{H}_{\mathrm{II}}} \frac{\mathscr{E}_V}{\mathscr{D}_V}\,.$$

Circular Couette flow is stable to arbitrary periodic (in x) axisymmetric disturbances when

$$A>0 \quad \text{and} \quad \Omega_1>\Omega_2\,, \tag{40.15}$$

and

$$\mathscr{E}_V(0) < (\nu/G_V)^2\,. \tag{40.16}$$

When these conditions hold,

$$\frac{\mathscr{E}_V(t)}{[\nu - G_V\mathscr{E}_V^{1/2}(t)]^2} \leq \frac{\mathscr{E}_V(0)}{[\nu - G_V\mathscr{E}^{1/2}(0)]^2}\exp\left\{\frac{-\nu\hat{\Lambda}_V^{1/2}t}{(b-a)}\right\}. \tag{40.17}$$

The stability criterion $A>0$, $\Omega_1>\Omega_2$ was established for axisymmetric solutions of the linearised stability equations by Synge (1938 A, B). The present result (Joseph and Hung, 1971) is a nonlinear extension of Synge's result.

Proof: To prove the conditional stability theorem (II) for Couette flow, we shall need to establish the following evolution equation:

$$\frac{d\mathscr{E}_V}{dt} = -2B\left\langle\frac{wv^2}{r^3}\right\rangle - \nu\mathscr{D}_V\,. \tag{40.18}$$

This equation is the sum of A times (40.2) and Eq. (40.19) below. Eq. (40.19) follows from the integration of $\left\langle\frac{V}{r}v\,(40.1\,\mathrm{b})\right\rangle$ over a period cell; in carrying out the integration, we calculate, using $\nabla_2\cdot r\mathbf{u}=0$,

$$\left\langle\frac{V}{r}v(\mathbf{u}\cdot\nabla_2)v\right\rangle = B\left\langle\frac{wv^2}{r^3}\right\rangle,$$

$$\left\langle\frac{V}{r}v\left(\nabla_2^2-\frac{1}{r^2}\right)v\right\rangle = -\left\langle\frac{V}{r}|\nabla_2 v|^2\right\rangle + 2B\left\langle\frac{v^2}{r^4}\right\rangle - \left\langle\frac{V}{r}\frac{v^2}{r^2}\right\rangle,$$

and

$$2B\left\langle\frac{v^2}{r^4}\right\rangle = 2\left\langle\frac{V}{r}\frac{v}{r}\partial_r v\right\rangle.$$

Combining the last two equations we find that

$$\left\langle\frac{V}{r}v\left(\nabla_2^2-\frac{1}{r^2}\right)v\right\rangle = -\left\langle\frac{V}{r}(\partial_x v)^2\right\rangle - \left\langle\frac{V}{r}\left(r\partial_r\frac{v}{r}\right)^2\right\rangle,$$

and

$$\frac{1}{2}\frac{d}{dt}\left\langle\frac{V}{r}v^2\right\rangle + \left\langle\left(\frac{V}{r}+\frac{B}{r^2}\right)\frac{wv^2}{r}\right\rangle + 2A\left\langle\frac{V}{r}wv\right\rangle = -\nu\left\langle\frac{V}{r}(\partial_x v)^2 + \frac{V}{r}\left(r\partial_r\frac{v}{r}\right)^2\right\rangle. \tag{40.19}$$

We may simplify the computation by introducing variables

$$\frac{v}{r}=\phi\,,\quad \sqrt{A}w=\psi\,,\quad \sqrt{A}u=\gamma\,.$$

In these variables (40.18) becomes

$$\frac{1}{2}\frac{d}{dt}\langle\psi^2+\gamma^2+(rV)\phi^2\rangle+\frac{2B}{\sqrt{A}}\left\langle\frac{\psi\phi^2}{r}\right\rangle$$

$$=-\nu\left\langle|\nabla_2\psi|^2+|\nabla_2\gamma|^2+\left|\frac{\psi}{r}\right|^2+(rV)|\nabla_2\phi|^2\right\rangle.$$

Since

$$\frac{d}{dr}(rV)=\frac{d}{dr}(Ar^2+B)=2Ar>0,$$

we have (since $A>0$) that

$$b^2\Omega_2>rV>a^2\Omega_1.$$

Hence, we may continue the inequality (40.11) as

$$\langle\phi^4\rangle\leqslant\frac{\langle rV\phi^2\rangle\langle rV|\nabla_2\phi|^2\rangle}{2a^5\Omega_1^2}\left\{\frac{\alpha}{\pi}\frac{(b-a)}{\hat{\lambda}^{1/2}}+1\right\}$$

and the remainder of the proof leading to (40.17) follows along the path leading from (40.11) to (40.14).

Exercise 40.1: In the conditional stability theorems (I) and (II), we used two of the three energy identities (40.2), (40.3) and (40.18). Consider the functional formed as a linear combination of these three identities. Prove a conditional stability theorem for this functional. Form a variational problem for the "optimum" stability limit for the sum of the three energy identities. Formulate the problem of finding the values of the coupling constants which give the largest region of conditional stability with monotonic decay.

Exercise 40.2 (A maximum principle for the total angular momentum; Rabinowitz, 1973):
The operator

$$L\equiv\sum_{i,j=1}^{3}a_{ij}(\mathbf{x},t)\frac{\partial^2}{\partial x_i\partial x_j}+\sum_{i=1}^{3}b_i(\mathbf{x},t)\frac{\partial}{\partial x_i}-\frac{\partial}{\partial t}$$

is said to be uniformly parabolic in a four-dimensional $(\mathbf{x},t)$ domain ($\mathbf{x}\in\mathscr{V}$, $0<t<T$) if for fixed t there is one number $\mu>0$ such that

$$\sum_{i,j=1}^{3}a_{ij}(\mathbf{x},t)\xi_i\xi_j\geqslant\mu\sum_{i=1}^{3}\xi_i^2$$

for all triplets of real numbers (ξ_1,ξ_2,ξ_3).

A special case of the maximum principle for parabolic operators (Nirenberg, 1953; a convenient reference is Protter and Weinberger, 1967) is as follows: *Let u satisfy* $(L+h)u=0$ for $\mathbf{x}\in\mathscr{V}$ and $0<t<T$ *and suppose that the coefficients of L are bounded and that* $h(\mathbf{x},t)\leqslant 0$. *Further, suppose* $-m\leqslant u\leqslant M$ *at* $t=0$ *and on the boundary* $\partial\mathscr{V}$ *of* $\mathscr{V}$. *Then,* $-m\leqslant u\leqslant M$ for $\mathbf{x}\in\mathscr{V}\cup\partial\mathscr{V}$ and $0\leqslant t\leqslant T$.

Use the result just stated to show that the total angular momentum

$$r^2\Omega=Ar^2+B$$

of Couette flow plus disturbance (rv where $v(r,x,t)$ is periodic in x) is bounded between the maximum and minimum of the values

$$a^2\Omega(a,x,t), \quad b^2\Omega(b,x,t), \quad r\Omega(r,x,0).$$

Other applications of the maximum principle are to be found in Exercises 40.3, 51.2, and in § 71.

Exercise 40.3: Show that the vorticity of two-dimensional flow and the weighted vorticity ζ/r of axisymmetric flow obey the maximum principle for parabolic operators.

§ 41. The Optimum Energy Stability Boundary for Axisymmetric Disturbances of Couette Flow

Our task now is to find the stability limits $v_{\mathscr{E}}(\lambda)$. Here $\lambda>0$ is a free parameter and we can select it to obtain the largest region for certain stability to axisymmetric disturbances. We call the value $v_{\mathscr{E}}(\tilde{\lambda})=\tilde{v}_{\mathscr{E}}$ which maximizes the region of stability,

$$\tilde{v}_{\mathscr{E}}=\min_{\lambda>0} v_{\mathscr{E}}(\lambda), \tag{41.1}$$

the "optimum stability boundary". It turns out that the value of $\tilde{\lambda}$ can be selected so that $\tilde{v}_{\mathscr{E}}$ and the linear limit v_L for Taylor vortices are virtually indistinguishable when $b^2\Omega_2<a^2\Omega_1$ and $\Omega_2/\Omega_1>0$.

The following result will help to characterize the optimizing value $\lambda=\tilde{\lambda}$.

Lemma:

$$\tilde{\lambda}+1=\frac{-B\left\langle \dfrac{w\phi}{r^2}\right\rangle}{A\langle w\phi\rangle}. \tag{41.2a}$$

When $b^2\Omega_2<a^2\Omega_1$ *and* $\Omega_2/\Omega_1>0$, *there exists* $\bar{r}$, $a\leqslant\bar{r}\leqslant b$, *such that* $\left\langle \dfrac{w\phi}{r^2}\right\rangle=\langle w\phi\rangle/\bar{r}^2$. *Then*

$$\tilde{\lambda}+1=\frac{\eta^2(1-\Omega_2/\Omega_1)}{(\bar{r}/b)^2(\eta^2-\Omega_2/\Omega_1)}>1\,. \tag{41.2b}$$

The formula (41.2b) expresses the requirement that $\lambda=\tilde{\lambda}$ when $\partial v_{\mathscr{E}}/\partial\lambda=0$. Here, the limit $v_{\mathscr{E}}(\lambda)$ which solves (40.5) is most conveniently found as the principal eigenvalue of Euler's equations for (40.5),

$$\left\{\frac{B}{r^2\sqrt{\lambda}}-\frac{(\lambda-1)A}{\sqrt{\lambda}}\right\}\phi+v_{\mathscr{E}}\left(\nabla_2^2-\frac{1}{r^2}\right)w=\partial_r p, \tag{41.3a}$$

$$\left\{\frac{B}{r^2\sqrt{\lambda}}-\frac{(\lambda-1)A}{\sqrt{\lambda}}\right\}w+v_{\mathscr{E}}\left(\nabla_2^2-\frac{1}{r^2}\right)\phi=0, \tag{41.3b}$$

$$\nu_{\mathscr{E}}\nabla_2^2 u = \partial_x p, \tag{41.3c}$$

$$\partial_r(rw) + \partial_x(ru) = 0, \tag{41.3d}$$

and

$$w = \phi = u = 0, \quad \text{at} \quad r = a, b. \tag{41.3e}$$

Eqs. (41.3) can be reduced to ordinary differential equations for the Fourier coefficient $\hat{w}(r)$ and $\hat{\phi}(r)$; set $w(r, x) = \hat{w}(r)\cos(\alpha x)$,

$$\phi = \hat{\phi}(r)\cos(\alpha x), \quad \hat{u} = u(r)\sin(\alpha x), \quad p = \hat{p}(r)\cos(\alpha x),$$

and eliminate $\hat{u}(r)$ and $\hat{p}(r)$ from the resulting set. The ratio on the right of (41.2a) reduces to

$$-\frac{B}{A}\frac{\int_a^b (\hat{w}(r)\hat{\phi}(r)/r^2) r dr}{\int_a^b \hat{w}(r)\hat{\phi}(r) r dr}. \tag{41.4}$$

To establish the existence of the mean value $\bar{r}$, it will suffice to show that $\hat{w}(r)$ and $\hat{\phi}(r)$ are one-signed when $\lambda = \tilde{\lambda}$.

Consider the set of $\lambda = \lambda(\hat{r})$,

$$\lambda + 1 = \frac{-B}{A\hat{r}^2}, \quad a \leqslant \hat{r} \leqslant b. \tag{41.5}$$

For these λ,

$$\frac{B}{r^2} - (\lambda - 1)A = 2A + B\left[\frac{1}{r^2} + \frac{1}{\hat{r}^2}\right] = \frac{2\Omega_1}{1-\eta^2}\left\{\frac{\Omega_2}{\Omega_1}\left(1 - \frac{\eta^2}{\rho^2}\right) - \eta^2\left(1 - \frac{1}{\rho^2}\right)\right\} \geqslant 0,$$

where $\dfrac{1}{\rho^2} = \dfrac{b^2}{2}\left\{\dfrac{1}{r^2} + \dfrac{1}{\hat{r}^2}\right\}$ and $\eta \leqslant \rho \leqslant 1$. It is clear from this inequality and Eqs. (41.3) that the ordinary differential equations which govern the Fourier coefficients $\hat{w}(r)$ and $\hat{\phi}(r)$ can be converted into integral equations with oscillatory kernels. It follows that both $\hat{w}(r)$ and $\hat{\phi}(r)$ are one-signed when $a \leqslant r \leqslant b$ (see Appendix *D*). Hence, for these functions (41.4) is equal to

$$-\frac{B}{A\bar{r}^2(\hat{r})}, \quad a < \bar{r} < b.$$

Now, since

$$\frac{1}{b^2} \leqslant \frac{1}{\bar{r}^2(\hat{r})} \leqslant \frac{1}{a^2}$$

when $a \leqslant \bar{r} \leqslant b$, there must be at least one value of $\hat{r}$ for which $1/\bar{r}^2 = 1/\hat{r}^2$. For this $\hat{r}$, $\lambda = \tilde{\lambda}$.

§ 42. Comparison of Linear and Energy Limits

A most interesting aspect of the results of this section follows from comparing the linear and energy stability boundaries. Taylor's original calculation of the linear stability limit and subsequent ones assume exchange of stability as well as axial symmetry. The linear limit is then found as, say, the principal eigenvalue ν_L of

$$2\left\{A+\frac{B}{r^2}\right\}\frac{\phi}{\sqrt{\lambda}}+\nu_L\left(\nabla_2^2-\frac{1}{r^2}\right)w=\partial_r p, \tag{42.1a}$$

$$\nu_L\nabla_2^2 u=\partial_x p, \tag{42.1b}$$

$$-2A\sqrt{\lambda}w+\nu_L\left(\nabla_2^2-\frac{1}{r^2}\right)\phi=0, \tag{42.1c}$$

subject to (41.3d,e).

From these equation it is easy to prove that within the mean-radius approximation[5]

$$\nu_L=\tilde{\nu}_{\mathscr{E}}.$$

Proof: In the mean-radius approximation, we replace r with its arithmetic mean in the term

$$\frac{B}{r^2}=\frac{4B}{(a+b)^2}$$

but not elsewhere. Then each of the systems (41.3) and (42.1) can be combined into a sixth-order problem of the form

$$-F^2\partial_{xx}^2 w+\nu^2\left(\nabla_2^2-\frac{1}{r^2}\right)^3 w=0, \tag{42.2a}$$

where

$$w=\partial_r w=\left(\nabla_2^2-\frac{1}{r^2}\right)^2 w=0 \quad\text{at}\quad r=a,\, b. \tag{42.2b}$$

For the energy problem,

$$F^2=F_{\mathscr{E}}^2=\left(\frac{4B}{(a+b)^2\sqrt{\lambda}}-\frac{(\lambda-1)A}{\sqrt{\lambda}}\right)^2 \quad\text{and}\quad \nu=\nu_{\mathscr{E}}. \tag{42.3a}$$

---

[5] The application to the linear stability problem (40.1) of the mean-radius approximation in the narrow-gap limit has been thoroughly discussed by Chandrasekhar (1961, pp. 299—315). This approximation is good when $|\Omega_2/\Omega_1-1|$ is small. He finds that the errors introduced by the approximation do not exceed one percent.

For the linear problem,

$$F^2 = -4A\left(A + \frac{4B}{(a+b)^2}\right) \quad \text{and} \quad \nu = \nu_L . \tag{42.3b}$$

Denote the smallest eigenvalue of (42.2) by

$$\frac{F^2}{\nu^2} = \Lambda^2 .$$

This is some fixed number.

Consider the energy problem (42.3a) and search for the optimum value $\lambda = \tilde{\lambda}$

$$\left(\frac{1}{\tilde{\nu}_{\mathscr{E}}}\right)^2 = \Lambda^2 \max_{\lambda > 0} F_{\mathscr{E}}^{-2} = \left(\frac{1}{\nu_{\mathscr{E}}(\tilde{\lambda})}\right)^2 .$$

To find this value, set $\partial F_{\mathscr{E}}/\partial\lambda = 0$. This leads us again to the relation (41.2b)

$$\tilde{\lambda} = -\left\{\frac{4B}{(a+b)^2} + A\right\}/A > 0$$

which, when inserted back into F, gives

$$\left(\frac{1}{\tilde{\nu}_{\mathscr{E}}}\right)^2 = \frac{\Lambda^2}{-4A\left(A + \dfrac{4B}{(a+b)^2}\right)} = \left(\frac{1}{\nu_L}\right)^2 , \tag{42.4}$$

as asserted. Eq. (42.4) is an exact result in the rotating plane Couette flow limit ($\Omega_V = a(\Omega_1 - \Omega_2)/(b-a)$ fixed, $\eta \to 1$).

The agreement which (42.4) asserts when the mean-radius approximation is valid fails when Ω_2/Ω_1 is too negative. But (42.4) already exhibits the main features of the numerical results. These numerical results are given graphically in Figs. 37.1 a, b for the parameter values which are usually studied in the laboratory. For these parameter values, the only branch of the optimum stability boundary which exists has $A < 0$ and $B > 0$. The numerical results are computed from the eigenvalue problem (41.3) by the Runge-Kutta method. In the first quadrant, the numerically computed values on the optimum stability boundary nearly coincide with the Taylor boundary. The numerically computed values of $\tilde{\lambda}$ are given with good accuracy by (41.2b) with $\bar{r}$ taken near the radius $\bar{r}/b = (1+\eta)/2$ when $\Omega_2/\Omega_1 > 1$ (see Joseph and Hung (1971) for numerical results).

In Figs. 37.1 a, b we have also compared the present result with Serrin's. The energy $\mathscr{E} = \frac{1}{2}\langle u^2 + v^2 + w^2 \rangle$ of every disturbance of Couette flow must decay monotonically when (Ω_1, Ω_2) are in Serrin's band. Outside this band one can find an axisymmetric disturbance whose energy increases initially. The conditional stability theorem for Couette flow guarantees monotonic decay of the energy $\frac{1}{2}\langle u^2 + \lambda v^2 + w^2 \rangle$ of axisymmetric disturbances of restricted size.

This completes the energy analysis for axisymmetric disturbances of Couette flow. There is, however, yet another interesting energy analysis which can be constructed for circular Couette flow. In this analysis two-dimensional disturbances are again assumed from the start but axial symmetry is not assumed. The problem can then be cast into the following form: Find the spiral angle of the disturbance which makes a weighted energy increase at the smallest value of R. This problem is taken up in the more general context of spiral Couette flow. For Couette flow with $\eta \to 1$, the answer is that the extremalizing disturbance is a Taylor vortex when Ω_1 and Ω_2 are such that $R < 177.2$. For the other values of (Ω_2, Ω_1) the extremalizing disturbance is a two-dimensional wave (of the Harrison-Orr type with $u \equiv 0$) whose energy increases initially when $R > 177.2$. These waves which have no circumferential vorticity are sometimes seen in flow between cylinders as a transient (see Fig. 36.1c).

Chapter VI

Global Stability of Spiral Couette-Poiseuille Flows

In this chapter we consider the stability of spiral Couette-Poiseuille flow induced by the shear of rotating and sliding the cylinders relative to one another and by a uniform pressure gradient along the pipe axis.

In the analysis, we shall be concerned with problems of spiral flow in which Spiral Couette-Poiseuille flow bifurcates into secondary cellular motions of the Taylor vortex type. These "Taylor vortices" are spirals rather than toroidal rings and under certain conditions the spiral may develop into flute-like vortices, little cylindrical vortices whose axes are nearly parallel to cylinder generators (see Figs. 46.3, 4, 5). The bifurcating spiral vortices do not appear for all values of the parameters; as in the classical case of Taylor vortices, the spiral vortices are dominated by rotational forces associated with a Rayleigh discriminant.

The plan of this chapter is as follows: The velocity field for spiral flows, dimensionless variables and spiral flow stability parameters are given in § 43. In § 44 we discuss the spectral problem and the energy variational problem for spiral flow in general terms; a distinguished spiral direction for the critical eigenvector of energy theory is delimited; procedures for obtaining eigenvalue bounds for the parameters associated with the spectral problem are reviewed in a sequence of directed exercises. In § 45 we derive conditions for the nonexistence of subcritical instabilities. The conditions are applied in § 46 to the problem of rotating Poiseuille flow; the rotation lowers the threshold for instability to energy-like values and the observed instabilities have a double spiral structure with one set of vortices winding up the inner cylinder and another set winding down the outer cylinder. The stability problem for spiral Couette flow is considered in §§ 47—52. Many of the main stability features of the spiral Couette flow are already present when the ratio of the radius of the inner to outer cylinder ($a/b=\eta$) is near to unity. The limit $\eta\to 1$ will be called a rotating plane Couette flow (RPCF) and it is to be regarded as representing the limit $\eta\to 1$ of Couette flow between rotating and sliding cylinders.

In the limit of narrow gap, a 'modified' energy theory is constructed. This theory exploits the consequences of assuming the existence of a preferred spiral direction along which disturbances do not vary. The flow is also analyzed from the viewpoint of linearized theory. Both problems depend strongly on the sign of Rayleigh's discriminant, $-2\Omega\zeta$. Here Ω is the component of angular velocity, and ζ is the component of total vorticity of the basic flow in the direction perpendicular to the spiral ribbons on which the disturbance is constant. When the discriminant is negative, there is evidently no instability to infinitesimal disturbances, and the spiral disturbance whose energy increases at the smallest R is a roll whose axis is perpendicular to the stream. This restores and generalizes Orr's (1907) results among disturbances having a preferred spiral direction. When the discriminant is positive, the critical disturbance of linear theory and the modified energy theory are spiral vortices. The differences between the energy and linear limits can be made smaller in the restricted class of disturbances with coincidence achieved for axisymmetric disturbances in the rotating cylinder problem in the limit of narrow gap. For the sliding-rotating case, the critical disturbance of the linear theory appears as a periodic wave in a coordinate system fixed on the outer cylinder. This wave has a dimensionless frequency equal to $-\frac{1}{2}a\sin(\chi-\psi)$, where a is the wave-number, χ is the angle between the pipe axis and the direction of motion of the inner cylinder relative to the outer one, and ψ is the disturbance spiral angle.

§ 43. The Basic Spiral Flow. Spiral Flow Angles

Two concentric cylinders of radii a and b ($b>a$) rotate around their common axis with angular velocity Ω_1 and Ω_2, respectively. The cylinders are pulled axially so that the difference in axial speed is U_c, and the coordinate system is chosen so that the inner and outer cylinders have axial speeds U_c and zero, respectively. An axial pressure gradient $\hat{P}$ is maintained so that the maximum excess axial speed above that induced by sliding is U_p. The physical problem and coordinates are sketched in Fig. 43.1.

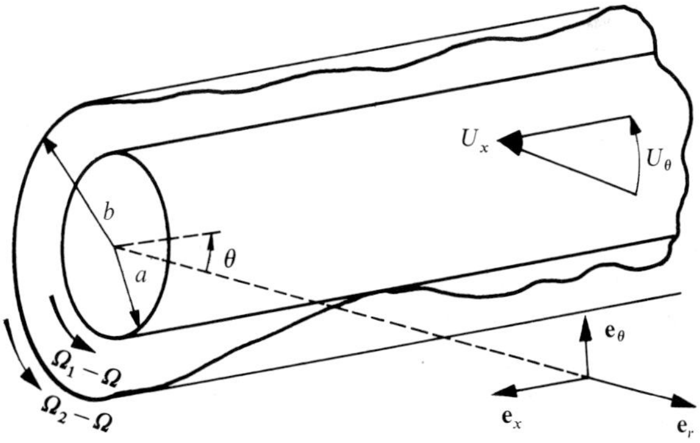

Fig. 43.1: Spiral flow between rotating-sliding cylinders as seen in a rotating coordinate system

(i) The Basic Flow

Let (r, θ, x) be the polar cylindrical coordinates and $a\leqslant r\leqslant b$, $\eta=a/b$. A basic solution of the Navier-Stokes equations

$$\hat{P}\mathbf{e}_x+\nu\nabla^2\mathbf{U}=0\,,\qquad \nabla\cdot\mathbf{U}=0$$

and boundary conditions

$$U_x(a)=U_c, \quad U_x(b)=U_r(a)=U_r(b)=0, \quad \hat{U}_\theta(a)=a\Omega_1, \quad \hat{U}_\theta(b)=b\Omega_2$$

is

$$\mathbf{U}=(U_r,\hat{U}_\theta,U_x)=\left(0, Ar+\frac{B}{r}, U_c\frac{\ln r/b}{\ln\eta}+U_p\left\{\frac{[1-(r/b)^2]\ln\eta^2+(\eta^2-1)\ln(r/b)^2}{\ln\eta^2+(\eta^2-1)\{1+\ln[(\eta^2-1)/\ln\eta^2\}}\right\}\right) \tag{43.1}$$

where

$$A=\frac{b^2\Omega_2-a^2\Omega_1}{b^2-a^2}, \quad B=\frac{-a^2b^2(\Omega_2-\Omega_1)}{b^2-a^2}.^{1} \tag{43.2}$$

It will sometimes be convenient to express the velocity components relative to a coordinate system which rotates with a constant angular velocity Ω. In this rotating system, U_x is unchanged but the tangential component of velocity $\hat{U}_\theta$ is replaced with

$$U_\theta=(\Omega_2-\Omega)r+\frac{\eta a(\Omega_2-\Omega_1)\left[\left(\frac{r}{b}\right)^2-1\right]}{(1-\eta^2)r/b}=(A-\Omega)r+\frac{B}{r}. \tag{43.3}$$

When $\Omega_2=\Omega_1$, $B=0$ and the tangential motion is a rigid body rotation $\hat{U}_\theta=\Omega_2 r$. $U_\theta=0$ if the coordinate system is rotating with angular velocity $\Omega_2=\Omega_1$.

It is easily verified that A is half the axial vorticity of the tangential velocity $\mathbf{e}_\theta\hat{U}_\theta(r)$. When $A=0$, $\hat{U}_\theta(r)$ is a potential flow. For this flow the angular momentum $r\hat{U}_\theta(r)=B$ is a constant.

The flow described by (43.1) will be called Spiral Poiseuille-Couette flow. Our interest is mainly with the special cases:

(43.1 a) *Spiral Couette flow:* $U_p=0$
(43.1 b) *Rotating Couette flow:* $U_p=\Omega_2-\Omega_1=0$, $\Omega_1\neq 0$
(43.1 c) *Sliding Couette flow:* $U_p=\Omega_2=\Omega_1=0$
(43.1 d) *Circular Couette flow:* $U_p=U_c=0$
(43.1 e) *Spiral Poiseuille flow:* $U_c=0$
(43.1 f) *Rotating Poiseuille flow:* $U_c=\Omega_2-\Omega_1=0$, $\Omega_1\neq 0$
(43.1 g) *Annular Poiseuille flow:* $U_c=\Omega_2=\Omega_1=0$
(43.1 h) *Spiral Parabolic Poiseuille flow:* U_p and U_c are related so that $(U_r, \hat{U}_\theta, U_x)=(0, Ar+B/r, U_m(b^2-r^2)/(b^2-a^2))$
(43.1 i) *Rotating Parabolic Poiseuille flow:* Spiral Parabolic-Poiseuille flow with $\Omega_1=\Omega_2\neq 0$
(43.1 j) *Parabolic Poiseuille flow:* $\Omega_1=\Omega_2=0$ in (43.1 h)

---

[1] In the notation of (17.5), $U_p=U_m$.

(43.1 k) *Hagen-Poiseuille flow:* Parabolic Poiseuille flow when there is no inner cylinder

(43.1 l) *Rotating Hagen-Poiseuille flow:* (43.1 k) when the pipe rotates about its axis with constant angular velocity Ω

(43.1 m) *Plane Poiseuille flow* (see Exercise 17.1): The $\eta \to 1$ limit of Annular Poiseuille flow (43.1 g)

(43.1 n) *Rotating Plane Couette flow* (see § 53): The $\eta \to 1$ limit of Spiral Couette flow (43.1 a)

(43.1 o) *Plane Couette flow:* The $\eta \to 1$ limit of sliding Couette flow.

It will be convenient to use dimensionless variables. The units of length, time, velocity and angular velocity are $[b-a, (b-a)/\hat{U}, \hat{U}, \hat{U}/(b-a)]$ where

$$\hat{U} = \sqrt{U_c^2 + U_p^2 + a^2|\Omega_2 - \Omega_1|^2}\,. \tag{43.4}$$

The same symbols $\mathbf{u}, u, v, w, r, \theta, x, \alpha, \beta$ are used for dimensionless and dimensional quantities. The dimensionless velocity components of the basic flow are designated as

$$(0,\ V(r),\ U(r)) = (0,\ U_\theta/\hat{U},\ U_x/\hat{U}) \tag{43.5}$$

and

$$\Omega = \Omega(b-a)/\hat{U}\,.$$

The Reynolds number for spiral flow is defined as

$$R = \hat{U}(b-a)/\nu\,.$$

The study of the stability of spiral flow is simplified by the fact that the dimensionless description of the velocity field of the spiral flow is independent of the Reynolds number.

We have already considered (in Chapters III, IV and V) problems of stability and bifurcation for special cases of spiral flow in which the basic flow is purely axial or purely circumferential. In this chapter we shall study problems of spiral flow where circumferential and axial motions combine. Linear stability theories for spiral flows have been given by Chandrasekhar (1961), Datta (1965), Howard and Gupta (1962), Hughes and Reid (1968) and Krueger and DiPrima (1964). These authors restrict their study to axi-symmetric disturbances. The basic spiraling properties of the instability were first revealed by the linear inviscid analysis of Ludwieg (1961) and, more decisively, in an ingenious experiment (Ludwieg, 1964; see § 52). Ludwieg's experiment put to rest the then current belief that rigid rotation always stabilizes. No studies of bifurcation of spiral flow, rigorous or formal, have yet appeared.

(ii) Distinguished Spiral Flow Directions

The most striking feature about the secondary flows which bifurcate from spiral flow are the spiral vortices. To describe these it is important to introduce certain distinguished directions; these are described by angles measured from cylinder generators in planes tangent to the cylinder of radius r. The unit vectors $(\mathbf{e}_r, \mathbf{e}_\theta, \mathbf{e}_x)$ form an orthogonal triad in the direction $\mathbf{e}_x$ of the advancing inner cylinder (Fig. 43.1).

(a) *Basic flow spiral angle* $\chi(r)$:

$$\begin{aligned} U_\theta(r) &= (U_\theta^2 + U_x^2)^{1/2} \sin\chi \\ U_\theta(r)/U_x(r) &= V(r)/U(r) = \tan\chi\,. \end{aligned} \tag{43.6}$$

(b) *Disturbance spiral flow angle* $\psi\,(r)$: Let p be a disturbance pressure and let x' be a spiral coordinate on the cylinder at r making a constant angle ψ between x' and the axial direction $x(\mathbf{e}_{x'}\cdot\mathbf{e}_x = \cos\psi)$. The angle ψ is a disturbance spiral flow angle if at r,

$$u',\ v',\ w \text{ and } p \text{ are independent of } x'\,. \tag{43.7a}$$

Here u' and v' are components in the orthogonal directions x' and y'. It is well to observe that the critical disturbances of linear and energy theory necessarily possess a preferred spiral direction. The direction is determined by selection of critical wave numbers for the axis (α) and in the azimuth (n). Thus, we have

$$0 = \frac{\partial p(r,\theta,x)}{\partial x'} = \cos\psi\frac{\partial p}{\partial x} + \sin\psi\frac{1}{r}\frac{\partial p}{\partial\theta} = i\hat{p}\left(\alpha\cos\psi + \frac{n}{r}\sin\psi\right)$$

for $p = \hat{p}(r)\exp(i\alpha x + in\theta)$. This leads to

$$\alpha\cos\psi + \frac{n}{r}\sin\psi = 0\,. \tag{43.7b}$$

The critical disturbances of energy and linear theory are spiral cell disturbances which are sometimes called spiral vortices;

$\psi_{\mathscr{E}}(r)$ is the spiral flow angle of the critical disturbance of energy theory,
$\psi_L(r)$ is the spiral flow angle of the critical disturbance of linear theory.

(c) *Energy spiral angle* $\beta_{\mathscr{E}}(r)$
Note that

$$rD(V/r) = 2D_{r\theta}\,, \qquad DU = 2D_{rx} \tag{43.8}$$

where $D_{r\theta}$ and D_{rx} are the only non-zero components of the stretching tensor $\mathbf{D}[\mathbf{U}]$ of the basic spiral flow. We define $\rho_\beta(r)$ and $\beta_{\mathscr{E}}(r)$ by

$$\begin{aligned} 2D_{r\theta} &= \rho_\beta \sin\beta_{\mathscr{E}}\,, \\ 2D_{rx} &= \rho_\beta \cos\beta_{\mathscr{E}}\,. \end{aligned} \tag{43.9}$$

§ 44. Eigenvalue Problems of Energy and Linear Theory

The analysis starts from the nonlinear Navier-Stokes equation for the difference motion written relative to a system of coordinates rotating with a steady angular velocity $\boldsymbol{\Omega}=\mathbf{e}_x\Omega$:

$$\frac{\partial \mathbf{u}}{\partial t}+2\boldsymbol{\Omega}\wedge\mathbf{u}+\mathbf{u}\cdot\nabla\mathbf{u}+(\mathbf{u}\cdot\nabla)\,\mathbf{U}+(\mathbf{U}\cdot\nabla)\mathbf{u}=-\nabla p+\frac{1}{R}\nabla^2\mathbf{u}\,,\qquad \nabla\cdot\mathbf{u}=0\,,\quad (44.1\,\text{a, b})$$

in the cylindrical annulus and

$$\mathbf{u}=0\quad\text{at}\quad r=\tilde{a},\tilde{b}\quad\text{where}\quad \tilde{a}=\eta/(1-\eta)\,,\qquad \tilde{b}=1/(1-\eta)\qquad (44.1\,\text{c})$$

and $\mathbf{u}(r,\theta,x)$ is periodic in θ with period 2π and almost periodic in x. By $\mathbf{U}$ we understand the spiral-flow problem $\mathbf{U}=\mathbf{U}(r)$ of § 43. The nonlinear problem (44.1) determines the border of stability and instability of spiral flow. It determines the critical Reynolds number R_G below which all disturbances eventially decay.

(i) Eigenvalue Problem of Energy Theory

The critical value R_G for global stability is not known except in certain special cases when $R_{\mathscr{E}}=R_L$ where $R_{\mathscr{E}}$ and R_L are the first critical values of energy and linear theory, respectively. In general, $R_{\mathscr{E}}<R_G<R_L$. The first critical value of energy theory is determined as

$$\frac{1}{R_{\mathscr{E}}}=\max_{\mathbf{H}_x}\Lambda[\mathbf{u}]\qquad (44.2)$$

where $\mathbf{H}_x$ is the set of vectors defined in § 20 and

$$\Lambda[\mathbf{u}]=\mathscr{I}[\mathbf{u}]/\mathscr{D}[\mathbf{u}]=-\langle\mathbf{u}\cdot\nabla\mathbf{U}\cdot\mathbf{u}\rangle/\langle|\nabla\mathbf{u}|^2\rangle\,.\qquad (44.3)$$

The energy stability limit may be obtained as the maximum eigenvalue of the Euler equation

$$\mathbf{u}\cdot\mathbf{D}[\mathbf{U}]=-\nabla p+\Lambda\nabla^2\mathbf{u}\,,\qquad \mathbf{u}\in\mathbf{H}_x\qquad (44.4)$$

where $\mathbf{D}[\mathbf{U}]$ is the stretching tensor for $\mathbf{U}$ and

$$\mathbf{u}\cdot\mathbf{D}[\mathbf{U}]=\tfrac{1}{2}\left[\mathbf{e}_r(vrD(V/r)+uDU\right]+\mathbf{e}_\theta wrD(V/r)+\mathbf{e}_x wDU]$$

where $V(r)$ and $U(r)$ are the components of spiral flow of § 43. The eigenfunction $\mathbf{u}$ belonging to the eigenvalue $\Lambda=1/R_{\mathscr{E}}$ gives the form of the disturbance whose energy increases at the smallest R. We now show that this disturbance should vary only slightly along the energy-spiral lines.

Using the spiral angle $\beta_{\mathscr{E}}$ defined by (43.9) we may write (44.2) as

$$\frac{1}{R} = \max\left[-\langle \rho_\beta w u'' \rangle / \langle |\nabla \mathbf{u}|^2 \rangle\right] \tag{44.5 a}$$

where

$$u'' = u\cos\beta + v\sin\beta \tag{44.5 b}$$

and the maximum is over solenoidal fields $\mathbf{u} = (u, v, w)$ which vanish on $r = \tilde{a}, \tilde{b}$.

In general, (44.5 a) will have its largest value among functions which allow one to raise the value of $|wu''|$ on each cylinder $r = \text{const.}$ without increasing the value of the dissipation denominator $|\nabla \mathbf{u}|^2$ too sharply. The value of the dissipation cannot be easily controlled among functions w since w and $\partial w/\partial r$ both vanish at the walls. To maximize (44.5) we need to give u'' the greatest freedom to take on the values which increase $|wu''|$ at the smallest dissipation cost. Of course, u'' is not free when w is given since the vectors $\mathbf{u}$ which compete for the maximum have $\operatorname{div}\mathbf{u} = 0$. If

$$\mathbf{e}_{x''} = \sin\beta_{\mathscr{E}}\mathbf{e}_\theta + \cos\beta_{\mathscr{E}}\mathbf{e}_x, \qquad \mathbf{e}_{y''} = \cos\beta_{\mathscr{E}}\mathbf{e}_\theta - \sin\beta_{\mathscr{E}}\mathbf{e}_x \tag{44.6}$$

are orthogonal energy-spiral directions on the cylinder of radius r the only derivative of u'' which appears in $\operatorname{div}\mathbf{u} = 0$ is $\partial u''/\partial x''$. If, on some cylinder, $r = \bar{r}$,

$$\psi_{\mathscr{E}}(\bar{r}) = \beta_{\mathscr{E}}(\bar{r}) \tag{44.7}$$

then $\partial u''/\partial x'' = 0$ and $|wu''|$ can hold to its largest value on each energy-spiral line $y'' = y' = \text{const.}$ on the cylinder $\bar{r}$. The argument implies that the critical disturbance of energy theory should vary only slightly along energy-spiral lines. (44.7) can be proved when $\eta = a/b \to 1$ (see § 51) and it is supported by numerical analysis for general values of a/b (Joseph and Munson, 1970).

Exercise 44.1 (Almost periodic solutions of (44.2)): All vectors $\mathbf{u} \in \mathbf{H}_x$ may be represented as

$$\mathbf{u}(r,\theta,x) \sim \sum_{n=-\infty}^{\infty} \sum_{\substack{j=-\infty \\ j\neq 0,\, \alpha_j \neq 0}}^{\infty} \mathbf{u}_{j,n}(r)\, e^{i(n\theta + \alpha_j x)}, \tag{44.8}$$

where, since $\mathbf{u}$ is real-valued,

$$\alpha_j = \alpha_{-j} \quad \text{and} \quad \mathbf{u}_{j,n} = \bar{\mathbf{u}}_{-j,-n}.$$

Here, the overbar means complex conjugate, and

$$\mathbf{u}_{j,n} = \frac{1}{2\pi} \lim_{L\to\infty} \frac{1}{2L} \int_0^{2\pi} d\theta \int_{-L}^{L} dx\, \mathbf{u}(r,\theta,x)\, e^{-i(n\theta + \alpha_j x)}$$

are Fourier coefficients. Show that

$$\frac{1}{R_{\mathscr{E}}} = \max_{j,n} \Lambda(\alpha_j, n) = \max_{\alpha,n} \Lambda(\alpha, n) = \Lambda(\alpha_{\mathscr{E}}, n_{\mathscr{E}}), \tag{44.9}$$

where, below, $\mathbf{u}=\mathbf{u}_{j,n}(r)$ and

$$\Lambda(\alpha,n)=-\langle\mathbf{u}\cdot\mathbf{D}[\mathbf{U}]\cdot\mathbf{u}\rangle/\mathscr{D}_n[\mathbf{u},\alpha]$$

where, since we averaging over functions of r alone,

$$\langle\cdot\rangle=\frac{2(1+\eta)}{1-\eta}\int_{\tilde{a}}^{\tilde{b}}\cdot\, r\,dr$$

$$\mathscr{L}=\frac{1}{r}D(rD)-(n^2+1)/r^2-\alpha^2,\qquad D=\frac{d}{dr},\qquad L=\mathscr{L}+1/r^2, \tag{44.10}$$

and

$$\begin{aligned}\mathscr{D}_n[\mathbf{u};\alpha]&=-\langle w\mathscr{L}\overline{w}\rangle-\langle v\mathscr{L}\overline{v}\rangle-\langle uL\overline{u}\rangle-2in(\langle\overline{v}w/r^2\rangle-\langle\overline{w}v/r^2\rangle)\\&=\langle|Dw|^2+|Dv|^2+|Du|^2+\alpha^2(|w|^2+|v|^2+|u|^2)\rangle\\&\quad+n^2\left\langle\left|\frac{u}{r}\right|^2\right\rangle+\left\langle\left|\frac{nw+iv}{r}\right|^2+\left|\frac{nv-iw}{r}\right|^2\right\rangle.\end{aligned} \tag{44.11}$$

Hint: Follow the proof of 20.7.

Exercise 44.2 (Euler equations for the Fourier coefficients): Show that the Euler equations for (44.9) may be reduced to two equations for the unknown coefficients $u(r)$ and $f=rw(r)$ where $\mathbf{U}_{j,n}(r)=\mathbf{u}(r)=(u(r),v(r),w(r))$:

$$n^2f(DU)/r=-2\Lambda\{i\alpha rDLf-(n^2+\alpha^2r^2)Lu-2\alpha^2rDu\}+\alpha nfrD(V/r), \tag{44.12a}$$

$$\begin{aligned}&-r^{-1}DU\{(n^2+\alpha^2r^2)u+i\alpha rDf\}-i2\Lambda\,\alpha rf\,D\left(\frac{1}{r}DU\right)\\&+\left\{\frac{\alpha}{n}(n^2+\alpha^2r^2)urD(V/r)-\frac{in}{r}f\,D(rD(V/r))-\frac{in}{r}(2n^2+\alpha^2r^2)DfrD(V/r)\right\}\\&=2\Lambda(L^2f+4i\alpha Lu)\end{aligned} \tag{44.12b}$$

At the boundaries, $r=\tilde{a},\tilde{b}$,

$$u=f=Df=0. \tag{44.12c}$$

(ii) The Spectral Problem

The spectral problem may be obtained from (44.1) by substituting solutions proportional to $e^{-\sigma t}$, $\sigma=\hat{\xi}+i\hat{\omega}$, into the equation which is obtained from (44.1) by linearization,

$$-\sigma\mathbf{u}+2\mathbf{\Omega}\wedge\mathbf{u}+\mathbf{u}\cdot\nabla\mathbf{U}+\mathbf{U}\cdot\nabla\mathbf{u}=-\nabla p+\lambda\nabla^2\mathbf{u} \tag{44.13}$$

where $\mathbf{u}\in\mathbf{H}_x$. The critical values of $\lambda=\lambda_L$ are those for which $\hat{\xi}(\lambda)=0$. Spiral Couette flow is stable by the criteria of the spectral problem when $\hat{\xi}(\lambda)>0$ for all eigenvalues $\sigma(\lambda)$. Spiral flow loses stability when R exceeds $R_L=\lambda_L^{-1}$ where λ_L^{-1} is the first critical Reynolds number of the spectral problem.

The equations for the Fourier coefficients of $\mathbf{u}$ and p in the spectral problem (44.13) are

$$-\hat{\xi}w+i\mathscr{S}w-2\frac{\mathbb{V}}{r}v+Dp-\lambda\left\{\mathscr{L}w-\frac{2in}{r^2}v\right\}=0, \tag{44.14a}$$

$$-\hat{\xi}v+i\mathscr{S}v+\frac{w}{r}D(\mathbb{V}r)+\frac{inp}{r}-\lambda\left\{\mathscr{L}v+\frac{2in}{r^2}w\right\}=0, \tag{44.14b}$$

$$-\hat{\xi}u+i\mathscr{S}u+wDU+i\alpha p-\lambda Lu=0, \tag{44.14c}$$

$$\frac{1}{r}D(rw)+\frac{in}{r}v+i\alpha u=0, \tag{44.14d}$$

$$w=v=u=0 \quad \text{at} \quad r=\tilde{a},\tilde{b}. \tag{44.14e}$$

where $\mathscr{L}$ and L are defined by (44.10) and

$$\mathscr{S}=\frac{n}{r}V+\alpha U-\hat{\omega}, \qquad \mathbb{V}(r)\equiv V(r)+\Omega r. \tag{44.15}$$

The first critical value of linear stability theory may be obtained from the principal eigenvalue of (44.14)

$$\frac{1}{R_L}=\max_{\alpha,n}\lambda(\alpha,n)=\lambda(\alpha_L,n_L). \tag{44.16}$$

Some interesting features of (44.14) are most easily detected by examination of energy integrals for (44.14). These are formed as follows:

$$\begin{aligned}0&=\langle\overline{w}(44.14\text{a})\rangle+\langle\overline{v}(44.14\text{b})\rangle+\langle\overline{u}(44.14\text{c})\rangle\\&=i\langle\mathscr{S}|\mathbf{u}|^2\rangle-2\left\langle\frac{\mathbb{V}}{r}\overline{w}v\right\rangle+\left\langle\frac{w\overline{v}}{r}D(r\mathbb{V})\right\rangle+\langle DU\overline{u}w\rangle\\&\quad+\lambda\mathscr{D}_n[\alpha;\mathbf{u}]-\hat{\xi}\langle|\mathbf{u}|^2\rangle.\end{aligned} \tag{44.17}$$

The real part of (44.17) is

$$-\hat{\mathscr{I}}[\mathbf{u}]+\lambda\mathscr{D}_n[\alpha;\mathbf{u}]-\hat{\xi}\langle|\mathbf{u}|^2\rangle=0 \tag{44.18}$$

and the imaginary part of (44.17) gives

$$\langle(\overline{u}w-u\overline{w})DU\rangle+\left\langle(w\overline{v}-\overline{w}v)\frac{D(r^3\mathbb{V})}{r^4}\right\rangle+2i\langle\mathscr{S}|\mathbf{u}|^2\rangle=0. \tag{44.19}$$

It is clear from the form (44.15) of $\mathscr{S}$ that the frequency $\hat{\omega}$ can be regarded as a wave on the spiral $\mathscr{S}=$const.

Eq. (44.18) can be made the basis for estimates of the decay constant $\hat{\xi}$ when α, n and $\boldsymbol{\lambda}$ are given; when $\hat{\xi}=0$

$$\boldsymbol{\lambda}=\hat{\mathscr{I}}[\mathbf{u}]/\mathscr{D}_n[\alpha;\mathbf{u}]\leqslant 1/R_{\mathscr{E}}\,.$$

Eq. (44.19) can be used to estimate wave speeds (Exercise (44.10)).

The problem (44.14) simplifies considerably when the flow is straight and parallel ($\mathbb{V}\equiv 0$). Then, assuming axisymmetric disturbances ($n=0$), we may reduce (44.14) to the cylindrical Orr-Sommerfeld problem (34.4). In considering the Orr-Sommerfeld problem it is more general to imagine arbitrary parallel flows $U(r)$.[2] Some parts of the Orr-Sommerfeld theory can be developed rigorously at this level of generality; in particular, estimates of the critical values of the Reynolds number and of wave speeds readily follow from the application of isoperimetric inequalities to energy integrals for the Orr-Sommerfeld problem (Exercises 44.4—44.10).

Exercise 44.3: Eliminate p and u from (44.14) and show that

$$\begin{aligned}\boldsymbol{\lambda}\left\{L^2 f-4L\left[\frac{Df+inv}{r}\right]\right\}&=i\mathscr{S}Lf+iD\mathscr{S}Df-\frac{2i\mathscr{S}}{r}(Df+inv)-\frac{in}{r^2}D(r\mathbb{V})Df\\&\quad-i\alpha DU\left(Df-\frac{2f}{r}\right)+\frac{2(n^2+\alpha^2r^2)}{r^2}\mathbb{V}v\end{aligned}\tag{44.20}$$

and

$$\begin{aligned}L\left[(n^2+\alpha^2r^2)\frac{v}{r}\right]&=inL\left(\frac{Df}{r}\right)-\frac{2in\alpha^2 f}{r^2}+\frac{2\alpha^2}{r}D(rv)\\&\quad+\frac{\boldsymbol{\lambda}^{-1}}{n}\left\{\frac{\mathscr{S}}{r}(n^2+\alpha^2r^2)(Df+inv)-\alpha^2 r\mathscr{S}Df-\frac{\alpha n^2 DU}{r}f+\alpha^2 n\left[\frac{D(r\mathbb{V})}{r}\right]f\right\},\end{aligned}\tag{44.21}$$

where

$$Df=f=v=0\quad\text{at}\quad r=\tilde{a},\tilde{b}\,.\tag{44.22}$$

---

[2] When there is no imposed rotation, the mechanism of instability associated with Rayleigh's discriminant cannot easily operate. Then, as in parallel flow, a different cause for the instability can be found. Two different mechanisms are commonly thought to be central in the instability of parallel flows. One of these is associated with the presence of a point of inflection (more exactly, with a vorticity maximum in the velocity profile). Only if a point of inflection is present, could there be instability in a fluid with no viscosity. Many parallel flows, like Poiseuille flow, have no inflection point. These flows are unstable to small disturbances at finite Reynolds numbers because of the destabilizing effects of viscosity.

Arbitrary parallel flows can be generated as exact solutions of the Navier-Stokes equations with appropriately defined body forces. This justification for considering arbitrary parallel flows merely moves the question; not many body force fields, much less arbitrary ones, are physically realizable. Many persons motivate the consideration of arbitrary parallel flows by an appeal to the theory of "nearly" parallel flows. The theory of parallel flow is thought to be an important subject because there are many important flows which are "nearly" parallel. The sense in which "nearly" parallel flows, like jets, are nearly parallel is still debatable; it is questionable whether the Orr-Sommerfeld theory is a valid approximation to all flows which are thought to be nearly parallel (see Appendix E).

Exercise 44.4 (Joseph and Munson, 1970): (a) Show that the solution of the minimum problem

$$\underline{\lambda}^2(l,\eta)=\min\frac{\int_\eta^1 r(du/dr)^2\,dr}{\int_\eta^1 r^{l+1}u^2\,dr}$$

among functions $u(r)$ which vanish at $r=\eta,1$ is $(l\neq -2)$

$$\underline{\lambda}^2(l,\eta)=\left(\frac{2+l}{2}\right)^2\underline{\lambda}^2(0,\ \eta^{(2+l)/2}),$$

where $\underline{\lambda}(0,\eta)$ is the first positive root of the Bessel function equation

$$0=J_0(\underline{\lambda}\eta)\,Y_0(\underline{\lambda})-J_0(\underline{\lambda})\,Y_0(\underline{\lambda}\eta)\,, \tag{44.23}$$

(see Fig. 44.1) and

$$\underline{\lambda}^2(-2,\eta)=\pi^2/(\log\eta)^2\,.$$

(b) Show that

$$\frac{1}{R}\leqslant\min\left[a_1+a_2+a_3,(b_1a_1+b_2a_2+b_3a_3)(\alpha^2+(1-\eta)^2n^2)^{1/2}\right], \tag{44.24}$$

where R is any positive critical value of (44.14) when $\mathrm{re}(\sigma)=0$. Find a_i and b_i $(i=1,2,3)$; show that b_i does not depend on α or n and that $a_i\to 0$ as $\alpha^2+n^2\to\infty$, and a_i is finite for $\alpha^2+n^2=0$. Consider Couette flow between rotating cylinders $(U\not\equiv 0)$ and show that (44.24) reduces to

$$\begin{aligned}\frac{1}{R}\leqslant&\left[\frac{2\eta}{1+\eta}\right]\Big/\left[\frac{\pi^2(1-\eta)^2}{(\ln\eta)^2}+\alpha^2\eta^2+(1-\eta)^2(n-1)^2\right]\\&\times\min\left[\tfrac{1}{2},\frac{1}{2(1-\eta)}\left[\frac{\pi^2}{2(\ln\eta)^2}+\tfrac{1}{2}\right]^{-1/2}(\alpha^2+(1-\eta)^2n^2)^{1/2}\right].\end{aligned} \tag{44.25}$$

Show that $R\to\infty$ for small and large α^2+n^2.

Exercise 44.5 (Synge, 1938A): Consider the problem (44.14) for parallel flow $(\mathbb{V}=0)$ and axisymmetric disturbances $(n=0)$. Let $\hat{\xi}=-\alpha c_i$, $\hat{\omega}=\alpha c_r$. Show that

$$c_i(\mathcal{I}_1^2+\alpha^2\mathcal{I}_0^2)=\frac{i}{2}\langle DU(wD\overline{w}-\overline{w}Dw)\rangle-\mathcal{D}/\alpha R \tag{44.26a}$$

and

$$c_r(\mathcal{I}_1^2+\alpha^2\mathcal{I}_0^2)=\langle U(|Dw|^2+|w/r|^2+\alpha^2|w|^2)+\tfrac{1}{2}\Lambda|w|^2\rangle \tag{44.26b}$$

where

$$\Lambda=r^3D\{(1/r)DU\}\,,$$

$$\mathcal{I}_0^2=\langle|w|^2\rangle=\int_a^b|w|^2\,r\,dr\,,$$

$$\mathcal{I}_1^2=\langle|Dw|^2+|w/r|^2\rangle\,,$$

$$\mathcal{I}_2^2=\langle|D^2w|^2\rangle+3\left\langle\left|\frac{1}{r}Dw\right|-\left|\frac{w}{r^2}\right|^2\right\rangle$$

and

$$\mathcal{D}=\mathcal{I}_2^2+2\alpha^2\mathcal{I}_1^2+\alpha^4\mathcal{I}_0^2\,.$$

Exercise 44.6 (Joseph 1968, Carmi, 1970); Show that the ratios

$$\mathcal{I}_1^2/\mathcal{I}_0^2, \quad \mathcal{I}_2^2/\mathcal{I}_1^2, \quad \mathcal{I}_2^2/\mathcal{I}_0^2$$

defined in Exercise (44.5) are bounded below for functions w which have $w = Dw = 0$ at $r = \tilde{a}, \tilde{b}$.

Find the minimum values $\underline{\lambda}_1^2, \underline{\lambda}_2^2, \underline{\lambda}_3^2$, respectively, of these functionals (Fig. 44.1). Show that $\underline{\lambda}_1, \underline{\lambda}_2, \underline{\lambda}_3$ are principal eigenvalues of the Euler equations

$$\tilde{\mathcal{L}}w + \underline{\lambda}_1^2 w = 0, \quad \tilde{\mathcal{L}}^2 w + \underline{\lambda}_2^2 \tilde{\mathcal{L}}w = 0, \quad \tilde{\mathcal{L}}^2 w - \underline{\lambda}_3^2 w = 0$$

where

$$\tilde{\mathcal{L}} = \frac{1}{r}D(rD) - 1/r^2$$

and the boundary conditions $w = 0$ (for $\underline{\lambda}_1$) and $w = Dw = 0$ (for $\underline{\lambda}_2$ and $\underline{\lambda}_3$).

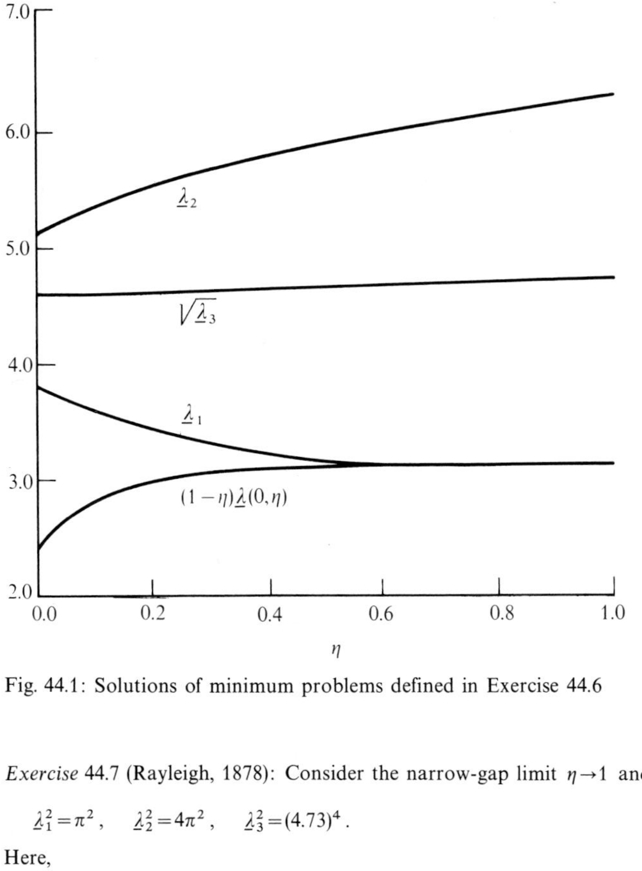

Fig. 44.1: Solutions of minimum problems defined in Exercise 44.6

Exercise 44.7 (Rayleigh, 1878): Consider the narrow-gap limit $\eta \to 1$ and show that

$$\underline{\lambda}_1^2 = \pi^2, \quad \underline{\lambda}_2^2 = 4\pi^2, \quad \underline{\lambda}_3^2 = (4.73)^4. \tag{44.27}$$

Here,

$$\underline{\lambda}_3^2 = \min\left[\int_0^1 |d^2 w/dz^2|^2\, dz / \int_0^1 w^2\, dz\right]$$

for w such that $w(z) = dw/dz = 0$ at $z = 0,1$ with a similar characterization of $\underline{\lambda}_1^2$ and $\underline{\lambda}_2^2$.

Exercise 44.8 (Joseph, 1968): Let $U(r)$ be a parallel motion in the annulus and let

$$q = \max_r |DU|, \quad \tilde{a} < r < \tilde{b}.$$

This motion is monotonically and globally stable against axisymmetric, two-dimensional disturbances provided that

$$\alpha R q < f(\alpha;\eta) = \max[M_1, M_2],$$

where (44.28)

$$M_1 = \underline{\lambda}_3\underline{\lambda}_2 + 2^{3/2}\alpha^3, \qquad M_2 = \underline{\lambda}_3\underline{\lambda}_2 + 2\alpha^2\underline{\lambda}_1.$$

Hint: Consider (44.26) with $c_r = 0$. Find the minimum value of R over kinematically admissible w. How does this relate to the original nonlinear problem for axisymmetric disturbances?

Exercise 44.9 (Georgescu, 1970): For plane parallel flow, the estimates of (44.27) and (44.28) apply. Show that

$$\alpha R q < \max[M(\alpha), M_1]$$

where

$$M_1 = 2\pi(4.73)^2 + 2^{3/2}\alpha^3 \quad \text{and} \quad M(\alpha)$$

is the envelope of the family

$$M(a,\alpha) = 2\pi(4.73)^2 + a\alpha^2\pi + 2\alpha^3\sqrt{2-a} \leqslant \frac{\mathscr{I}_2^2 + 2\alpha^2\mathscr{I}_1^2 + \mathscr{I}_0^2}{\mathscr{I}_1\mathscr{I}_0}.$$

Using the envelope idea, show that

$$\min_\alpha Rq = 74.28$$

improves the earlier result (Joseph, 1968)

$$\min_\alpha Rq = 72.26, \qquad a = 2.$$

Another approach to improving these values is given by Yih (1969).

Exercise 44.10 (Joseph, 1969): Let $\sigma = i\alpha(c_r + ic_i)$ be any eigenvalue of (44.14) when the basic motion is parallel and the disturbance is two-dimensional and axisymmetric. Then,

$$\alpha c_i \leqslant \frac{q}{2} - \frac{\underline{\lambda}_1^2(\underline{\lambda}_2^2 + 2\alpha^2) + \alpha^4}{R(\underline{\lambda}_1^2 + \alpha^2)}$$

and

$$\left(\hat{\Psi}(r) = rD\left(\frac{1}{r}DU\right)\right). \tag{44.29a}$$

$$\text{(a)}\ \hat{\Psi}_{\min} \geqq 0,\ U_{\min} \leqslant c_r \leqslant U_{\max} + \hat{\Psi}_{\max}/2(\underline{\lambda}_1^2 + \alpha^2); \tag{44.29b}$$

$$\text{(b)}\ \hat{\Psi}_{\min} \leqq 0 \leqq \hat{\Psi}_{\max}, \quad U_{\min} + \hat{\Psi}_{\min}\, 2(\underline{\lambda}_1^2 + \alpha^2) \leqslant c_r \leqslant U_{\max} + \hat{\Psi}_{\max}/2(\underline{\lambda}_1^2 + \alpha^2). \tag{44.29c}$$

$$\text{(c)}\ \hat{\Psi}_{\max} \leqq 0, \quad U_{\min} + \hat{\Psi}_{\min}/2(\underline{\lambda}_1^2 + \alpha^2) \leqslant c_r \leqslant U_{\max}, \tag{44.29d}$$

where maximum and minimum values are on the range $a \leqslant r \leqslant b$.

Eqs. (44.29) restrict the complex wave speed to a parabola in the complex c plane. These estimates leave open the possibility that the wave speed lies outside the range of U. This possibility is realized for some members of the family of flows in diverging wedge-shaped channels, the Jeffery-Hamel flows (Eagles, 1966).

Exercise 44.11: Consider statistically stationary turbulent Couette flow between rotating and sliding cylinders. Find the equations which govern the mean motion and the fluctuations. Find a first integral of the equations of the mean motion. Eliminate the mean motion from the energy functional and identify the response functional for the spiral flow system.

§ 45. Conditions for the Nonexistence of Subcritical Instability

The problem of spiral flow is characterized by an axial speed, an angular-velocity difference, an angular velocity and the radius ratio. It is possible to specify, *a priori*, an explicit relation among these four parameters to bring the instability and stability limits into coincidence or near coincidence.

The optimal adjustment of parameters can be obtained by applying the following criterion of comparison of energy and linear theory.

Suppose $\mathbf{u}$ *is a vector field which solves* (44.4) *and a vector* $\boldsymbol{\Omega}$ *exists such that*

$$\operatorname{curl}\{(\mathbf{U}\cdot\nabla)\mathbf{u}+2\boldsymbol{\Omega}\wedge\mathbf{u}-\tfrac{1}{2}(\mathbf{u}\wedge\operatorname{curl}\mathbf{U})\}=0\,. \tag{45.1}$$

Then $\mathbf{u}$ *is an eigensolution of* (44.13) *with eigenvalue* $\sigma(\Lambda)=0$.

Proof: Since $\mathbf{u}$ and $\mathbf{U}$ are solenoidal,

$$\operatorname{curl}\{\mathbf{u}\cdot(\nabla\mathbf{U}-\mathbf{D}[\mathbf{U}])\}=-\tfrac{1}{2}\operatorname{curl}(\mathbf{u}\wedge\operatorname{curl}\mathbf{U}).$$

Then if (45.1) holds

$$(\mathbf{U}\cdot\nabla)\mathbf{u}+2\boldsymbol{\Omega}\wedge\mathbf{u}+\mathbf{u}\cdot(\nabla\mathbf{U}-\mathbf{D}[\mathbf{U}])=-\nabla\phi\,,$$

and we can write (44.4) as

$$2\boldsymbol{\Omega}\wedge\mathbf{u}+\mathbf{u}\cdot\nabla\mathbf{U}+\mathbf{U}\cdot\nabla\mathbf{u}=-\nabla(p+\phi)+\Lambda\nabla^2\mathbf{u}\,. \tag{45.2}$$

Comparison of (45.2) with (44.13) shows that $\mathbf{u}$ is an eigenfunction of the latter with eigenvalue $\sigma(\Lambda)=0$.

Eq. (45.1) gives a procedure for finding the parameters of deepest instability. Suppose $\mathbf{U}$ is a spiral flow and $\boldsymbol{\Omega}=\mathbf{e}_x\Omega$. In polar cylindrical coordinates (r,θ,x), one can write (45.1) as

$$\begin{aligned}0=\operatorname{curl}\Bigg\{&\mathbf{e}_i\left(U\frac{\partial u_i}{\partial x}+\frac{V}{r}\frac{\partial u_i}{\partial\theta}\right)\Bigg\}+\mathbf{e}_i\left\{\frac{1}{2r}DU\frac{\partial u_i}{\partial\theta}-\left[2\Omega+\frac{V}{r}+\frac{1}{2r}D(rV)\right]\frac{\partial u_i}{\partial x}\right\}\\&+\tfrac{1}{2}\left\{-\mathbf{e}_\theta wrD\left(\frac{1}{r}DU\right)+\mathbf{e}_x wD(V/r)\right\}\end{aligned} \tag{45.3}$$

where

$$\mathbf{e}_i u_i=\mathbf{e}_r w+\mathbf{e}_\theta v+\mathbf{e}_x u\,.$$

To derive (45.3) we use the fact that $V(r)$ is the tangential component of velocity in spiral flow

$$D\left[\frac{1}{r}D(rV)\right]=0$$

and the continuity equation

$$\frac{1}{r}\frac{\partial(rw)}{\partial r}+\frac{1}{r}\frac{\partial v}{\partial\theta}+\frac{\partial u}{\partial x}=0.$$

Eq. (45.3) can be satisfied exactly in two cases: (a) for a suitably adjusted rotating plane Couette flow (43.1n) see Exercise 49.1, and (b) for rapidly rotating parabolic Poiseuille flow (43.1i) see Exercise 46.1. Ordinarily (45.3) could not hold for all $a\leqslant r\leqslant b$. But if the variable coefficients of $\mathbf{e}_r$, $\mathbf{e}_\theta$ and $\mathbf{e}_x$ are slowly varying, then (45.3) can be used to approximate the values of the parameters for which the linear and energy limits are closest.

Exercise 45.1 (Joseph and Munson, 1970): Assume that (44.7) holds. Then show how (45.3) suggests the following statements:

The critical energy disturbance is constant or slowly varying along the energy spiral. The linear and energy limits are closest when the disturbance spiral, the energy spiral, and the basic flow spiral coincide. The parameters can be adjusted to bring these angles into close agreement. For these parameters the critical disturbance of linear theory is steady or nearly steady in a suitably chosen rotating system.

Exercise 45.2 (Joseph and Munson, 1970; an application of (45.3) to Couette flow between rotating cylinders): Show that the energy bands shown in Figs. 37.1 are closest to Taylor's boundary when

$$\frac{\Omega_1}{\Omega_2}=\frac{1+2\eta-\eta^2}{\eta^2(\eta^2+2\eta-1)}.$$

§ 46. Global Stability of Poiseuille Flow between Cylinders which Rotate with the Same Angular Velocity

Consider the effect, on the stability of Poiseuille flow between cylinders, of rotating the cylinders about their symmetry axis with a constant angular velocity Ω. Large rates of rotation are very destabilizing. When Ω is large the Coriolis forces $\mathbf{u}\wedge\boldsymbol{\Omega}$ acting on the disturbance push the fluid in the directions perpendicular to $\mathbf{u}$ and $\boldsymbol{\Omega}$; in this way the disturbance $\mathbf{u}$ can acquire longitudinal vorticity.

Note that rotating the pipe as a whole cannot alter the form of the stretching tensor $\mathbf{D}$ for Poiseuille flow and, therefore, cannot alter the results of energy analysis. It follows from this that the energy analyses which were given in §§ 21 and 22 apply uniformly, independent of Ω. Rotation has a big effect on the instability of rotating Poiseuille flow, but the energy stability limit and the most energetic initial condition are independent of the rate of rotation.

(i) Rotating Parabolic Poiseuille Flow

A concise statement of the effect of rotation can be framed for parabolic Poiseuille flow. In dimensionless variables, with $\hat{U}=U_m$, $U(r)=\{1-(1-\eta)^2r^2\}/(1-\eta^2)$. Computations indicate that when $\Omega=0$, parabolic Poiseuille flow is absolutely stable ($R_L\to\infty$) to infinitesimal disturbances for all values of $\eta\leqslant 1$. Experiments show that there is instability when $U_m(b-a)/\nu>R_G$ where R_G is approximately 2000. Independent of Ω, $R_{\mathscr{E}}\simeq 100$.

Suppose that the most energetic initial disturbance of parabolic Poiseuille flow which rotates as a rigid body about its axis of symmetry x is x-independent. Then, in the limit $\Omega\to\infty$,

$$R<R_{\mathscr{E}}(\eta) \tag{46.1}$$

is both necessary and sufficient for global stability.

The hypothesis that the critical energy disturbance is x-independent is supported by numerical results (see Fig. 22.1). The criterion (46.1) is obviously sufficient for stability; we need to prove that the criterion is necessary. To do this we could show that

$$\lim_{\Omega\to\infty} R_L(\eta,\Omega)=R_{\mathscr{E}}(\eta)\,. \tag{46.2}$$

The relation (46.2) is implied by the comparison criterion (45.3). We leave the direct proof of (46.2) as Exercise (46.1) for the reader[3].

(ii) Rotating Poiseuille Flow

We now turn away from the special problem of parabolic Poiseuille flow and consider the problem of stability of laminar Poiseuille flow (17.5) between cylinders which rotate around their axis of symmetry with a common angular velocity. This comparison does not yield necessary and sufficient conditions for stability as it does for parabolic Poiseuille flow. Nonetheless, in the limit $\Omega\to\infty$ where α is finite, the energy and linear stability limits are close. The linearized equations of stability in this limit are obtained from (44.20, 21, 22). Since the basic flow is in rigid rotation $V(r)=0$, $\mathbb{V}=r\Omega$ and $\mathscr{S}=\alpha U(r)-\hat{\omega}=-\hat{\omega}$ ($\alpha\to 0$ if $\Omega\to\infty$ and $\alpha\Omega$ is finite). Then using $i\alpha ru=-(Df+inv)$ to eliminate v we pass to the limit and find that

$$\begin{aligned} \lambda L^2 f &= -i\hat{\omega}Lf-2n(\alpha\Omega)u\,,\\ \lambda Lu &= -i\hat{\omega}u-\left(\frac{2(\alpha\Omega)}{n}-\frac{DU}{r}\right)f \end{aligned} \tag{46.3}$$

---

[3] Pedley (1969) following the earlier work of Ludwieg (1961) on rotating Couette flow, found instability for rotating Hagen-Poiseuille flow when $\Omega\to\infty$ and $R>82.88$. His solution is just the one given in § 21 and it also leads to x-independent rolls whose axes are parallel to the pipe axis. However, this is not the most energetic initial disturbance of Hagen-Poiseuille flow and, hence, $R_{\mathscr{E}}\neq R_L$ in the limit $\Omega\to\infty$.

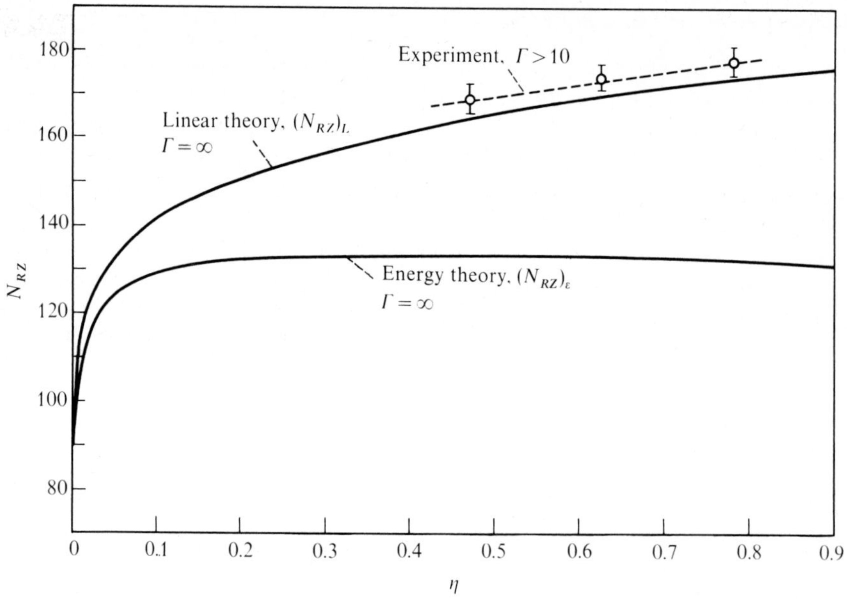

Fig. 46.1: Stability limits for Poiseuille flow in a rapidly rotating annulus. The experimental results are due to Nagib (1972)

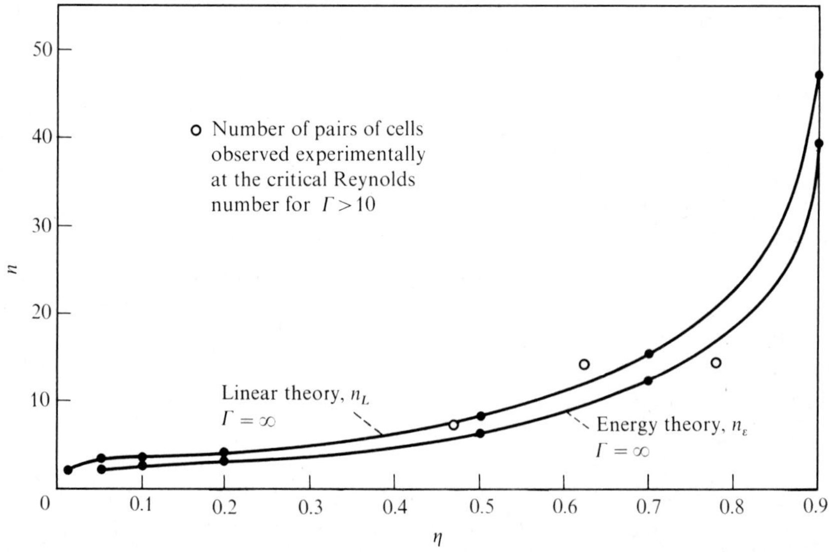

Fig. 46.2: Azimuthal periodicity of the critical energy and linear eigenfunctions. Black dots are computed; white dots are experimental (Nagib, 1972)

and at $r=\tilde{a}, \tilde{b}$,

$$Df=f=u=0.$$

Computed values of the linear and energy stability limits are shown graphically in Figs. 46.1 and 46.2. Linear stability results are taken from the paper of Joseph and Munson (1970); energy stability results are taken from the calculation described in § 22. The experimental results are due to Nagib (1972). The symbols used in Fig. 46.1 and Fig. 46.2 are defined below.

The stability of rotating Poiseuille flow in annular passages has been studied in experiments initiated by Nagib, Lavan, Fejer and Wolf (1971) and completed by H. Nagib (1972). The stability parameters used in the experiments are

$$N_{R\theta}=\frac{2\Omega(b-a)b}{\nu}$$

and

$$N_{RZ}=\frac{2(b-a)\bar{U}}{\nu}$$

where $\bar{U}=\langle U\rangle$ is the mass-flow average of the axial velocity (17.8). For Hagen-Poiseuille flow, $a=0$ and $\bar{U}=\frac{1}{2}U_m$. For plane Poiseuille flow, $a/b\to 1$ and $\bar{U}=\frac{2}{3}U_m$. The stability parameters are related to (Ω, R) as follows:

$$N_{R\theta}=\frac{2\Omega R}{1-\eta}, \qquad N_{RZ}=\frac{2\bar{U}R}{U_m}.$$

The swirl ratio Γ is defined as

$$\Gamma=\frac{N_{R\theta}}{N_{RZ}}=\frac{\Omega b}{\bar{U}}=\frac{U_m\Omega}{\bar{U}(1-\eta)}.$$

Comparison of our analysis with the experiments is possible because:

(1) The asymptotic ($\Gamma\to\infty$) value of N_{RZ} is nearly attained in the experiments at swirl ratios $\Gamma>5$.

(2) When the swirl ratio is large the energy and linear stability limits are close.

The following points of agreement between theory and experiment deserve attention:

(1) The observed stability limits when Γ is large (>10) are very close to the theoretical values for $\Gamma=\infty$ which are predicted by the linear theory (see Fig. 46.1).

(2) The critical eigenfunctions of energy and linear ($\Gamma=\infty$) theory are infinitely elongated x-independent rolls. The slow variation with x of the secondary flow when Γ is large (but not infinite) is shown clearly in Fig. 46.3 ($\Gamma=14.18$) and 46.4 ($\Gamma=5.88$).

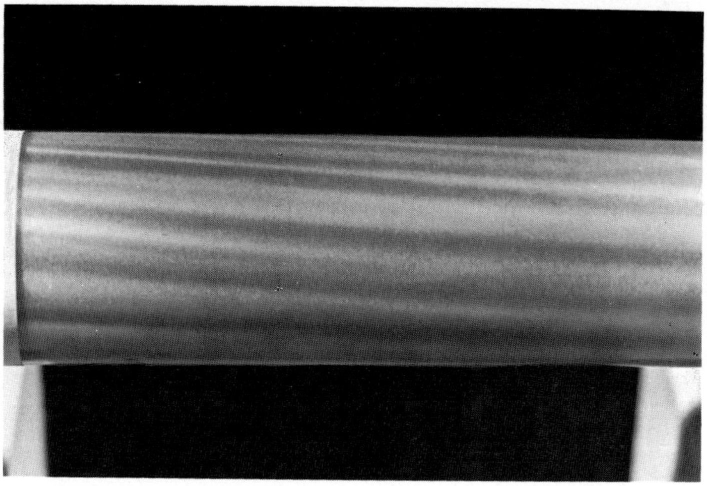

Fig. 46.3: Spiral vortices in rotating Poiseuille flow down the annulus with $\eta=0.625$. The flow is from right to left. The operating conditions for the flow are $N_{R\theta}=3120$, $N_{RZ}=220$ and $\Gamma=14.18$ (Nagib, 1972)

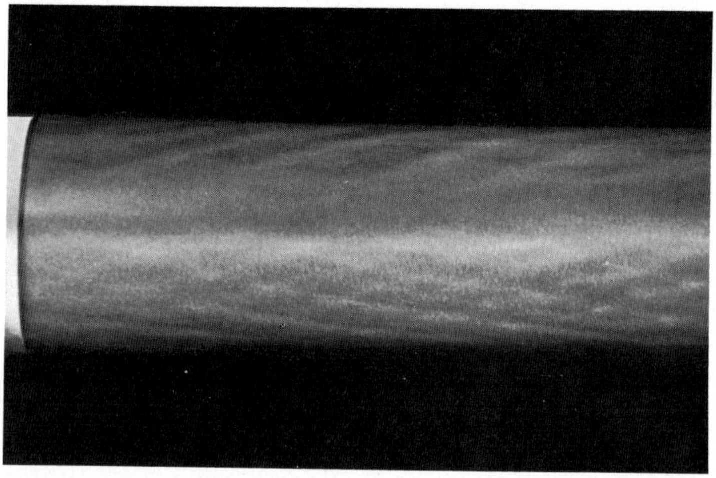

Fig. 46.4: Spiral vortices in rotating Poiseuille flow down the annulus. This figure shows two sets of vortices: an outer set and an inner set. The outer set is near to the outer cylinder and it dominates this photograph but less strongly than in Fig. 46.3. The spiral vortices near the inner wall are also visible. The two sets appear to make an equal and opposite angle with the axis. The flow is from right to left. The operating conditions for the flow are $N_{R\theta}=2000$, $N_{RZ}=340$ and $\Gamma=5.88$ (Nagib, 1972)

We also note that while the number of cells ($2n$) observed in the experiment is in good agreement with the linear theory for $\eta=0.47$, the experimental observations and the theoretically predicted values of n are not in perfect agreement

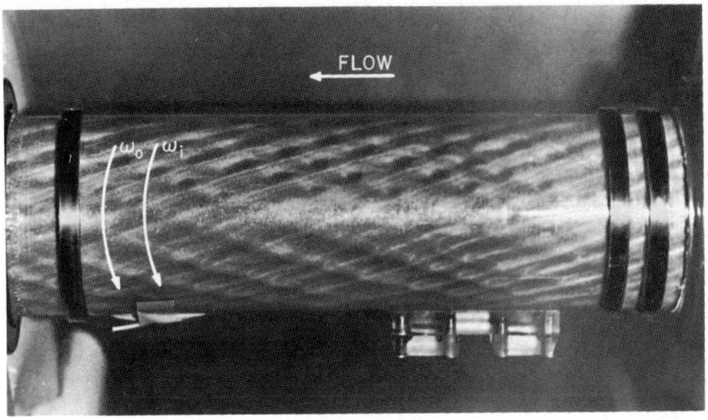

Fig. 46.5: Secondary flow observed for cylinders rotating in same direction; $(N_{R\theta})_O = 1795$, $(N_{R\theta})_I = 1670$, $N_{RZ} = 240$ and $\eta = 0.77$ (Nagib, 1972)

for $\eta = 0.625$ and 0.78 (see Fig. 46.2). It should also be noted that with $R > R_L$ the linear theory shows that spiral Poiseuille flow is unstable to more and more disturbances having values of n near to the critical value. Hence, observed secondary motions with $R > R_L$ need not have a unique value of n; *a* range of values to be determined by stability analysis of the secondary motion is expected.

A most interesting aspect of the experiments on rotating Poiseuille flow is the existence of the two spirals shown in Fig. 46.4. Nagib reports that these spirals are stationary in coordinates which rotate with the cylinders (see Exercise 46.2). The existence of the two spirals is probably associated with the fact that DU changes sign in the annulus; this sign change implies (see § 38) that the Rayleigh discriminant $\Omega\zeta$ can have the right sign for instability on both walls only if the spiral disturbance angle is of opposite sign on each wall. The same sign change in DU is ultimately responsible for the sign change in the longitudinal component of the disturbance velocity (Fig. 23.1).

The stability of rotating plane Poiseuille flow with $\Omega_1 = \Omega_2$ has been considered by Lezius and Johnston (1976). They reduce the spectral problem to a sixth order differential equation. This sixth order equation and the boundary conditions is adjoint to the problem governing the stability of Couette flow to Taylor vortices studied by Chandrasekhar (1961, p. 299). Counter-rotating vortices are shown in the stream function diagrams given by Lezius and Johnston.

The double spiral structure can be intensified further by differential rotation of the cylinder walls as in Fig. 46.5. When the cylinders rotate at different speeds the bifurcating spirals appear to be unsteady in every coordinate system. When the pipe rotates rapidly the solutions which are observed in the experiments seem to be like the ones computed from the Eqs. (46.3) of the linearized theory of stability. This suggests that spiral vortex flow bifurcates supercritically from spiral Poiseuille flow; spiral Poiseuille flow loses stability and spiral vortex flow gains stability when $R > R_L$. Linear stability results have not yet been given for finite values of Ω and when the cylinder walls rotate at different rates. The precise role of the Rayleigh discriminant in the mechanics of the double spiral structure needs further explanation. It would be interesting to see how this structure arises from a stability and bifurcation analysis of rotating Poiseuille flow.

Exercise 46.1: Assume that the critical energy disturbance for parabolic Poiseuille flow is x-independent. Use (45.3) to demonstrate that this disturbance also solves the spectral problem of the linear theory when the rotation parameter is properly adjusted. Hint: Put $\mathbf{e}_\theta \cdot \mathbf{u} = 0$ and replace $\partial/\partial\theta$ and $\partial/\partial x$ with in and $i\alpha$ to find that

$$0 = \operatorname{curl}\left\{\frac{1-(1-\eta)^2 r^2}{1-\eta^2}\alpha\mathbf{u}\right\} - \mathbf{u}\left\{n\frac{(1-\eta)}{(1+\eta)} + 2\alpha\Omega\right\}. \tag{46.4}$$

Now consider the limit $\alpha \to 0$, $\Omega \to \infty$ where $\alpha\Omega$ is finite and not zero.

Exercise 46.2: Prove that every solution of (46.3) must have

$$\hat{\omega} = 2n(\alpha\Omega)\frac{\operatorname{im}\langle u\bar{f}\rangle}{\langle|Df|^2\rangle} = \left\{2\frac{\alpha\Omega}{n} - \operatorname{im}\langle u\bar{f}\rangle + \operatorname{im}\left\langle\frac{DU}{r}u\bar{f}\right\rangle\right\}/\langle|\mathbf{u}|^2\rangle. \tag{46.5}$$

Show that $\hat{\omega} = 0$ when $U(r)$ is parabolic Poiseuille flow and (46.4) holds.

§ 47. Disturbance Equations for Rotating Plane Couette Flow

For definiteness we shall choose $\boldsymbol{\Omega} = -\Omega_2\mathbf{e}_x$ where $\Omega_2 > 0$ is the angular velocity of the outer cylinder. (We have here changed the sign of Ω.) Then the coordinates rotate with the outer cylinder. In the rotating system the outer cylinder is stationary and the inner cylinder moves with a circumferential speed $a(\Omega_1 - \Omega_2)$ and slides forward at the rate U_c (see Fig. 47.1).

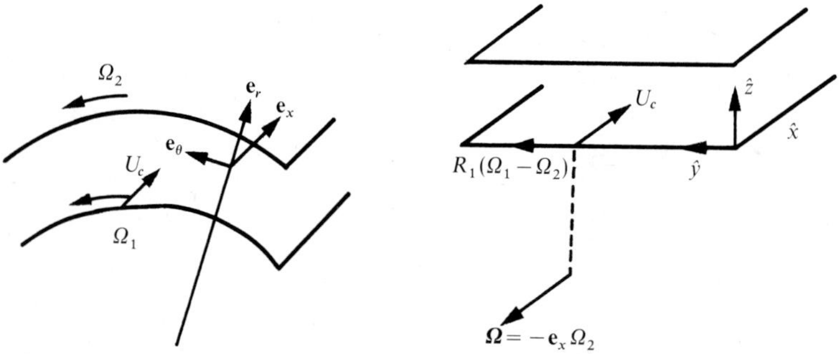

Fig. 47.1: Rotating sliding cylinders in a coordinate system rotating with the outer cylinder in the narrow-gap (rotating plane Couette flow) limit

In the limit $\eta \to 1$ with $b-a$ fixed one may find from (43.3) that

$$U_\theta = \frac{\eta a(\Omega_2 - \Omega_1)[(\hat{r}/b)^2 - 1]}{(1-\eta^2)\hat{r}/b} \to \frac{a(\Omega_1 - \Omega_2)}{b-a}[(b-a) - \hat{z}], \tag{47.1 a}$$

and from (43.1) (with $U_p = 0$)

$$U_x = U_c \ln(\hat{r}/b)/\ln\eta \to \frac{U_c}{b-a}[(b-a)-\hat{z}], \tag{47.1b}$$

where we have set $\hat{r} = a + \hat{z}$ $(0 \leqslant \hat{z} \leqslant b-a)$.

Since we are considering the limit $\eta \to 1$, we must allow for the possibility that with a fixed rate of shear,

$$b - a \to 0.$$

Then we should want

$$\frac{a(\Omega_2 - \Omega_1)}{b-a} = \Omega_V, \quad \frac{U_c}{b-a} = \Omega_U \tag{47.2}$$

to be bounded. The boundedness of Ω_V and Ω_U when $\eta \to 1$ implies that $\Omega_2 - \Omega_1 \to 0$. Hence, apart from the uninteresting case in which $\Omega_1 \to 0$, $\Omega_2 \to 0$, we are here restricted to the problem in which Ω_2 and Ω_1 have the same sense.

In dimensionless variables $z = \hat{z}/(b-a)$, we have

$$V = \frac{U_\theta}{\hat{U}} = (1-z)\sin\chi, \quad U = (1-z)\cos\chi, \tag{47.3a, b}$$

where

$$\hat{U} = \sqrt{a^2(\Omega_1 - \Omega_2)^2 + U_c^2} = (b-a)\sqrt{\Omega_U^2 + \Omega_V^2},$$

$$\sin\chi = \Omega_V\,(b-a)/\hat{U} = \Omega_V/\sqrt{\Omega_V^2 + \Omega_U^2}, \quad \cos\chi = \Omega_U/\sqrt{\Omega_V^2 + \Omega_U^2},$$

and

$$\Omega = \Omega_2/\sqrt{\Omega_U^2 + \Omega_V^2}, \quad \mathbf{\Omega} = -\mathbf{e}_x \Omega.$$

Disturbances of rotating plane Couette flow necessarily satisfy the following equations:

$$\frac{\partial u}{\partial t} + (\mathbf{u} + \mathbf{U})\cdot\nabla u - w\cos\chi = -\partial_x p + \lambda\nabla^2 u, \tag{47.4a}$$

$$\frac{\partial v}{\partial t} + (\mathbf{u} + \mathbf{U})\cdot\nabla v + (2\Omega - \sin\chi)w = -\partial_y p + \lambda\nabla^2 v, \tag{47.4b}$$

$$\frac{\partial w}{\partial t} + (\mathbf{u} + \mathbf{U})\cdot\nabla w - 2\Omega v = -\partial_z p + \lambda\nabla^2 w, \tag{47.4c}$$

where

$$\mathbf{U}=\mathbf{e}_x U+\mathbf{e}_y V \quad \text{and} \quad R=\lambda^{-1}=\hat{U}(b-a)/\nu\,,$$

and

$$\mathbf{u}\in\mathbf{H}\equiv\{\mathbf{u}:\operatorname{div}\mathbf{u}=0,\ \mathbf{u}|_{z=0,1}=0,\ \mathbf{u}\in AP(x,y)\}\,. \tag{47.4d}$$

§ 48. The Form of the Disturbance Whose Energy Increases at the Smallest R

The energy problem for rotating plane Couette flow can be formed directly from (47.4) or as the $\eta\to 1$ limit of (44.2). In either case

$$\frac{1}{R_{\mathscr{E}}}=\max_{\mathbf{H}}\frac{\sin\chi\langle wv\rangle+\cos\chi\langle wu\rangle}{\langle|\nabla\mathbf{u}|^2\rangle}=\max_{\mathbf{H}}\Lambda[\mathbf{u};\chi]\,. \tag{48.1}$$

Here there is stability when

$$R=\frac{(b-a)\sqrt{U_c^2+a^2(\Omega_1-\Omega_2)^2}}{\nu}=\frac{(b-a)^2\sqrt{\Omega_V^2+\Omega_U^2}}{\nu}\leqslant R_{\mathscr{E}}\,.$$

When $\eta=1$, we find, using (43.9) and (47.3), that

$$\tan\beta_{\mathscr{E}}=\frac{rD(V/r)}{DU}\to\tan\chi\,,\qquad \rho_\beta=-1\,. \tag{48.2}$$

We next change coordinates so that x' lies in the direction of the stream at an angle χ from x

$$u'=u\cos\chi+v\sin\chi\,,$$

$$v'=-u\sin\chi+v\cos\chi\,,$$

and $w'=w$ to find that

$$\frac{1}{R_{\mathscr{E}}}=\max_{\mathbf{H}}\frac{\langle u'w'\rangle}{\langle|\nabla\mathbf{u}'|^2\rangle}\,. \tag{48.3}$$

The hypothesis of § 44 is that the maximizing function for (48.3) does not vary along the direction x'. Orr (1907) had a different hypothesis: the maximizing function for (48.3) is among fields $\mathbf{u}'$ such that $v'=0$, $\partial\mathbf{u}'/\partial y'=0$. He found that there is monotonic stability to disturbances of this type when

$$R<177\cdot 2\,.$$

Orr's assumption about the form of the disturbance which increases at the smallest R is not correct since we shall see that the energy of an x'-independent disturbance can increase when

$$R > 2\sqrt{1708}\,.$$

The energy problem (48.3) is independent of rigid rotation (Ω_2). Exactly the same problem is generated by plane Couette flow with $\Omega_2 = \Omega_1 = 0$. In this flow the motion is driven by the forward velocity U_c of the lower plate. With this understanding we can use the equations of § 44 in the present case.

It will be convenient to imagine now that the coordinates were initially aligned in the prime system and then drop the primes. We shall also change the origin of the coordinate z so that $-\frac{1}{2} < z < \frac{1}{2}$.

The Euler equations for (48.3) are

$$2\Lambda\nabla^2\mathbf{u} + \mathbf{e}_x w + \mathbf{e}_z u = \nabla p\,.$$

Elimination of v and p using the continuity equation, followed by a Fourier reduction, leads to

$$2\Lambda L^2 w - a^2 u - i\alpha D w = 0\,, \tag{48.4a}$$

$$2\Lambda\{a^2 L u - i\alpha D L w\} + \beta^2 w = 0\,, \tag{48.4b}$$

where $L = D^2 - a^2$, $D = d/dz$, $a^2 = \alpha^2 + \beta^2$ and

$$u = w = Dw = 0 \quad \text{at} \quad z = \pm\tfrac{1}{2}\,.$$

The sixth order problem

$$4\Lambda^2 L^3 w - 4\Lambda a i\tau L D w + a^2(1-\tau^2) w = 0\,, \tag{48.5}$$

where

$$\tau = \alpha/a \quad \text{and}\,, \quad \text{at } z = \pm\tfrac{1}{2}\,,$$

$$w = Dw = L^2 w = 0\,, \tag{48.6}$$

is obtained by elimination from (48.4) and is equivalent to (48.4).

In the general case, (48.5, 6) is a formidable problem. But when $\tau = 0$, it reduces to the Bénard problem and $\Lambda^{-1} = R = R_{\mathscr{E}} = 2\sqrt{1708}$, and $a = 3\cdot 117$. It is clear that Orr's limit $R = 177\cdot 2$ is more than twice as large as the value $R = 82.6$, which is the limit for energy stability when $\tau = \alpha = 0$. For this lower limit, the disturbance is independent of x'.

The result of numerical integration of (48.5, 6) is given in Table 48.1. One first seeks the field of minimum eigenvalues $\Lambda^{-1}(a, \tau)$ and then the smallest of these for the fixed τ; that is,

$$\tilde{R}(\tau) = \Lambda^{-1}(a(\tau), \tau) = \min_{a>0} \Lambda^{-1}(a, \tau)\,. \tag{48.7}$$

It will be recalled that $\tau=\alpha/a$ is a normalized wave number in the stream (x') direction. When $\tau=1$, $\alpha=a$ and $\beta=0$; that is, the disturbance varies along the stream and not transverse to it. This is the disturbance which Orr thought would solve the problem

$$R_{\mathscr{E}}=\min_{0\leqslant\tau\leqslant 1}\tilde{R}(\tau).$$

But it is clear from the results just given that the value $\tau=0$, corresponding to a disturbance which is independent of x', solves (48.7) and gives the form of the disturbance which increases initially at the smallest R.

Table 48.1: The principal eigenvalues $\tilde{R}(\tau)$ and the minimizing wave number $a(\tau)$ for (48.5, 6, 7)

τ	$a(\tau)$	$(\tilde{R}/2)^2$	$\tilde{R}$
0.0	3.1155	1707.76	82.65
0.05	3.1161	1711.22	82.73
0.10	3.1170	1721.69	82.99
0.15	3.1187	1739.42	83.41
0.20	3.1206	1764.86	84.02
0.25	3.1229	1798.68	84.82
0.30	3.1265	1841.80	85.82
0.35	3.1308	1895.48	87.07
0.40	3.1373	1961.40	88.57
0.45	3.1440	2041.81	90.37
0.50	3.1533	2139.77	92.51
0.55	3.1654	2259.44	95.07
0.60	3.1773	2406.60	98.11
0.65	3.1915	2589.54	101.77
0.70	3.2182	2820.41	106.21
0.75	3.2436	3117.65	111.67
0.80	3.2842	3510.71	118.50
0.85	3.3305	4048.71	127.26
0.90	3.4100	4818.49	138.83
0.95	3.4294	5982.70	144.70
1.00	3.7800	7851.42	177.22

Busse (1972 B) has shown that the numerical computations needed to find the disturbance whose energy increases at the smallest R may be reduced to the calculation of only three extreme values. The proof of his result is of independent interest[4].

Problem: Find

$$\frac{1}{R_{\mathscr{E}}}=\max_{\mathbf{H}}\frac{\langle u'w'\,\hbar(z)\rangle}{\langle|\nabla\mathbf{u}'|^2\rangle} \tag{48.8}$$

[4] The proof given here differs in details from the original one given by Busse.

where $\mathscr{h}(z)=\mathscr{h}(-z)$ (for Couette flow $\mathscr{h}=1$). Here the domain of integration is the fluid layer bounded at $z=\pm\frac{1}{2}$. A Fourier series reduction like that given in § 20 leads us to the result

$$\frac{1}{R_{\mathscr{E}}}=\max_{\alpha,\beta}\max_{\mathbf{H}_B}\frac{\langle \mathrm{re}(w'u')\mathscr{h}\rangle}{\langle |Du'|^2+|Dv'|^2+|Dw'|^2+a^2(u'^2+v'^2+w'^2)\rangle} \tag{48.9}$$

where

$$\mathbf{H}_B\equiv\{\mathbf{u}=\mathbf{u}(z):\mathbf{u}(\pm\tfrac{1}{2})=0,\ i(\beta v'+\alpha u')+Dw=0\}$$

and

$$a^2=\alpha^2+\beta^2\,.$$

The solution of the problem (48.9) necessarily involves the determination of the two wave numbers α and β and these determine a unique direction $\mathbf{e}_y=\mathbf{e}_{y'}\cos\psi-\mathbf{e}_{x'}\sin\psi$ along which derivatives $\partial_y=i(\alpha\,\partial_y x'-\beta\,\partial_y y')=i(\alpha\sin\psi+\beta\cos\psi)=0$.

Parameters in the primed and unprimed coordinates are related by:

$$\begin{aligned}
w'&=w\,,\\
u'&=u\cos\psi+v\sin\psi\,,\\
v'&=-u\sin\psi+v\cos\psi\,,\\
\beta&=-a\sin\psi\,,\qquad \alpha=a\cos\psi\,.
\end{aligned} \tag{48.10}$$

We are going to show that $\psi=\pi/2$ and $\mathbf{e}_y=-\mathbf{e}_{x'}$.

Rephrasing (48.9) in the new variables (48.10) leads to

$$\frac{1}{R_{\mathscr{E}}}=\max_{\substack{a>0\\0\leq\psi\leq\pi}}\frac{1}{R(a,\psi)} \tag{48.11}$$

where

$$\frac{1}{R(a,\psi)}=\max_{\mathbf{H}_B}\frac{\cos\psi\langle\mathscr{h}\,\mathrm{re}(wu)\rangle+\sin\psi\langle\mathscr{h}\,\mathrm{re}(wv)\rangle}{\langle|Dw|^2+|Dv|^2+|Du|^2\rangle+a^2\langle|u|^2+|v|^2+|w|^2\rangle} \tag{48.12}$$

and $\mathbf{H}_B$ is as before except that the continuity equation now becomes

$$Dw+iau=0\,. \tag{48.13}$$

Suppose now that (u, v, w) are maximizing functions for (48.12). We may decompose these functions into even and odd parts; e.g., $w=w_e+w_0$ where

$2w_e = w(z) + w(-z)$, etc. Elimination of u in (48.12) using (48.13) followed by this decomposition leads one to the following:

$$\frac{1}{R(a,\psi)}$$

$$=\frac{-\cos\psi(\mathscr{I}_1[w_e, Dw_0] + \mathscr{I}_1[w_0, Dw_e]) + \sin\psi(\mathscr{I}_2[w_e, v_e] + \mathscr{I}_2[w_0, v_0])}{\cos^2\psi\mathscr{D}_2[w_e] + \cos^2\mu\mathscr{D}_2[w_0] + \sin^2\psi\mathscr{D}_2[w_e] + \mathscr{D}_1[v_e] + \sin^2\mu\mathscr{D}_2[w_0] + \mathscr{D}_1[v_0]},$$

where $0 \leqslant \mu \leqslant \pi/2$, (48.14)

$$\mathscr{I}_1[w, Dw] = a\langle \hbar\,\mathrm{re}(iwDw)\rangle,$$

$$\mathscr{I}_2[w, v] = a^2\langle \hbar\,\mathrm{re}(wv)\rangle,$$

$$\mathscr{D}_1[v] = a^2\langle |Dv|^2\rangle + a^4\langle |v|^2\rangle,$$

and

$$\mathscr{D}_2[w] = \langle |D^2 w|^2 + 2a^2|Dw|^2 + a^4|w|^2\rangle.$$

Lemma: *consider the maximum problem* (48.12) *with* $a > 0$, $0 \leqslant \psi \leqslant \pi/2$. *Let* $\hbar(z)$ *be a symmetric function. Then*

$$\frac{1}{R(a,\psi)} \leqslant \max\left[\frac{1}{\cos\mu\, R(a,0)}, \frac{1}{R(a,\pi/2)}, \frac{\sin\psi}{\sin\mu}\frac{1}{R_0(a,\pi/2)}\right] \tag{48.15}$$

where

$$\frac{1}{R_0(a,\pi/2)} = \max_{w_0, v_0}\left\{\frac{\mathscr{I}_2[w,v]}{\mathscr{D}_2[w] + \mathscr{D}_1[v]}\right\},$$

and w_0 *and* v_0 *are odd functions satisfying the boundary conditions.*

Theorem: *Suppose that for some* μ, $0 < \mu < \pi/2$

$$\frac{1}{R(a,\pi/2)}$$

is the largest of the three values in (48.15). *Then*

$$\frac{1}{R_{\mathscr{E}}} = \frac{1}{R(\tilde{a},\tilde{\psi})} \equiv \max_{\substack{a>0\\ 0\leqslant\psi\leqslant\pi/2}} \frac{1}{R(a,\psi)} = \frac{1}{R(\tilde{a},\pi/2)}. \tag{48.16}$$

Under the hypothesis of the theorem $\tilde{\psi} = \pi/2$ and the disturbance whose energy increases initially at the smallest R is independent of x'. For plane Couette flow, $\hbar = 1$ and the hypothesis of the theorem holds:

$$\max_{a\geqslant 0}\frac{1}{R(a,\pi/2)} = \frac{1}{2\sqrt{1708}} \quad \text{(Joseph (1966))},$$

$$\max_{a\geqslant 0}\frac{1}{R(a,0)} = \frac{1}{177{\cdot}2} \quad \text{(Orr (1907))},$$

and

$$\max_{a\geq 0}\frac{1}{R_0(a,\pi/2)}=\frac{1}{2\sqrt{17610.4}}$$ (see Chandrasekhar (1961) p. 38, Table 1).

Proof: Suppose the lemma holds and $\psi=\tilde{\psi}$ where

$$\frac{1}{R(a,\tilde{\psi})}=\max_{0\leq\psi\leq\pi/2}\frac{1}{R(a,\psi)}. \tag{48.17}$$

Then by (48.17) and (48.15), respectively,

$$\frac{1}{R(a,\pi/2)}\leq\frac{1}{R(a,\tilde{\psi})}\quad\text{and}\quad\frac{1}{R(a,\tilde{\psi})}\leq\frac{1}{R(a,\pi/2)}.$$

These two inequalities hold simultaneously only with equality. This proves the theorem. To prove the lemma we note that application of the inequality

$$\frac{\sum a_n}{\sum b_n}\leq\max_n\frac{a_n}{b_n},\qquad a_n\geq 0,\qquad b_n\geq 0$$

to (48.14) implies that

$$\frac{1}{R(a,\psi)}\leq\max\left[\frac{A}{\cos\mu},\ B,\ \frac{\sin\psi}{\sin\mu}C\right],$$

where

$$A=\frac{\mathscr{I}_1[w_e\cos\psi,Dw_0\cos\mu]+\mathscr{I}_1[w_0\cos\mu,Dw_e\cos\psi]}{\mathscr{D}_2[w_e\cos\psi]+\mathscr{D}_2[w_0\cos\mu]}$$

$$\leq\max\frac{\mathscr{I}_1[w,Dw]+\mathscr{I}_1[w,Dw]}{\mathscr{D}_2[w]}, \tag{48.18a}$$

$$B=\frac{\mathscr{I}_2[w_e\sin\psi,v_e]}{\mathscr{D}_2[w_e\sin\psi]+\mathscr{D}_1[v_e]}\leq\max\frac{\mathscr{I}_2[w,v]}{\mathscr{D}_2[w]+\mathscr{D}_1[v]} \tag{48.18b}$$

and

$$C=\frac{\mathscr{I}_2[w_0\sin\mu,v_0]}{\mathscr{D}_2[w_0\sin\mu]+\mathscr{D}_1[v_0]}\leq\max\frac{\mathscr{I}_2[w_0,v_0]}{\mathscr{D}_2[w_0]+\mathscr{D}_1[v_0]}. \tag{48.18c}$$

It is important to note that the maximizing functions for (48.13) are necessarily such that the numerators in (48.18a, b, c) are each non-negative. To show this, suppose the opposite is true; say (48.18a) is negative and (48.18b, c) are positive. Now change the sign of w_0 and v_0. This leaves the absolute value of the three ratios unaltered but makes all three non-negative. Our supposition (one negative, two positive) could, therefore, not give (48.13) its maximum value. In the same way, by changing the sign of two functions, we may verify that only non-negative numerators maximize (48.13). Since these numerators are not negative, they are admissible as non-negative numbers in (48.16). This proves Busse's theorem.

§ 49. Necessary and Sufficient Conditions for the Global Stability of Rotating Plane Couette Flow

The equations which follow upon linearizing (47.4) may be combined into the following pair:

$$\left(\frac{d}{dt}-\lambda\nabla^2\right)\nabla^2 w-2\Omega\phi=0, \tag{49.1a}$$

$$\left(\frac{d}{dt}-\lambda\nabla^2\right)\phi+\cos\chi\,\partial^2_{yx}w+(2\Omega-\sin\chi)\,\partial^2_{xx}w=0, \tag{49.1b}$$

where

$$\phi=(\partial^2_{xx}+\partial^2_{yy})v+\partial^2_{zy}w,$$

and

$$\frac{d}{dt}=\frac{\partial}{\partial t}+(1-z)\sin\chi\,\partial_y+(1-z)\cos\chi\,\partial_x.$$

At the boundary

$$\phi=w=\partial_z w=0. \tag{49.1c}$$

Assuming disturbances in the form of Fourier series with terms proportional to $\exp(\sigma t+i\alpha x+i\beta y)$, we find the following equations for the Fourier coefficients $\hat{w}(z)$ and $\hat{\phi}(z)$:

$$(\mathrm{re}(\sigma)+i\hat{\mathscr{S}}-\lambda L)L\hat{w}-2\Omega\hat{\phi}=0, \tag{49.2a}$$

$$(\mathrm{re}(\sigma)+i\hat{\mathscr{S}}-\lambda L)\hat{\phi}-\alpha(\beta\cos\chi+\alpha(2\Omega-\sin\chi))\hat{w}=0, \tag{49.2b}$$

$$\hat{\phi}=\hat{w}=D\hat{w}=0 \quad \text{at} \quad z=0,1, \tag{49.2c}$$

where

$$L=D^2-a^2,$$

$$\hat{\mathscr{S}}=c-az\sin(\chi-\psi),$$

$$c=\mathrm{im}(\sigma)+a\sin(\chi-\psi),$$

and

$$\sin(\chi-\psi)=\frac{\beta}{a}\sin\chi+\frac{\alpha}{a}\cos\chi\,, \tag{49.3}$$

where

$$\alpha=-a\sin\psi\,,\qquad \beta=a\cos\psi\,. \tag{49.4}$$

Here, ψ is the angle between x and x' where x is parallel to $\boldsymbol{\Omega}=-\Omega_2\mathbf{e}_x$ and x' is the direction along which disturbances are constant (Fig. 49.1).

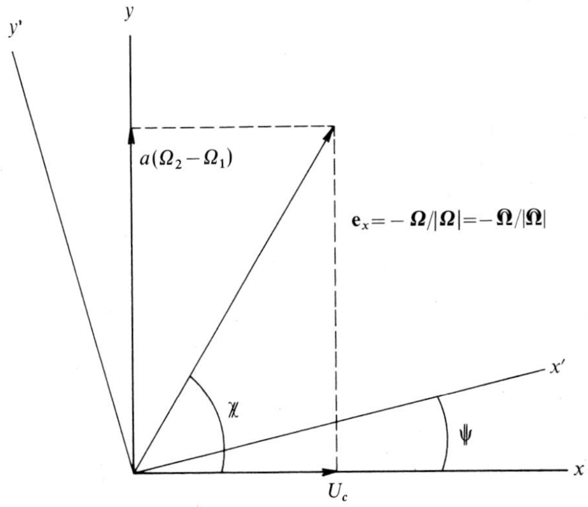

Fig. 49.1: Spiral flow angles

Eqs. (49.2a, b) combine into

$$(\mathrm{re}(\sigma)+i\hat{\mathscr{S}}-\lambda L)^2 L\hat{w}+a^2\mathbb{F}\hat{w}=0\,, \tag{49.5}$$

which is to be solved along with

$$\hat{w}=D\hat{w}=(\mathrm{re}(\sigma)+i\hat{\mathscr{S}}-\lambda L)L\hat{w}=0 \quad \text{at} \quad z=0,1\,. \tag{49.6}$$

Here,

$$\mathbb{F}(\psi,\chi,\Omega)=2\Omega\sin\psi(\cos(\chi-\psi)-2\Omega\sin\psi)\,, \tag{49.7}$$

and

$$-\infty\leqslant\mathbb{F}\leqslant\tfrac{1}{4}\cos^2(\chi-\psi)\,.$$

The linear stability limit R_L is the smallest critical value of R (for these, $\mathrm{re}(\sigma)=0$) minimized with respect to the wave number radius a and spiral angle ψ.

A special solution of (49.5) can be found when $\hat{\mathscr{S}}=0$. It will not ordinarily be possible to put $\hat{\mathscr{S}}(z)=0$, but this is possible when $\chi=\psi$. Then the spiral angle χ which is always equal to the energy spiral angle $\beta_{\mathscr{E}}$ (see Eq. 48.2) is also the spiral angle $\psi_{\mathscr{E}}$ along which disturbances do not vary. In this case the problem $L^3\hat{w}+\lambda^{-2}a^2\mathbb{F}\hat{w}=0$ and (49.6) define the Bénard problem

$$\lambda^{-2}\mathbb{F}(\chi,\chi,\Omega)=g(a^2)\,,\qquad \min_{a^2} g(a^2)=g((3.12)^2)=1708\,. \tag{49.8}$$

Eqs. (49.8) were first given by Kiessling (1963). They imply that

$$\lambda^{-2}=1708/\mathbb{F}(\chi,\chi,\Omega)\,.$$

Of course, the special solution (49.8) could hold only so long as $\Omega(-\infty<\Omega<\infty)$ is assigned so that

$$\mathbb{F}(\chi,\chi,\Omega)=2\Omega\sin\chi(1-2\Omega\sin\chi)>0\,.$$

The values of χ and Ω which make $\mathbb{F}=\frac{1}{4}$ have a special importance: *Suppose that* $\Omega=1/(4\sin\chi)$. *Then* $\mathbb{F}=\frac{1}{4}$, *and the criterion*

$$R=\frac{(b-a)^2}{\nu}\sqrt{\Omega_U^2+\Omega_V^2}<2\sqrt{1708} \tag{49.9}$$

is both necessary and sufficient for all periodic disturbances to decay monotonically.

The proof of this theorem follows by comparing the linear solution with the unmodified energy problem (Busse, 1970 B). It follows that the criterion $R<2\sqrt{1708}$ is both necessary and sufficient for stability and that the most persistent infinitesimal disturbance is just the one whose energy increases initially at the smallest R.

Exercise 49.1 (Joseph and Munson, 1970): Show that the criterion (49.9) follows directly from (45.1). Show that (45.1) can be written in the dimensionless variables of this section as

$$\operatorname{curl}(\partial_{x''}\mathbf{u})-\{2\Omega_{x''}\partial_{x''}+(2\Omega_{y''}+\tfrac{1}{2})\partial_{y''}\}\mathbf{u}=0\,,\qquad \mathbf{u}=\hat{\mathbf{u}}(z)\exp i\{\alpha''x''+\beta''y''\}\,, \tag{49.10}$$

where x'' is the direction of the motion of the bottom plate relative to the top. The energy eigenfunction $\mathbf{u}$ is independent of x'' (see § 48) and

$$\Omega_{y''}=-\tfrac{1}{4}\,.$$

Show that (49.10) holds in the "fast rotation" limit when $\Omega_{x''}\to\infty$ and

$$\alpha''\Omega_{x''}=\text{const}\,.$$

independent of the value of $\Omega_{y''}$. What is the value of β'' and of the constant in this limit?

§ 50. Rayleigh's Criterion for the Instability of Rotating Plane Couette Flow, Wave Speeds

In the general case, the energy spiral angle χ and the disturbance spiral angle ψ do not coincide. Then $\hat{\mathscr{S}}$ cannot vanish everywhere.

For the case $\hat{\mathscr{S}} \equiv 0$, McIntyre and Pedley (Pedley, 1969) have shown that a necessary and sufficient condition for the existence of an inviscid $(R \to \infty)$ solution of (49.2a, b) is that $\mathbb{F} > 0$. Furthermore, they show that $-\mathbb{F}$ is the product of the overall angular velocity and the total vorticity; that is, $\mathbb{F}$ is Rayleigh's discriminant.

In the general case $\chi \neq \psi$, $\mathbb{F}$ is still Rayleigh's discriminant; that is,

$$\mathbb{F}(\psi, \chi, \Omega) = -2\Omega_{y'}\zeta_{y'}, \tag{50.1}$$

where

$$\Omega_{y'} = \mathbf{e}_{y'} \cdot \boldsymbol{\Omega}$$

and

$$\zeta_{y'} = \mathbf{e}_{y'} \cdot (2\boldsymbol{\Omega} + \operatorname{curl} \mathbf{U})$$

are components of the overall angular velocity ($\boldsymbol{\Omega}$) and of the total vorticity ($2\boldsymbol{\Omega} + \operatorname{curl} \mathbf{U}$) in the direction y' normal to the direction x' in which disturbances do not vary. To verify (50.1) note that

$$\boldsymbol{\Omega} = -\Omega \mathbf{e}_x, \qquad \mathbf{U} = (\mathbf{e}_x \cos\chi + \mathbf{e}_y \sin\chi)(1 - z)$$

and use the geometry of Fig. 49.1.

Numerical analysis of (49.5, 6) gives solutions whenever $\mathbb{F} > 0$ and not otherwise (see Table 51.1). Problem (49.5, 6) contains plane Couette flow ($\mathbb{F} = 0$) as a special case.

The determination of the wave speed $\operatorname{im}(\sigma)$ of the most persistent small disturbance has a particularly simple solution. The answer is that

$$\operatorname{im}(\sigma) = -\frac{a}{2}\sin(\chi - \psi). \tag{50.2}$$

The argument leading to (50.2) starts with the observation that every solution of (49.5, 6) or the equivalent problem (49.2a, b, c) has

$$\langle \hat{\mathscr{S}}[|2\Omega\hat{\phi}|^2 + a^2\mathbb{F}(|D\hat{w}|^2 + a^2|\hat{w}|^2)] \rangle = 0. \tag{50.3}$$

To prove (50.3) set $f = -2\Omega\hat{\phi}/a\mathbb{F}^{1/2}$ and introduce $\mathbb{F}$ of (49.7) into (49.2a, b). This leads us to

$$[\operatorname{re}(\sigma) + i\hat{\mathscr{S}}]L\hat{w} + a\mathbb{F}^{1/2}f = \lambda L^2\hat{w} \tag{50.4a}$$

and

$$[\mathrm{re}(\sigma)+i\hat{\mathscr{S}}]f - a\mathbb{F}^{1/2}\hat{w} = \lambda Lf. \tag{50.4b}$$

Form $\langle \bar{\hat{w}}(50.4\,\mathrm{a})\rangle - \langle \bar{f}(50.4\,\mathrm{b})\rangle$ to produce

$$\langle(\mathrm{re}(\sigma)+i\hat{\mathscr{S}})(\bar{\hat{w}}L\hat{w}-|f|^2)\rangle = \lambda\langle |L\hat{w}|^2+|Df|^2+a^2|f|^2\rangle - a\mathbb{F}^{1/2}\langle f\bar{\hat{w}}+\bar{f}\hat{w}\rangle \tag{50.5}$$

where the overbar designates the complex conjugate. Subtraction of the complex conjugate of (50.5) from (50.5) gives

$$0=\langle\hat{\mathscr{S}}(\bar{\hat{w}}L\hat{w}+\hat{w}L\bar{\hat{w}}-2|f|^2)\rangle = -2\langle\hat{\mathscr{S}}(|D\hat{w}|^2+a^2|\hat{w}|^2+|f|^2)\rangle,$$

where the term

$$\langle D\hat{\mathscr{S}}(\bar{\hat{w}}D\hat{w}+\hat{w}D\bar{\hat{w}})\rangle = 0$$

which arises from integration by parts vanishes because $D\hat{\mathscr{S}}$ is a constant.

Eq. (50.3) shows that $\hat{\mathscr{S}}(z)$ must change sign on (0,1). In fact, $\hat{\mathscr{S}}(\frac{1}{2})=0$. To see this, choose $c=\mathrm{im}(\sigma)+a\sin(\chi-\psi)=\dfrac{a}{2}\sin(\chi-\psi)$ so that $\hat{\mathscr{S}}(\frac{1}{2})=0$. Then write (50.3) and (50.4) in the variable $z'=z-\frac{1}{2}$, $-\frac{1}{2}\leqslant z'\leqslant\frac{1}{2}$. The form of (50.3,4) is unchanged by the variable change but

$$\hat{\mathscr{S}}(z') = -z'a\sin(\chi-\psi). \tag{50.6}$$

Next, decompose f and $\hat{w}$ into even and odd parts; insert these representations into (50.4) and identify the even and odd parts of the resulting equations; for example, from (50.4b) and (50.6) we find

$$\mathrm{re}(\sigma)\hat{f}_0 - iz'\sin(\chi-\psi)a\hat{f}_e = \lambda L\hat{f}_0 + a\mathbb{F}^{1/2}\hat{w}_0.$$

This and the other three equations show that we may take

$$\hat{w}=w_e(z')+iw_0(z'),\qquad \hat{f}=f_e(z')+if_0(z'), \tag{50.7}$$

where w_e, f_e, w_0 and f_0 are real functions. Now using (50.5) and (50.7), the condition (50.3) is satisfied identically.

Exercise 50.1: Show that

$$2\lambda\langle|L\hat{w}|^2+|Df|^2+a^2|f|^2\rangle = 2a\sqrt{\mathbb{F}}\langle\bar{f}w+f\bar{w}\rangle - a\sin(\chi-\psi)\langle\tau\rangle \tag{50.8}$$

where $\tau=-i(\bar{w}Dw-wD\bar{w})$. This equation gives $\lambda(a,\psi)$ for eigensolutions of (50.4) and the boundary conditions. Find the eigenvalue problem which is adjoint to (50.4) and show that the same eigenvalue (a,ψ) is given by

$$\lambda\langle Lw^*L\bar{w}+Df^*D\bar{f}+a^2f^*\bar{f}\rangle = a\sqrt{\mathbb{F}}\langle\bar{w}^*f+\bar{f}^*w\rangle + a\sin(\chi-\psi)\langle i\bar{w}^*Dw\rangle \tag{50.9}$$

where w^* and f^* are the adjoint eigenfunctions. Using the result of Exercise (B4.6) or otherwise show that the disturbance angle which leads to the lowest critical value of R is a root of the equation

$$a\langle \overline{w^* f + \bar{f}^* w}\rangle \frac{\partial\sqrt{\mathbb{F}}}{\partial\psi} = a\cos(\chi-\psi)\langle i\overline{w^* Dw}\rangle . \tag{50.10}$$

Show that when $\chi-\psi=\pi/2$ the disturbance angle which minimizes R also maximizes Rayleigh's discriminant. Does a formula like (50.10) follow from (50.8)?

§ 51. The Energy Problem for Rotating Plane Couette Flow when Spiral Disturbances are Assumed from the Start

A distinguished spiral direction ($\mathbf{e}_{x'}$) is found as a part of the solution of the linear stability problem. This direction is determined by the wave number which gives the largest critical value λ for a neutral solution ($\text{re}(\sigma)=0$) of the spectral problem. The spiral disturbance is also observed in experiments. It is, therefore, reasonable to examine the consequence of assuming the preferred direction from the start. The aim here is a nonlinear analysis which takes advantage of the presumed spiral form for the disturbance.

It will be convenient to decompose the motion along the normal to the direction x' in which u, v, w and p do not vary; for example, $u=u(y', z, t)$

$$\begin{aligned}\mathbf{U}(z) &= (1-z)[(\cos\chi\cos\psi+\sin\chi\sin\psi)\mathbf{e}_{x'}+(\sin\chi\cos\psi-\sin\psi\cos\chi)\mathbf{e}_{y'}]\\ &= (1-z)[\cos(\chi-\psi)\mathbf{e}_{x'}+\sin(\chi-\psi)\mathbf{e}_{y'}],\end{aligned} \tag{51.1 a}$$

and

$$-\boldsymbol{\Omega} = \Omega\cos\psi\,\mathbf{e}_{x'} - \Omega\,\mathbf{e}_{y'}\sin\psi = \Omega\,\mathbf{e}_x . \tag{51.1 b}$$

The governing equations for the x'-independent disturbances are

$$\frac{\partial u}{\partial t} + (\mathbf{u}+\mathbf{U})\cdot\nabla u + (2\Omega\sin\psi - \cos(\chi-\psi))w = \lambda\nabla_2^2 u , \tag{51.2a}$$

$$\frac{\partial v}{\partial t} + (\mathbf{u}+\mathbf{U})\cdot\nabla v + (2\Omega\cos\psi - \sin(\chi-\psi))w = -\frac{\partial p}{\partial y'} + \lambda\nabla_2^2 v \tag{51.2b}$$

$$\frac{\partial w}{\partial t} + (\mathbf{u}+\mathbf{U})\cdot\nabla w - 2\Omega(v\cos\psi + u\sin\psi) = -\frac{\partial p}{\partial z} + \lambda\nabla_2^2 w , \tag{51.2c}$$

where

$$\nabla_2^2 = \frac{\partial^2}{\partial y'^2} + \frac{\partial^2}{\partial z^2}$$

and

$$\frac{\partial v}{\partial y'}+\frac{\partial w}{\partial z}=0. \tag{51.2d}$$

The boundary conditions are

$$\mathbf{u}=0 \quad \text{at} \quad z=0,1 \tag{51.2e}$$

and $\mathbf{u}$ is almost periodic in y'.

The disturbance velocity component u cannot be driven by a disturbance pressure gradient because the assumption that $\partial\mathbf{u}/\partial x'=0$ implies that $\partial^2 p/\partial x'^2=0$. Then $\partial p/\partial x'=K(y',z)$ and since p is almost periodic in x', it is bounded as $x'^2\to\infty$ and it follows that $K\equiv 0$.

There are several consequences of the independence of p upon x'. One consequence is the existence of *two* energy identies: one governing the energy of the longitudinal component of velocity

$$\tfrac{1}{2}\frac{d}{dt}\langle u^2\rangle+(2\Omega\sin\psi-\cos(\chi-\psi))\langle wu\rangle=-\lambda\langle|\nabla_2 u|^2\rangle, \tag{51.3a}$$

and one governing the evolution of the energy of the transverse components,

$$\tfrac{1}{2}\frac{d}{dt}\langle w^2+v^2\rangle-\sin(\chi-\psi)\langle wv\rangle-2\Omega\sin\psi\langle wu\rangle=-\lambda\langle|\nabla_2 w|^2+|\nabla_2 v|^2\rangle. \tag{51.3b}$$

Eqs. (51.3) are the subject of analysis of this section. We form the sum

(51.3 b) + λ(51.3 a)

with $\lambda>0$ and let $\phi=\sqrt{\lambda}u$ to obtain

$$\begin{aligned}\tfrac{1}{2}\frac{d}{dt}&\langle w^2+v^2+\phi^2\rangle-\sin(\chi-\psi)\langle wv\rangle-\sqrt{\lambda}\cos(\chi-\psi)\langle w\phi\rangle\\ &-2\Omega\sin\psi\left(\frac{1}{\sqrt{\lambda}}-\sqrt{\lambda}\right)\langle w\phi\rangle=-\lambda\langle|\nabla_2 w|^2+|\nabla_2 v|^2+|\nabla_2\phi|^2\rangle.\end{aligned} \tag{51.4}$$

This is the equation governing the evolution of the disturbance energy when spiral disturbances are assumed from the start; it can be written as

$$\frac{d\mathscr{E}}{dt}=-\mathscr{H}_\lambda-\lambda\mathscr{D}=-\mathscr{D}\left(\lambda-\frac{(-\mathscr{H}_\lambda)}{\mathscr{D}}\right)\leq-\mathscr{D}(\lambda-\Lambda_\lambda), \tag{51.5}$$

where

$$\mathscr{E}=\langle w^2+v^2+\phi^2\rangle/2\,,\qquad \mathscr{D}=\langle|\nabla_2 w|^2+|\nabla_2 v|^2+|\nabla_2\phi|^2\rangle\,,$$

$$-\mathscr{H}_\lambda=-2\Omega\sin\psi\left(\sqrt{\lambda}-\frac{1}{\sqrt{\lambda}}\right)\langle w\phi\rangle+\sqrt{\lambda}\cos(\chi-\psi)\langle w\phi\rangle+\sin(\chi-\psi)\langle wv\rangle\,,$$

and

$$\Lambda_\lambda=\max_{\mathbf{H}_2}\frac{-\mathscr{H}_\lambda}{\mathscr{D}}\,, \tag{51.6}$$

where $\mathbf{H}_2$ is the set of x'-independent kinematically admissible vectors; that is, vectors $\mathbf{u}$ satisfying $\operatorname{div}\mathbf{u}=0$ and boundary and periodicity conditions.

The energy inequality (51.5) integrates to

$$\mathscr{E}(t)\leqslant\mathscr{E}(0)\exp\{-2\hat{\Lambda}(\lambda-\Lambda_\lambda)t\}\,, \tag{51.7}$$

where $\mathscr{D}>2\hat{\Lambda}\mathscr{E}$ for all $\mathbf{u}\in\mathbf{H}_2$ and provided that $R<R_\lambda\equiv\Lambda_\lambda^{-1}$. The criterion here is independent of the size of the initial disturbances and applies globally. It is clear that the vector $\mathbf{u}\in\mathbf{H}_2$ which solves (51.6) is also the form of the x'-independent disturbance whose energy increases initially at the smallest R.

Again $\lambda>0$ is a free parameter which is selected to maximize the interval $0<R<R_\lambda$ on which global monotonic stability can be assured. Thus,

$$\tilde{\tilde{R}}(\chi,\psi,\Omega)=\max_{\lambda>0}R_\lambda\,. \tag{51.8}$$

If $R<\tilde{\tilde{R}}$, rotating Couette flow is monotonically and globally stable to x'-independent disturbances making an angle ψ with x ($\mathbf{e}_x=-\mathbf{\Omega}/|\mathbf{\Omega}|$).

Of course, one cannot know at the start whether nature will select a single direction x' along which disturbances are constant. Moreover, even if such a direction is selected, it will not be possible to specify its angle ψ with x at the start. However, one can seek the angle $\psi=\psi_{\mathscr{E}}$ for which

$$R_{\mathscr{E}}(\chi,\Omega)=\tilde{\tilde{R}}(\chi,\psi_{\mathscr{E}}(\chi,\Omega),\Omega)=\min_{0\leqslant\psi\leqslant 2\pi}\tilde{\tilde{R}}\,. \tag{51.9}$$

If $R<R_{\mathscr{E}}$, RPCF *is monotonically and globally stable to all x'-independent disturbances.*

Two types of disturbances are especially important:

(a) Disturbances of Orr's type. These are of the type considered in the Orr-Sommerfeld theory. They are two-dimensional disturbances which do not vary in the direction perpendicular to the plane of the motion,

$$\chi-\psi=\pm\pi/2\,.$$

(b) Disturbances of Taylor's type. These disturbances are constant on straight lines formed in the intersection of the plane of the motion and the planes $z=\text{const.}$

$$\chi-\psi=0 \quad \text{or} \quad \pi .$$

The values of $\tilde{\tilde{R}}$ and R depend strongly on the sign of the Rayleigh discriminant. The first case to be considered is the case $\mathbb{F}\leqslant 0$. This corresponds to the situation in which the angular momentum increases outward in the rotating cylinder problem. This case includes plane Couette flow ($\mathbb{F}=0$). The following theorem holds:

Let $\mathbb{F}\leqslant 0$. *The x'-independent disturbance whose energy increases at the smallest value of $R(>R_{\mathscr{E}})$ is of Orr's type. Moreover,*

$$R_{\mathscr{E}}=177.22$$

is the value calculated by Orr (1907).

Proof: When $\mathbb{F}\leqslant 0$ one can find a value $\lambda=\tilde{\lambda}=-4\Omega^2\sin^2\psi/\mathbb{F}\geqslant 0$ which will make the coefficient of $\langle w\phi\rangle$ in (51.6) vanish. Any other choice of λ leads to a larger value Λ_λ. The maximum of (51.6) is not smaller than the maximum of (51.6) among the smaller class of vectors which have $\phi\equiv 0$. This same maximum (with $\phi\equiv 0$) is attained in the larger class in which ϕ is not zero from the outset and $\lambda=\tilde{\lambda}$. When $\lambda=\tilde{\lambda}$, we have

$$\frac{1}{\tilde{\tilde{R}}}=\max_{\mathbf{H}_2}\frac{\sin(\chi-\psi)\langle wv\rangle}{\langle|\nabla w|^2+|\nabla v|^2+|\nabla\phi|^2\rangle}, \tag{51.10}$$

where $\mathbf{H}_2$ is defined under (51.6). The maximizing vector for (51.10) has $\phi\equiv 0$. The maximum value of $\langle wv\rangle/\langle|\nabla w|^2+|\nabla v|^2\rangle$ in $\mathbf{H}_2$ is 1/177.22 (Orr, 1907). Hence,

$$\tilde{\tilde{R}}=177.22/\sin(\chi-\psi)$$

and

$$R=\min_\psi \tilde{\tilde{R}}=177.22$$

is attained when

$$\chi-\psi=\pi/2 .$$

This disturbance does not vary on lines perpendicular to the plane of the motion.

When $\mathbb{F}>0$ we cannot select a positive value of λ which will make the coefficient of $\langle w\phi\rangle$ in (51.6) vanish. Then the optimizing value for (51.8) $\lambda=\tilde{\lambda}$ is sought as the root of

$$0=\frac{\partial}{\partial\lambda}\Lambda_\lambda=\frac{1}{\mathscr{D}}\frac{\partial}{\partial\lambda}(-\mathscr{H}_\lambda).$$

Since **u** is a maximizing vector,

$$\frac{\partial}{\partial\lambda}(-\mathscr{H}_\lambda)=\frac{1}{2\sqrt{\lambda}}\frac{\partial(-\mathscr{H}_\lambda)}{\partial\sqrt{\lambda}}=\frac{1}{2\sqrt{\lambda}}\left\{-2\Omega\sin\psi\left(1+\frac{1}{\lambda}\right)+\cos(\chi-\psi)\right\}\langle w\phi\rangle .$$

Hence

$$\tilde{\lambda}=\frac{\Omega\sin\psi}{\frac{1}{2}\cos(\chi-\psi)-\Omega\sin\psi}=\frac{4\Omega^2\sin^2\psi}{\mathbb{F}}, \tag{51.11}$$

$$-\mathscr{H}_{\tilde{\lambda}}\equiv-\mathscr{H}=2\sqrt{\mathbb{F}}\,\langle w\phi\rangle+\sin(\chi-\psi)\langle wv\rangle , \tag{51.12}$$

and

$$\frac{1}{\tilde{\tilde{R}}}=\max_{\mathbf{H}_2}-\mathscr{H}/\mathscr{D}\equiv\max_{\mathbf{H}_2}\Lambda[\mathbf{u};\chi,\psi,\Omega] . \tag{51.13}$$

It is convenient to reformulate (51.13) as an eigenvalue problem with eigenvalues $\Lambda(\chi,\psi,\Omega)$:

$$\sqrt{\mathbb{F}}\,\phi+\sin(\chi-\psi)v/2+\Lambda\nabla_2^2 w=\partial_z p ,$$

$$\sin(\chi-\psi)w/2+\Lambda\nabla_2^2 v=\partial_{y'}p$$

and

$$\sqrt{\mathbb{F}}\,w+\Lambda\nabla_2^2\phi=0 .$$

Using the continuity Eq. (51.2d) and after normal-mode reduction to ordinary differential equations, one finds that

$$4\Lambda^2 L\hat{w}-4i\Lambda a\sin(\chi-\psi)L\hat{w}+\mathbb{F}a^2\hat{w}=0 \tag{51.14a}$$

where $L=D^2-a^2$ and

$$\hat{w}=D\hat{w}=L^2\hat{w}=0 \quad \text{at} \quad z=0,1 . \tag{51.14b}$$

The required stability limit is found as an eigenvalue of (51.14)

$$\tilde{\tilde{R}}=\min_a\Lambda^{-1}(\chi,\psi,\Omega,a) . \tag{51.15}$$

The criterion $R<\tilde{\tilde{R}}$ guarantees stability for all disturbances making an angle ψ with x. The criterion

$$R<\min_\psi\tilde{\tilde{R}}\equiv R_{\mathscr{E}} \tag{51.16}$$

suffices for stability to disturbances making any angle with x. The values $R_{\mathscr{E}}$ and the minimizing angles $\psi_{\mathscr{E}}$ are displayed in Table 51.1.

The stability criterion $R<\tilde{\tilde{R}}$ has an interesting consequence when the velocity of sliding is zero ($\chi=90°$) and the disturbances are axisymmetric ($\psi=90°$). In this case, referring first to the linear equations (49.5) and (49.6) we find that

Table 51.1: Values of the critical parameters of linear and energy stability theory for the stability of RPCF (Hung, Joseph and Munson, 1972)

	Ω	$a_{\mathscr{E}}$	$\psi_{\mathscr{E}}$	λ	$R_{\mathscr{E}}$	$\mathbb{F}_{\mathscr{E}}$	a_L	ψ_L	im(σ)	R_L	$\mathbb{F}_L$
	0.30	3.12	34.80°	0.6057	91.95	0.1936	2.925	20.915°	0.27693	111.19	0.1644
	0.50	3.12	25.85°	0.8290	85.68	0.2293	3.009	19.55°	0.24961	92.97	0.2180
	0.70	3.12	19.77°	0.9250	83.74	0.2424	3.061	17.10°	0.18917	86.62	0.2390
$\eta=1.0$	0.90	3.12	15.75°	0.9649	83.01	0.2474	3.10	14.50°	0.12161	84.08	0.2462
$\chi=10°$	1.44	3.12	10.00°	1.000	82.66	0.2500	3.12	10.00°	0	82.66	0.2500
	1.80	3.12	8.25°	1.0696	82.73	0.2495	3.114	8.10°	−0.05162	82.84	0.2497
	2.20	3.12	6.75°	1.0747	82.83	0.2489	3.101	6.63°	−0.091144	83.24	0.2491
	2.40	3.12	6.15°	1.0635	82.86	0.2486	3.097	6.20°	−0.10264	83.44	0.2485
	0.20	3.12	51.01°	0.4993	92.84	0.1936	3.07	35.70°	0.152456	100.11	0.1778
	0.30	3.12	43.02°	0.7246	85.48	0.2325	3.093	34.75°	0.12806	88.36	0.2239
	0.40	3.12	35.90°	0.8925	83.15	0.2466	3.108	32.60°	0.070494	83.79	0.2448
$\eta=1.0$	0.50	3.12	30.00°	1.000	82.66	0.2500	3.12	30.00°	0	82.66	0.2500
$\chi=30°$	0.60	3.12	25.40°	1.0678	82.92	0.2481	3.11	27.15°	−0.077317	83.46	0.2471
	0.70	3.12	21.73°	1.0998	83.43	0.2443	3.086	24.38°	−0.151106	85.43	0.2412
	0.80	3.12	18.90°	1.1193	84.06	0.2400	3.042	21.77°	−0.21773	88.13	0.2352
	0.90	3.12	16.65°	1.1280	84.65	0.2358	3.00	19.48°	−0.27392	91.23	0.2299
	1.00	3.12	14.81°	1.1265	85.20	0.2320	2.935	17.51°	−0.317375	94.55	0.2254
	0.10	3.12	75.87°	0.2526	105.83	0.1489	3.106	62.13°	0.05772	108.76	0.1454
	0.15	3.12	71.61°	0.4097	92.48	0.1978	3.105	62.21°	0.059868	93.91	0.1948
$\eta=1.0$	0.20	3.12	67.89°	0.5978	86.06	0.2297	3.110	61.50°	0.040705	86.67	0.2278
$\chi=60°$	0.28868	3.12	60.00°	1.000	82.66	0.2500	3.12	60.00°	0	82.66	0.2500
	0.40	3.12	47.35°	1.5192	86.14	0.2279	3.10	56.10°	−0.105424	88.49	0.2216
	0.50	3.12	35.54°	1.7669	92.52	0.1912	2.98	50.86°	−0.23668	108.33	0.1642
	0.60	3.12	26.40°	1.7823	98.95	0.1597	2.404	42.82°	−0.35504	176.31	0.1137

$\hat{\mathscr{S}}=\mathrm{im}(\sigma)$ and (49.7) then implies $\mathrm{im}(\sigma)=0$. This reduces (49.5) and (49.6) to the Bénard problem (see § 62) with minimum eigenvalue $R_L^2=1708$, $a=3.12$. On the other hand, with $\chi=\psi$, the energy equations (51.14a, b) imply $\tilde{\tilde{R}}^2\mathbb{F}=1708$, $a=3.12$. *Hence,* $\tilde{\tilde{R}}=R_L$ *and the criterion* $R<\tilde{\tilde{R}}$ *is both necessary and sufficient for global stability of Couette flow between rotating cylinders to axisymmetric disturbances in the limit* (47.1) *and* (47.2)[5].

The previous result applies to the RPCF of spiral Couette flow when the velocity of sliding is zero ($\chi=90°$) and the disturbances are axisymmetric. If axisymmetry is not assumed we must solve (51.9). Then numerical results of Hung, Joseph and Munson (1972) show that *the critical disturbance for* (51.9) *with* $\chi=90°$ *is either of Orr's type* ($\psi_{\mathscr{E}}=\pi/2$) *or of Taylor's type* ($\psi_{\mathscr{E}}=0$). The values of $\psi_{\mathscr{E}}(\Omega)$ are:

(a) $\Omega^*<\Omega<\Omega^{**}$, $\quad \psi_{\mathscr{E}}(\Omega)=90°$, $\quad 82.66\leqslant R_{\mathscr{E}}(\Omega)<177.2$,

(b) $\Omega<\Omega^*$, $\quad \Omega>\Omega^{**}$, $\quad \psi_{\mathscr{E}}(\Omega)=0$, $\quad R_{\mathscr{E}}=177.2$,

where $\Omega^*=0.028$ and $\Omega^{**}=0.472$.

[5] The equality of energy and linear limits here applies to all Ω for which $\mathbb{F}>0$. In contrast, if axial symmetry is not assumed from the start, the coincidence of the energy and linear limits occurs only when $\Omega=\frac{1}{4}$.

Exercise 51.1: Show that the maximum value of $\mathbb{F}(\chi, \psi, \Omega)$ considered as a function of Ω is

$$\tfrac{1}{4}\cos^2(\chi - \psi),$$

and

$$\tilde{\tilde{R}} \leqslant \max\ \tilde{\tilde{R}}(\chi, \psi, \Omega) = \tilde{R}(\sin(\chi - \psi))$$

where $\tilde{R}(\tau)$, $\tau = \sin(\chi - \psi)$ is given by the function $\tilde{R}(\tau)$ whose values are found in Table 48.1.

Exercise 51.2: Find the parabolic equation governed by

$$q(y', z, t) = u(y', z, t) + \{2\Omega \sin\psi - \cos(\chi - \psi)\}\, z\,,$$

where u is any solution of (51.2a) satisfying the boundary conditions (see Exercise 40.2).
Show that $q(y', z, t)$ lies between the maximum and minimum values of

$$[q(y', 0, t),\ q(y', 1, t),\ q(y', z, 0)]\,.$$

§ 52. Numerical and Experimental Results

In previous sections of this chapter we analyzed RPCF. This flow has been interpreted as a limiting case, $\eta \to 1$, of spiral Couette flow between rotating and sliding cylinders. The main qualitative properties of spiral Couette flow depend only weakly on the relative gap size. A brief summary of some of the results which hold when η is unrestricted is given in this section.

We first consider energy stability analysis for three-dimensional disturbances. Some of the results are given in Fig. 52.1. The energy stability limit $R(\eta, \chi)$ depends only weakly on χ and η. Numerical results reported by Joseph and Munson (1970) but not shown here show that there is a mean radius at which $\psi_{\mathscr{E}} = \beta_{\mathscr{E}}$ as was assumed by (44.7). An energy stability theory for x'-independent disturbances has not yet been given for general values of η.

Linear stability limits for $\eta = 0.5$ and $\eta = 0.8$ are given in Hung, Joseph and Munson (1972). The instability limit R_L, the wave speed $\mathrm{im}(\sigma)$ and the azimuthal periodicity η of the spiral disturbance are important experimental observables. The number of vortices seen in any given experiment should correspond to the number of zeros $(2n)$ of $\cos n\theta$ of the eigenfunction belonging to R_L.

We shall now consider experimental results for spiral flow between closely spaced cylinders. We compare these experiments with theory for rotating plane Couette flow. Here the linear theory gives instability when $\mathbb{F} > 0$ and stability when $\mathbb{F} \leqslant 0$. The equality may be thought to represent the situation in which either the angular momentum of the basic spiral flow is constant in planes parallel to the plane of the disturbance, or it can represent plane Couette flow in a non-rotating system $\Omega = \mathbb{F} = 0$. The condition $\mathbb{F} < 0$ is satisfied by rotating cylinders in which the outer cylinder rotates much faster than the inner one.

The linear theory of stability of plane Couette flow without rotation has never indicated anything other than absolute stability for this flow[6].

---

[6] L. Hopf (1914), C. Morawetz (1952), W. Wasow (1953), R. Grohne (1954), L. Dikii (1961), A. Gallagher and A. Mercer (1962), E. Riis (1962), J. Deardorff (1963), A. Davey (1973).

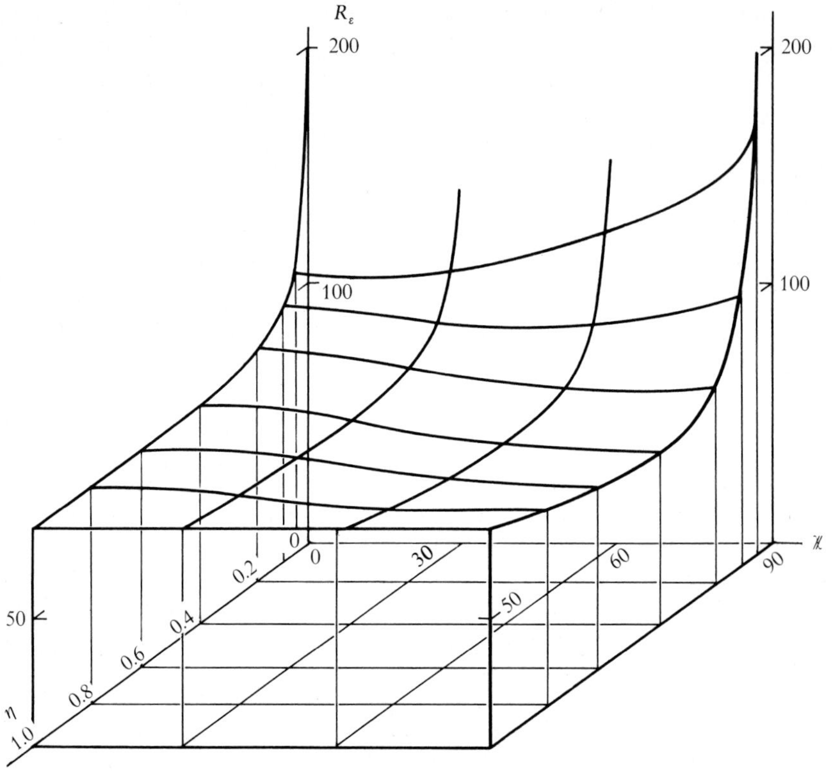

Fig. 52.1: Energy (stability) surface for Couette flow between rotating-sliding cylinders as a function of radius ratio, η, and the angle χ. Plane Couette flow in a rotating co-ordinate system is given by $\eta=1$; Couette flow in an annulus with no differential rotation is given by $\chi=0°$; and Taylor flow is given by $\chi=90°$. Because the circumferential wave-number, n, must be an integer, the above smooth surface is an approximation to the surface with discontinuous first derivatives. The smoothed-out version is barely distinguishable from the true surface (Joseph and Munson, 1970)

The experimental results of Schulz-Grunow (1958) for the allied problem of rotating cylinders with the inner one at rest do seem to indicate that if the amplitude of the disturbances is suppressed, then Couette flow is stable even at very large Reynolds numbers. On the other hand, for the same flow in natural circumstances, Couette (1890), Mallock (1888) and Taylor (1923) among others find a natural transition to turbulence. Reichardt (1956) claims to find that the plane Couette flow achieved in his experiments is stable when $U_c(b-a)/\nu<1500$.

When $\mathbb{F}\leqslant 0$ the observed instability cannot be explained by the linear theory of instability because the linear theory judges that RPCF with $\mathbb{F}\leqslant 0$ is absolutely stable. The energy theory gives sufficient conditions for stability and the form of the initial condition whose energy increases at the smallest R. For three-dimensional disturbances of RPCF, $R=2\sqrt{1708}$, $\frac{1}{20}$ of the 1500 observed in Reichardts experiment. For x'-independent disturbances $R=177.2$, $\frac{1}{10}$ of the aforementioned value 1500. The linear and energy theory are both in good agreement with experiment, independent of the sign of $\mathbb{F}$. Instability is always found when R is greater than the value given as critical by linear theory and RPCF is always stable when R is less than the critical value of energy theory. Unfortunately, such good agreements need not teach one much about the observed instability; when $\mathbb{F} \leqslant 0$ the energy result—stability with $R < 2\sqrt{1708}$ or even with Orr's $R < 177.2$—has very little to say about Reichardt's experiment and the linear theory, which probably gives instability only when $R \to \infty$, neither disagrees with observations nor does much to explain them.

The situation is greatly changed when $\mathbb{F} > 0$. Now there is a dynamic source for converting the energy of rotation into disturbance energy. The theshold of *instability* is lowered to energy-like values and when the rotation parameters are optimally adjusted, there is perfect or nearly perfect agreement between the energy and linear theory.

The comparison of the theoretical results with the experiment of Ludwieg (1964) is developed below. Ludwieg's apparatus is like a long sleeve bearing which is rotated around its axis at a fixed angular velocity and is geared to a shaft in the bearing in such a way that the shaft can be made to turn and slide relative to the rotating bearing. Since the clearance is small ($\eta = 0.8$), the flow develops almost instantly and is very nearly linear shear.

In Ludwieg's experiments, the parameter $\Omega_2(b-a)^2/\nu = 150$ is held fixed, and his results are expressed in terms of the parameters $\tilde{a}$, $\tilde{c}_z$ and $\tilde{c}_\phi$ which are related to the parameters of this chapter (see p. 161) as follows:

$$\tilde{a} = (1-\eta)/(1+\eta), \qquad R = 150/\Omega_2,$$
$$\Omega = (1 + \tilde{a}\tilde{c}_\phi)/\sqrt{\{(\tilde{a}+1)^2\tilde{c}_z^2 + (1-\tilde{c}_\phi)^2\}},$$

and

$$\chi = \arcsin\{(1-\tilde{c}_\phi)/\sqrt{\{(\tilde{a}+1)^2\tilde{c}_z^2 + (1-\tilde{c}_\phi)^2\}}\}.$$

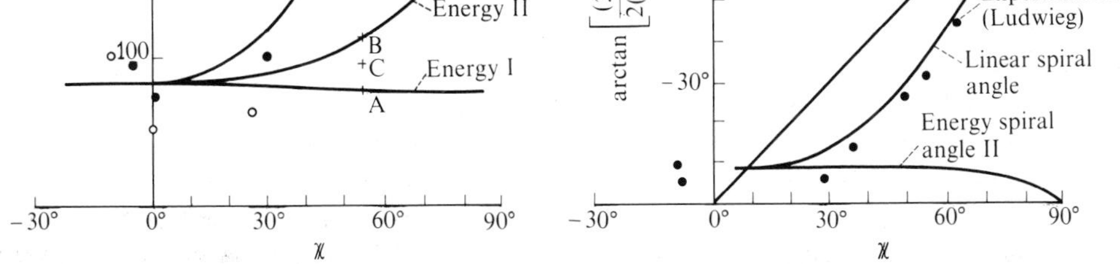

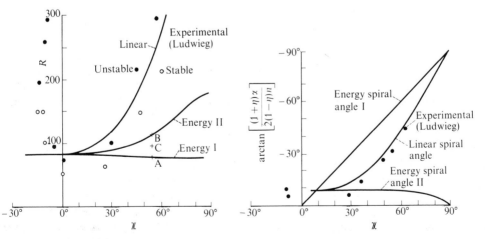

Fig. 52.2a, b: Comparison of theory and experiment (Ludwieg 1964) for $\eta = 0.8$, $(b-a)^2\Omega_2/\nu = 150$. Black dots are unable and white ones are stable (Hung, Joseph and Munson, 1972)

Since Ludwieg's data shows considerable scatter, a "mean" value of Ω is used to calculate the theoretical linear limit shown in Figs. 52.2a, b. The graphs marked "linear theory" are taken from the numerical integration for $\eta=0.8$ (Table 2; Hung, *et al.*, 1972). In Table 52.1 we have given values for Ludwieg's experiments.

Table 52.1: Spiral flow parameters for Ludwieg's experiment

χ (degrees)	Ω	R
60.5	0.572	262
54.1	0.81	185
52.4	1.09	138
45.0	1.49	100
29.4	2.13	70.4
12.7	2.42	62
0	2.33	64.5
−9.1	0.415	361
−10.2	0.514	292
−10.4	1.35	111
−10.9	0.68	220
11.5	0.22	163

The good agreement between the linear theory and Ludwieg's experiment shows that the instability being observed here can begin with an infinitesimal perturbation. There are four important observables in an experiment like Ludwieg's:

(1) the threshold limit,
(2) the spiral angle,
(3) the wave speed for the disturbances,
(4) the spiral vortex spacing.

Fig. 52.2a shows good agreement between the threshold limit R and the spiral angle χ. The wave speed $\mathrm{im}(\sigma)$ and the values n (given in Table 2, Hung *et al.*, 1972) which give the vortex spacing are not reported in Ludwieg's experiment.

The curves marked "Energy I" gives values associated with the disturbance whose energy increases at the smallest R. The curves marked energy "Energy II" are taken from the calculation of § 51 for the rotating plane Couette flow. The spiral angle ψ is in this case the limiting $\eta \to 1$ value.

In explanation of the two energy analyses, consider points A, B and C on Fig. 52.2a. At A we consider the critical energy disturbance which is a spiral vortex along energy spiral I. At B we consider the extremalizing solution of problem (51.16). This is also a spiral vortex whose axis lies along energy spiral II. At C the energy of disturbance A increases initially and the weighted energy of disturbance B decreases. The same weighted energy, but of A rather than B, decreases at a yet faster rate than B. Hence, the difference between the rate-of-change of the energy of A which is positive and the weighted energy which is negative is strongly positive.

To test the theoretical predictions of energy theory, it would be necessary to determine if the initial conditions whose energy will increase at values of $R > R_{\mathscr{E}}$ are sufficiently representative of physically realizable initial conditions. It would appear from experiments that even if such energetic disturbances are realizable, they are globally stable and *eventually* decay.

Exercise 52.1: Define a system of orthogonal spiral coordinates (r, ξ, η) where the surfaces $\xi = \text{const.}$ are swept out by radial vector $\mathbf{e}_r(t)$ which rotates and translates at constant speed. Write the IBVP for disturbances of spiral flow in these coordinates. Assume that along one spiral direction the disturbance does not vary. Form two energy identities for these spiral disturbances.

Chapter VII

Global Stability of the Flow between Concentric Rotating Spheres

In this chapter we shall consider the steady laminar motion of an incompressible viscous fluid contained between two concentric spheres which rotate about a common axis with fixed, but different, angular velocities (see Fig. 53.1). Some results are given for the flow between eccentric spheres (inner sphere displaced along the axis of rotation).

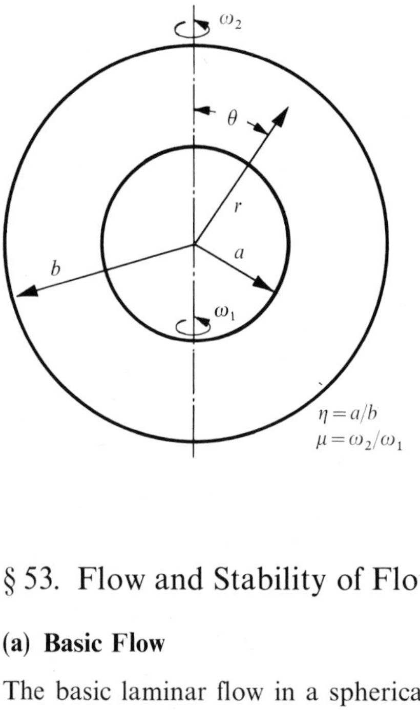

Fig. 53.1: Parameters for the flow between concentric spheres

§ 53. Flow and Stability of Flow between Spheres

(a) Basic Flow

The basic laminar flow in a spherical annulus is a function of two spatial variables, r and θ, and in non-dimensional form is strongly dependent on the Reynolds number. This Reynolds number dependence makes the study of flow in a spherical annulus different from, and perhaps more typical than, problems like spiral flow between cylinders which are independent of Reynolds number.

Since there is no simple solution for flow between spheres, various approximate solutions, valid over various ranges of the parameters, have been obtained.

These include (1) analytic perturbation solutions for small Reynolds numbers (Haberman, 1962; Ovseenko, 1963; and Munson and Joseph, 1971A), (2) Galerkin-type solutions for moderate Reynolds numbers (Munson and Joseph, 1971A), (3) finite-difference numerical solutions of the non-linear partial differential equations (C. Pearson, 1967; Greenspan, 1975) and (4) singular perturbation solutions (boundary layer, inviscid core) for large Reynolds numbers (Proudman, 1956). We stress the first two methods since they show most of the phenomena associated with flow in a sperical annulus and are in a form that is convenient for the stability analysis.

The basic flow is obtained from the governing Navier-Stokes equations which can be written in terms of a stream function in the meridian plane, ψ, and an angular velocity function, Ω, as follows (Rosenhead, 1963, p. 131)

$$-\frac{\psi_r\Omega_\theta-\psi_\theta\Omega_r}{r^2\sin\theta}=\frac{1}{R}\tilde{D}^2\Omega,$$

$$\frac{2\Omega}{r^3\sin^2\theta}[\Omega_r r\cos\theta-\Omega_\theta\sin\theta]-\frac{1}{r^2\sin\theta}[\psi_r(\tilde{D}^2\psi)_\theta-\psi_\theta(\tilde{D}^2\psi)_r]$$
$$+\frac{2\tilde{D}^2\psi}{r^3\sin^2\theta}[\psi_r r\cos\theta-\psi_\theta\sin\theta]=\frac{1}{R}\tilde{D}^4\psi, \tag{53.1}$$

where

$$\tilde{D}^2=\frac{\partial^2}{\partial r^2}+\frac{1}{r^2}\frac{\partial^2}{\partial\theta^2}-\frac{1}{r^2}\cot\theta\frac{\partial}{\partial\theta}$$

$$(\quad)_r=\partial/\partial r,$$

$$(\quad)_\theta=\partial/\partial\theta.$$

The non-dimensional velocities are related to ψ and Ω as follows:

$$U_r=\frac{\partial\psi/\partial\theta}{r^2\sin\theta},\qquad U_\theta=\frac{-\partial\psi/\partial r}{r\sin\theta},\qquad U_\phi=\frac{\Omega}{r\sin\theta}. \tag{53.2a}$$

The fluid volume is $\mathscr{V}=\{r,\theta,\phi:\eta\leqslant r\leqslant 1, 0\leqslant\theta\leqslant\pi, 0\leqslant\phi\leqslant 2\pi\}$, where the radius ratio $\eta=a/b$. Here a and b are the radii of the inner and outer spheres, respectively. The Reynolds number, R, is defined as $R=\omega_1 b^2/\nu$, where ν is the kinematic viscosity and ω_1 and ω_2 are the angular velocities of the inner and outer spheres.

The boundary conditions which complete the formulation of the problem are that $\psi=\partial\psi/\partial n=0$ on the boundaries, with n denoting the direction of the outward pointing unit normal, and that Ω is prescribed on $r=\eta$ and $r=1$. The governing dimensionless parameters are the radius ratio, η, the angular velocity ratio, $\mu=\omega_2/\omega_1$, and the Reynolds number, R.

The fluid motion consists of the "primary motion" about the axis of rotation given by U_ϕ (in terms of $\Omega(r,\theta)$) and the "secondary motion" in the meridian plane given by U_r and U_θ (in terms of $\psi(r,\theta)$). Although the secondary motion is small relative to the motion about the axis of rotation when the Reynolds number is small, it can become comparable to this primary motion for larger values of Reynolds number.

For small values of R, a solution to the governing equations, (53.1), can be obtained as a convergent series in powers of R (Munson and Joseph, 1971A):

$$\begin{aligned}\psi(r,\theta;R)&=R\psi_1(r,\theta)+R^3\psi_3(r,\theta)+R^5\psi_5(r,\theta)+\cdots\\ \Omega(r,\theta;R)&=\Omega_0(r,\theta)+R^2\Omega_2(r,\theta)+R^4\Omega_4(r,\theta)+\cdots\end{aligned} \tag{53.2b}$$

The component functions, $\psi_i(r,\theta)$ and $\Omega_i(r,\theta)$, can be obtained by a straightforward, but tedious, procedure. This consists of using Legendre polynomials in order to separate the r and θ dependence and solving the resulting ordinary differential equations. Haberman (1962) calculated the first term ("Stokes flow") of the series (see Exercise 53.2), Ovseenko (1963) calculated terms through R^3 and Munson and Joseph (1971A) calculated terms through R^7. The partial sum of seven terms provides an accurate solution for $R<80$ if $\eta=0.5$ and $\mu=\infty$.

For Reynolds numbers larger than those for which it is practical to obtain the perturbation solution, it is possible to use a Galerkin method to generate the basic flow. To do this (Munson and Joseph, 1971A) the stream function and angular velocity function (ψ and Ω) are approximated by a truncated series involving Legendre polynomials and various unknown functions of the radial coordinate, r. The unknown functions are governed by non-linear ordinary differential equations generated from (53.1). The nonlinear equations are integrated by numerical methods.

In Fig. 53.2 the angular velocity contours and streamlines of the secondary flow are plotted for $\eta=0.5$, $\mu=0$ and $R=100$. For low Reynolds numbers, the angular velocity of the primary swirling motion is approximately constant on

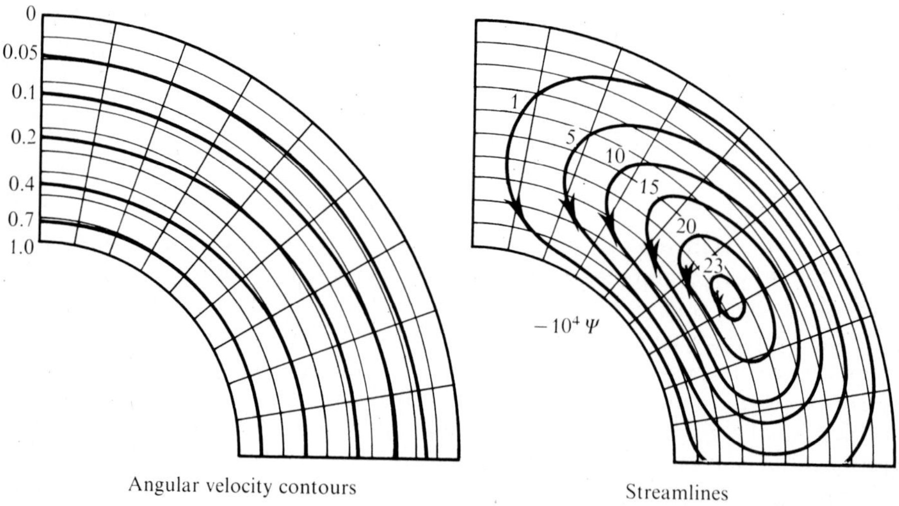

Fig. 53.2: Angular velocity contours and streamlines of the secondary flow ($\eta=0.5$, $\mu=0$, $R=100$) (Munson and Joseph, 1971A)

each sphere, while the secondary flow consists of a counter clockwise swirl driven by centripetal accelerations which force fluid away from the inner sphere near the equator. A sequence of photos clearly showing this motion is shown in the paper by Munson and Menguturk (1975).

The low Reynolds number flow generated when the outer sphere is rotated and the inner sphere is stationary is similar to the one just described except that the direction of the secondary flow is reversed. If both spheres are rotated in opposite directions, the secondary flow may consists of two swirls of opposite sense.

A typical solution for moderate R is shown in Fig. 53.3. Comparison of Figs. 53.2 and 53.3 shows that a change in R results in (1) a change in the speed ratio of secondary flow to primary flow, (2) a change in the shape of the contours of constant angular velocity and (3) a shift in the shape of the secondary flow contours.

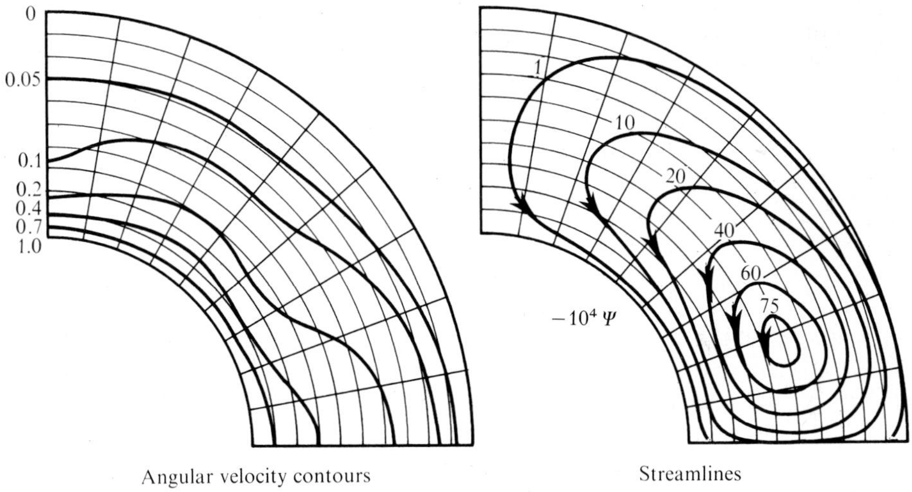

Fig. 53.3: Angular velocity contours and secondary flow for $\eta=0.5$, $\mu=0$, $R=400$ (Munson and Joseph, 1971A)

For large R, various phenomena associated with boundary layers and shear layers and other special effects become important (C. Pearson, 1967; Proudman, 1956).

The flow in an eccentric spherical annulus obtained by displacing the inner sphere along the axis of rotation can be constructed as a power series in two small parameters: the Reynolds number R and the eccentricity δ (Munson, 1974). Here $\delta=\varepsilon/b$, where ε is the distance between the center of the two spheres. The results indicate that the secondary flow swirls split about the equator of the inner sphere and that it takes more torque to rotate the spheres when they are not centered than when they are centered. Fig. 53.4 shows the secondary flow patterns for $\eta=0.44$, $\mu=0$, $\delta=0.3$ and $R=20$. It is interesting to note that as

Fig. 53.4: Basic flow for eccentric rotating spheres with $\eta=0.44$, $\mu=0$, $\delta=0.3$, $R=20$ (Munson and Menguturk, 1975)

R is increased, the flow eventually becomes turbulent, with turbulent flow confined to regions where the local Reynolds number, which is proportional to the local gap size, is larger than some critical value. Fig. 53.5 shows a situation for which the flow in the lower portion is turbulent while the upper portion is laminar.

(b) Stability Analysis

We turn now to the study of the stability of the basic laminar flow between rotating spheres. This problem is different from the problem of spiral flow considered in Chapters V and VI in that the flow is completely bounded and depends on two spatial variables and the Reynolds number R. These facts force the stability analysis to be more complex than for the simpler spiral flows.

We treat the global stability of flow between rotating spheres. The linear theory of stability is used to find a critical value $R=R_L$ such that when $R>R_L$ the basic laminar flow is unstable. The energy theory of stability is used to find a critical value $R=R_{\mathscr{E}}$ such that when $R<R_{\mathscr{E}}$ the basic laminar flow is definitely stable. The critical value R_L is determined as the first zero of the principal eigenvalue $\sigma(R_L)=0$ of the spectral problem of the linearized theory of stability

$$-\sigma\mathbf{u}+\mathbf{u}\cdot\nabla\mathbf{U}+\mathbf{U}\cdot\nabla\mathbf{u}+\nabla p-\frac{1}{R}\nabla^2\mathbf{u}=\nabla\cdot\mathbf{u}=0 \quad \text{in } \mathscr{V} \tag{53.3}$$

Fig. 53.5: Turbulent flow in the wide gap portion (bottom) of an eccentric spherical annulus with laminar flow in the narrow gap (top); $\eta=0.44$, $\mu=0$, $\delta=0.3$, $R=2200$ (Munson and Menguturk, 1975)

and $\mathbf{u}|_{\partial\mathscr{V}}=0$. The eigenvalues $\sigma(R)$ arise from substituting disturbances proportional to $e^{-\sigma t}$ into the Navier-Stokes equations linearized around the basic laminar flow between rotating spheres

$$\mathbf{U}=\mathbf{U}(r,\theta;\eta,\mu,R).$$

The eigenvalues $\sigma(R)$ depend on R because R appears in (53.3) explicity and implicity, through $\mathbf{U}$.[1]

The critical value $R_{\mathscr{E}}$ of the energy theory of stability is defined by

$$R_{\mathscr{E}}^{-1}=\max_H -\langle\mathbf{u}\cdot\nabla\mathbf{U}\cdot\mathbf{u}\rangle/\langle|\nabla\mathbf{u}|^2\rangle \tag{53.4a}$$

where H is the collection of smooth solenoidal vectors $\mathbf{u}$ which vanish on the boundary sphere, $\nabla\mathbf{U}$ is the velocity gradient of the basic flow and $\langle\ \rangle$ is a volume-averaged integral over the spherical annulus; $\langle c\rangle=c$ for any constant c.

[1] Munson and Menguturk (1975) assume that $\sigma(R)$ is real-valued at criticality; i.e., $\sigma(R_L)=0$. This assumption is supported by their experimental observations which indicate that the bifurcating flow is steady when R is slightly greater than R_L. They note that this assumption may not hold in all situations.

The Euler equation for (53.4a) is

$$\mathbf{D}[\mathbf{U}(R)]\cdot\mathbf{u}=\lambda\nabla^2\mathbf{u}-\nabla p \tag{53.4b}$$

where $\mathbf{u}$ is solenoidal and vanishes on the boundaries. $\lambda^{-1}(R)$ is the critical point of the functional (53.4a) and is an eigenvalue of (53.4a) is

The basic flow $U(\mathbf{x}; R)$ is stable if the minimum eigenvalue of (53.4b), λ^{-1}, is greater than the Reynolds number of the basic flow being considered. The condition for which $\lambda^{-1}=R_{\mathscr{E}}$ thus provides the energy stability limit. On the other hand, a basic flow is unstable if the eigenvalue $\sigma(R)$ of (53.3) is less than zero.

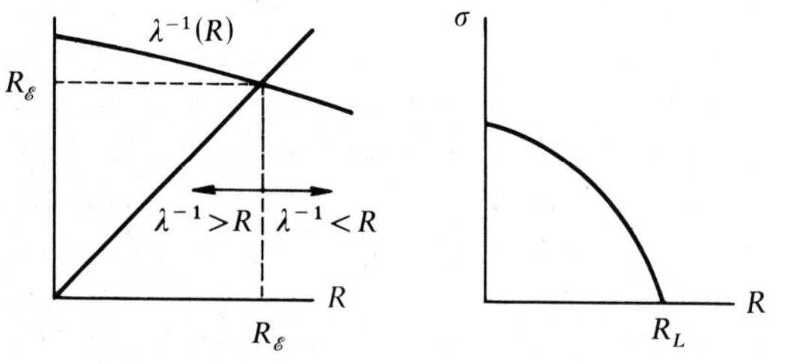

Fig. 53.6: Critical values of R for basic flows which depend on the Reynolds number

Munson and Joseph (1971B) and Munson and Menguturk (1975) use a Galerkin method to solve the stability problem of energy and linear theory. This method utilizes the accurate solution for the basic flow which was discussed in the preceding section and a representation of the disturbance in terms of the toroidal ($\mathbf{u}_2=\mathbf{T}$) and poloidal ($\mathbf{u}_1=\mathbf{S}$) decomposition given in Appendix B.6. The toroidal and poloidal vector fields are generated from toroidal and poloidal potentials

$$\mathbf{T}=\operatorname{curl}(\mathbf{r}\psi), \qquad \mathbf{S}=\operatorname{curl}^2(\mathbf{r}\chi)\,.$$

Details of this representation are given by Chandrasekhar (1961). The generating scalars $r\psi$ and $r\chi$ can be expanded in terms of spherical harmonics, $Y_l^m(\theta,\phi)$. This expansion leads to the following expression for the disturbance $\mathbf{u}$:

$$\mathbf{u}=\sum_l\sum_m\left\{\frac{l(l+1)}{r^2}S_l^mY_l^m,\ \frac{1}{r}S_l'^m\dot{Y}_l^m+\frac{im}{r\sin\theta}T_l^mY_l^m,\ \frac{im}{r\sin\theta}S_l'^mY_l^m-\frac{1}{r}T_l^m\dot{Y}_l^m\right\}$$

where (53.5)

$$(\quad)'\equiv d/dr \quad\text{and}\quad (\dot{\quad})\equiv d/d\theta\,.$$

The component functions, $T_l^m(r)$ and $S_l^m(r)$, must be obtained so that the disturbance given by (53.5) satisfies the appropriate eigenvalue problem ((53.3) for linear theory and (53.4) for energy theory).

Yakushin (1969, 1970) and Khlebutin (1968) have carried out approximate linear stability analyses for the narrow gap ($\eta \approx 1$) when the spheres rotate in the same direction at different angular velocities. They used a low order approximation for the basic flow ($\psi = R\psi_1, \Omega = \Omega_0$) and a Bubnov-Galerkin method to solve the linear stability equations. The results indicate that when the fluid layer is thin (narrow gap) the stability characteristics of the flow in the spherical annulus and the flow between rotating cylinders are similar (both qualitatively and quantitatively). This similarity is not unexpected since the portion of the spherical annulus near the equator looks very similar to the annulus between rotating cylinders (see Exercise 53.3).

When the gap is not small the flow in the spherical annulus is not locally similar to the flow in a cylindrical annulus and the low order approximation to the basic flow is not valid at the relatively large Reynolds numbers which are important in stability analysis. Bratukhin (1961) considered the linear theory for a large gap ($\eta = 0.5$) with the outer sphere stationary ($\mu = 0$). He used a low order approximation to represent the basic flow and a perturbation solution in powers of R for the eigenvalue $\sigma = \sigma(R)$. This approximate result indicates that $R_L = \omega_1 b^2/\nu \approx 400$. Munson and Menguturk (1975) have shown that R_L is much larger than 400 when a more accurate representation of the basic flow is used in the stability calculation.

Since the basic flow is not a function of ϕ, it is possible to consider disturbances for various values of m (the wave number in the circumferential direction) and to seek the minimum over integer values of m.

Although the physical interpretations and origins of the linear and energy theory are entirely different, the structure of the eigenvalue problem in each theory is superficially similar. This similarity is especially evident for eigenvalues $\sigma(R_L) = 0$ which were considered by Munson and Menguturk (1975). When $\sigma(R_L) = 0$, the methods used to solve for the critical values $R_{\mathscr{E}}$ and R_L are nearly identical. The toroidal-poloidal representation given by (53.5) is substituted into (53.3) and (53.4) and the coefficients of independent spherical harmonics are set to zero[2]. This leads to a set of coupled, linear ordinary differential equations with variable coefficients for the component functions $T_l^m(r)$ and $S_l^m(r)$. The ordinary differential equations are then solved by numerical methods.

Some typical results of the application of this method of analysis to the problem of stability of flow in a spherical annulus with $\eta = 0.5$ are shown in Fig. 53.7 and for other values of η in Fig. 53.8. The eigenfunction belonging to the eigenvalue $\sigma(R_L) = 0$ of linear theory is axisymmetric ($m = 0$) but the critical eigenfunction of energy theory is not axisymmetric ($m = 1$). Munson and Menguturk (1975) have noted that when the gap width it is necessary to include many functions in the truncated series describing the disturbance flow. For example for $\eta = 0.5$ it was necessary to use only seven functions in the series to obtain a reasonable convergence of the series (Munson and Joseph, 1971B),

[2] The identification of the independent spherical harmonics is complicated by the fact that the basic flow also depends on the angles and various products of the spherical harmonics must be reduced. It is best to proceed by using orthogonality properties of the toroidal-poloidal vectors. The algebra is easy but tedious (see Munson, 1970; Munson and Joseph, 1971B; Munson and Menguturk, 1974).

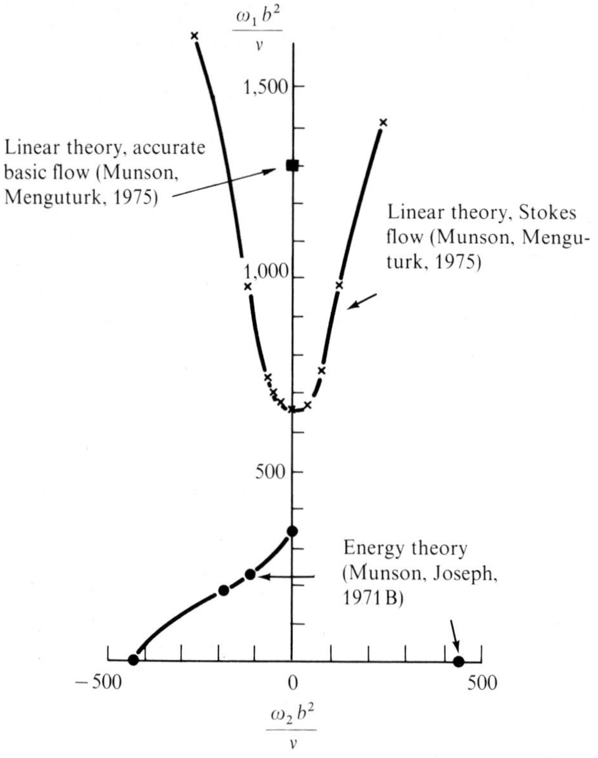

Fig. 53.7: Critical values of energy and linear theory for flow in a spherical annulus with $\eta = 0.5$. These values should be compared with the experimental results shown in Fig. 53.8 (Munson and Menguturk, 1975)

whereas for $\eta \approx 1$ many functions are needed. Yakushin (1970) used thirty functions for his Bubnov-Galerkin method in solving the linear theory equation for $\eta = 0.91$ and $\eta = 0.935$. When the gap is narrow the bifurcating flow near the equator is in the form of Taylor vortices and the structure of this flow is probably more naturally suited for analysis in cylindrical coordinates (see Exercise 53.3). In the next subsection the numerical results are compared with experiments.

(c) Experimental and Numerical Results

Several persons have carried out experimental studies of flow in a spherical annulus. The results depend strongly on the value of the radius ratio, η. Zierep and Sawatzki (1970), Khlebutin (1968), Munson and Menguturk (1975) and Morales-Gomez (1974) find that when the gap is small the stability properties are similar to those for Taylor flow for the same radius ratio. On the other hand, Sorokin, Khlebutin and Shaidurov (1966) and Khlebutin (1968) were unable to observe any transition in the spherical annulus flow when the gap is large ($\eta < 0.7$). Careful experiments by Munson and Menguturk (1975) did reveal various instabilities in the wide gap, but the bifurcating flows seem to have a structure different from those which are observed when the gap is small.

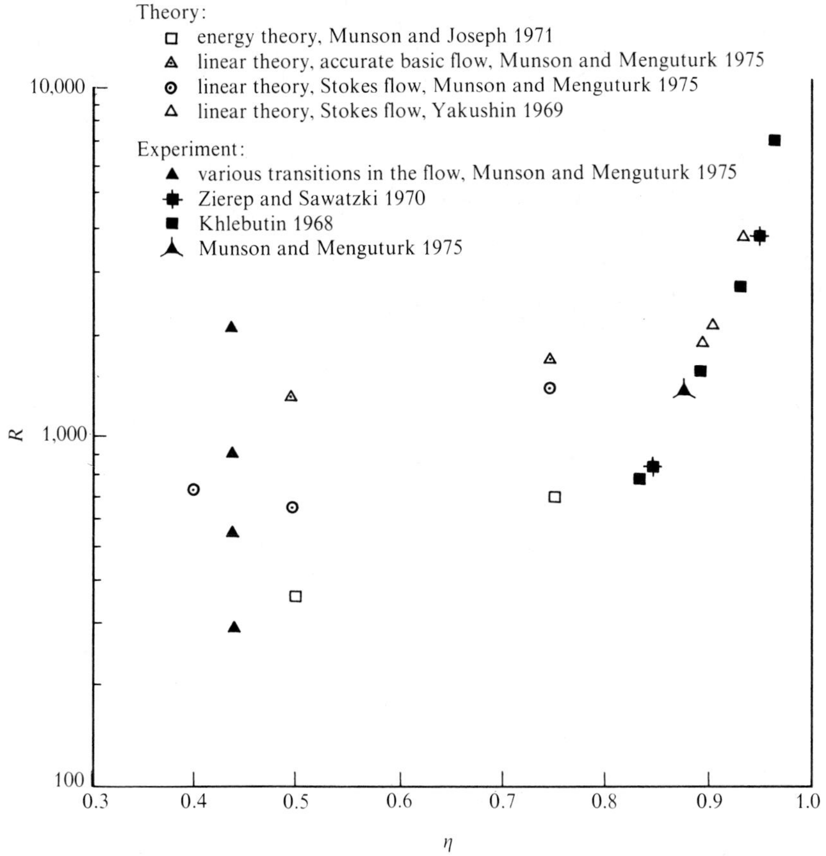

Fig. 53.8: Computed and observed critical values of R for flow in a spherical annulus with $\mu=0$ (Munson and Menguturk, 1975)

The apparatus shown schematically in Fig. 53.9 was used by Munson and Menguturk (1975) to obtain experimental results for flow in a spherical annulus. It includes (a) a clear outer spherical shell of diameter 12.95 cm, (b) an inner sphere rotated by a variable speed electric motor, (c) a torsion wire from which the plexiglass box containing the outer sphere is suspended, (d) a thermocouple to measure the temperature of the fluid in the annulus and (e) a means of measuring the angular rotation of the outer sphere (in order to determine the applied torque). Silicone oils with various viscosities were used as the fluid. Flow visualization was achieved by aluminum flakes suspended in the fluid or dye injection.

Flow visualization studies (Munson and Menguturk, 1975; Menguturk, 1974) confirm that the basic flow is as given by the theoretical results of subsection (a).

The theoretical torque, M, needed to rotate the sphere can be obtained once the velocity field is known by integrating the shear stress over the surface of the sphere. The dimensionless torque

$$\tilde{m}(R,\eta)=3M/(8\pi\nu\rho\omega_1 b^3)$$

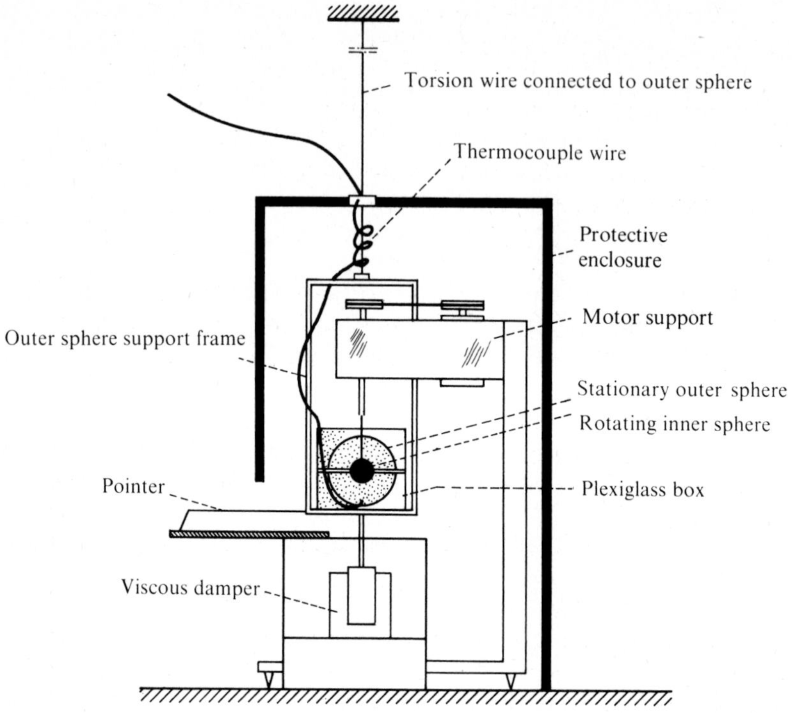

Fig. 53.9: Schematic diagram of the experimental apparatus for flow in a spherical annulus (Munson and Menguturk, 1975)

is a function of Reynolds number and has been calculated for low and moderate R by Munson and Joseph (1971A). The agreement between the theoretical and experimental results for $\eta = 0.44$ is shown in Fig. 53.10. For small R, where the secondary flow is negligible, $\tilde{m}$ is constant. For large R, the flow has a boundary

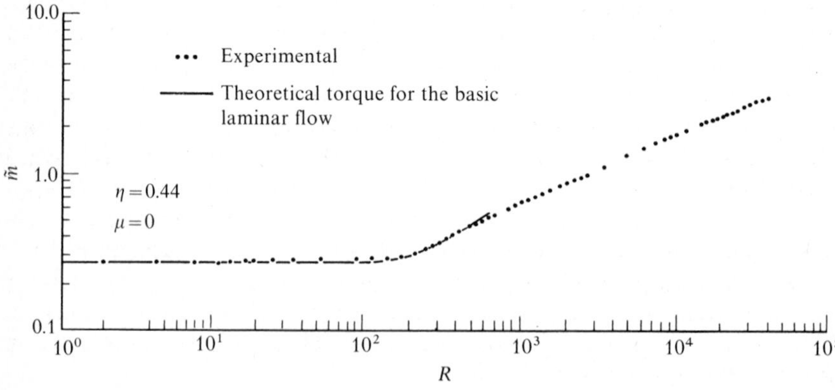

Fig. 53.10: Torque vs. Reynolds number. The response curve for flow between concentric rotating spheres (Munson and Menguturk, 1975)

layer structure and $\tilde{m} \sim R^n$ with $n=0.406$ for $\eta=0.44$. This dependence of $\tilde{m}$ on R is not unexpected but is suggested by boundary layer studies of related flows (Howarth, 1951; Sorokin, Khlebutin and Shaidurov, 1966).

Various critical values for flow in a spherical annulus can be determined from flow visualization experiments or torque characteristics. When the gap is small both of these methods work well, but when the gap is wide, the effects of a transition in the flow may be easily masked by experimental error or other features inherent in the flow. Typical stability experiment results are described below.

The initial instability in a spherical annulus with a small gap takes place as Taylor vortices in the region near the equator (see Fig. 53.12, 13). The instability occurs at Reynolds numbers corresponding very closely to the critical value R_L computed by Yakushin (1969). The physical dimensions of the vortices depends on the gap width (as in Taylor vortex flow) and the portion of the annulus that they occupy (i.e., how close they come to the poles) is also dependent on the ratio. It is noted that this Taylor vortex type instability is observed only if the gap width is sufficiently small ($\eta > 0.7$, or so). According to Munson and Menguturk (1975) the bifurcating flow which replaces the basic flow appears to be steady when R is slightly above the first critical value observed in the experiments.

For Reynolds numbers above the first critical value, various other modes of flow are observed for the narrow gap case. Zierep and Sawatzki (1970) catalog these flow configurations for two different narrow gap situations ($\eta=0.95$ and $\eta=0.848$) and present photographs of the flow. A somewhat similar situation for flow between rotating cylinders is well documented by Coles (1965).

Munson and Menguturk (1975) note that the various transitions in the flow are easily observed not only by flow visualization techniques, but also by the fact that the torque characteristics change quite noticeably when the gap is narrow. If the gap is wide, the transitions are much different and more difficult to observe.

It has been noted that the flow in a wide gap is considerably different from that in a narrow gap. In fact, several persons have studied the flow in a wide gap experimentally and were unable to detect any transitions by either torque measurements or flow visualization techniques. The apparent lack of transitions is due to the fact that the changes in the flow that occur due to a transition in the flow become much smaller as the gap size increases. Careful experiments (Munson and Menguturk, 1975) suggest that transitions in the flow do occur as the Reynolds number is increased, but their effects are not nearly as large for flow in a wide gap.

A series of experiments using the apparatus shown in Fig. 53.9 for $\eta=0.44$ showed that four break points (or transitions in the flow) occur between Reynolds numbers corresponding to the basic laminar flow and those for which complete turbulence was obtained. It was found that the flow in the wide gap is subcritically unstable and that the observed instability occurs at a Reynolds number close to the critical value of the energy theory, $R_{\mathscr{E}}$.

The experimental torque results ($\tilde{m}=\tilde{m}(R)$) for $\eta=0.44$ as shown in Fig. 53.10 seem to indicate a "crisis free" (no transition) situation since the curve appears to be smooth. However, upon a closer look at the torque data, several transitions are observed. Fig. 53.11 is a greatly enlarged portion of the torque curve in the neighborhood of $R=540$, the second transition point for this flow. Although the change in the torque curve is not nearly as large as those for flow in a narrow gap careful experiments show that the data is repeatable and free of hysteresis. The first and third break points, occurring at $R=290$ and $R=900$ for this radius ratio ($\eta=0.44$), produce similar small changes in the torque characteristics. The fourth and more pronounced, break occurs at $R=2100$, at which point the flow suddenly becomes turbulent.

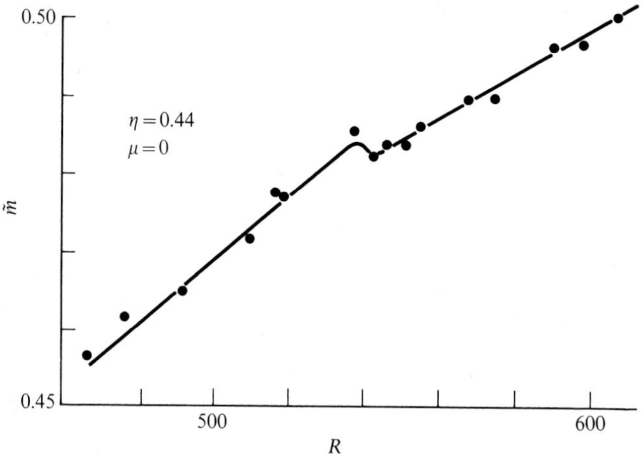

Fig. 53.11: Magnification of the response curve of Fig. 53.10 in the neighborhood of the second point of transition

According to energy theory (Munson and Joseph, 1971B), flow in a spherical annulus is stable to any disturbance (large or small) if $R<R_{\mathscr{E}}\approx 310$ when the $\eta=0.44$ and $\mu=0$. (This value is obtained by an extrapolation from the known results for $\eta=0.5$.) The close agreement between the critical value $R_{\mathscr{E}}\approx 310$ of energy theory and the first break point, $R=290$, is noteworthy. Since linear theory indicates that $R_L\approx 1300$ when a principle of exchange of stability is assumed, we may tentatively conclude that transition from the basic laminar flow is subcritical and, therefore, is caused by finite disturbances under conditions for which infinitesimal disturbances decay.

Careful observation of the flow field by using aluminum flakes suspended in the fluid was carried out in order to observe the nature of the various transitions in the flow. No instabilities were observed for Reynolds numbers near the first break point ($R=290$) of the torque curve. It is possible that the flow following this instability may occur in a form similar to the secondary motion of the basic flow and thus be indistinguishable from the basic flow, although it would produce a break in the torque curve.

At Reynolds numbers corresponding to the second break point ($R=540$), an instability appears in the form of small spots or puffs of turbulence. These

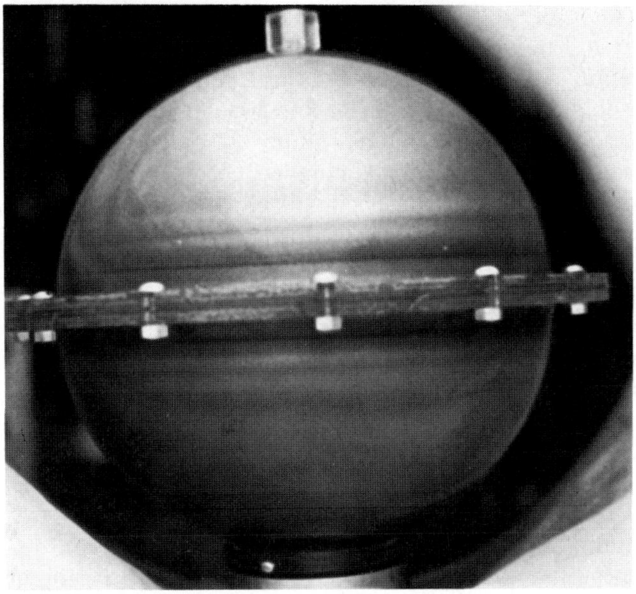

Fig. 53.12: Taylor vortices between rotating spheres. The inner, aluminum sphere ($a = 4.118$ inches) is rotating and the outer, plexiglass sphere ($b = 4.634$ inches) is stationary, $a^2\omega_1/\nu = 1078$ (Morales-Gomez, 1974)

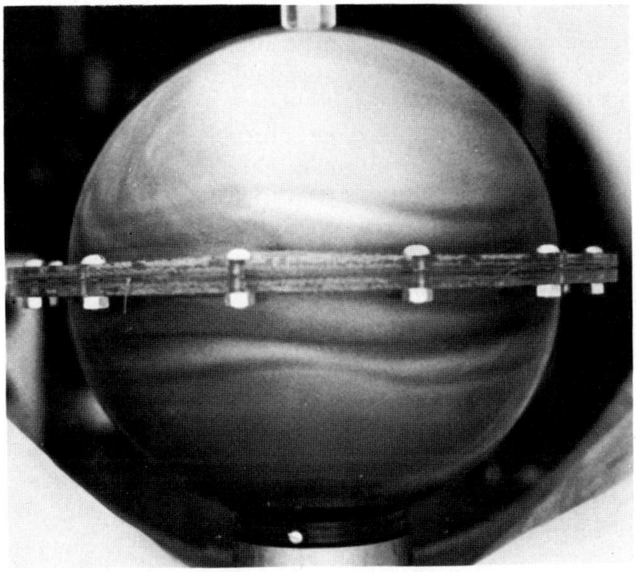

Fig. 53.13: Wavy Taylor vortices between rotating spheres. The axisymmetric Taylor vortices lose stability to wavy vortices when $a^2\omega_1/\nu \approx 1400$. In this photo $a^2\omega_1/\nu = 2696$ (Morales-Gomez, 1974)

turbulent spots occur at the center of the secondary basic flow swirl and rotate about the axis of rotation at a rate corresponding to the angular velocity of the primary flow at that location. At Reynolds numbers corresponding to the third break point ($R=900$) a slight waviness or unsteadiness is observed near the equator. This unsteadiness increases and spreads toward the poles as the Reynolds number is increased. Finally, for Reynolds numbers corresponding to the fourth break point ($R=2100$) the flow becomes completely turbulent.

For the first three break points the change in the dimensionless torque occurs as a decrease in $\tilde{m}$, whereas the fourth break (transition to turbulence) occurs as an increase in $\tilde{m}$. Munson and Menguturk note that the magnitude of these breaks in the torque curve for $\eta=0.44$ is considerably smaller than those which occur for larger values of η.

Thus the transitions involved in the wide gap appear to have a different character from those in the narrow gap (in particular there are no Taylor type vortices near the equator) and the instability is initiated as a subcritical one near the critical value of the energy theory. A summary of the various theoretical and experimental results is shown in Fig. 53.8.

Munson and Menguturk (1976) have obtained interesting stability results for the flow between eccentric rotating spheres. This flow was described briefly in subsection (a). The experiments of Munson and Menguturk indicate that the stability of the flow in the eccentric spheres is dependent upon the magnitude of the eccentricity and the radius ratio. When the gap is narrow, the Reynolds number for the initial transition of the basic flow (from the original laminar flow to Taylor type vortices) is insensitive to the amount of eccentricity. This result is not unexpected since the region near the equator (in which the flow first becomes unstable is similar to rotating cylinders whether the spheres are concentric or eccentric (see Figs. 53.12, 13).

When the gap is wide the various transitions depend strongly upon the eccentricity. This dependence on eccentricity is dramatically illustrated in Fig. 53.5. This figure shows that the flow in the region where the gap is small is laminar whereas the flow in the region where the gap is large is turbulent. The two flows are separated by a non-propagating laminar-turbulent interface.

Exercise 53.1: Consider the possibly unsteady axisymmetric motion of an incompressible viscous fluid contained between two concentric spheres which rotate about a common axis with different angular velocities. Let (r,θ,ϕ) be polar spherical coordinates. The velocity components and the pressure are independent of ϕ but are otherwise arbitrary. Show that the total angular momentum $rU_\phi \sin\theta$ is bounded above and below by the maximum and minimum values which this quantity takes on in the annulus at $t=0$ and at the boundary of the spheres for $t>0$.

Exercise 53.2 (Munson and Joseph, 1971A): Consider the series solution (53.2b). Show that to leading order

$$U_\phi \sim \Omega_0(r,\theta)/r\sin\theta = \sin\theta\, f(r)/r \tag{53.6}$$

where

$$f(r) = \frac{A}{r^2} + Br\,,$$

$$A = \frac{b^3a^3(\omega_1 a - \omega_2 b)}{b^3 - a^3}\,,$$

and

$$B = \frac{\omega_2 b^4 - \omega_1 a^4}{b^3 - a^3}.$$

Formulate and solve the perturbation problem for $\psi_1(r,\theta)$.

Exercise 53.3 (Rayleigh's discriminant): Define cylindrical coordinates $(\hat{r}, z, \phi)$

$$\hat{r} = r\sin\theta,$$

$$z = r\cos\theta,$$

$$\phi = \phi,$$

in the sphere. Show that instability according to the criterion of Rayleigh's discriminant is given by

$$\frac{d}{d\hat{r}}(\hat{r}^2 U_\phi) < 0$$

or

$$\Omega_z \zeta_z < 0 \tag{53.7}$$

where

$$\Omega_z = U_\phi / \hat{r}$$

is the total angular velocity and

$$\zeta_z = \zeta_r \cos\theta - \zeta_\theta \sin\theta$$

is the vorticity. Apply the criterion (53.7) to the Stokes flow (53.6) and show that

$$\Omega_z \zeta_z = \frac{f(r)}{r^3} \sin\theta \left[2\cos^2\theta \frac{f}{r} + \sin^2\theta \frac{df}{dr} \right]. \tag{53.8}$$

Show that Rayleigh discriminant is always *positive* near the poles $\theta = 0$, π and is *negative* at the equator whenever A and B are such

$$df^2/dr < 0.$$

Construct an argument, based on Rayleigh's discriminant to show that the flow (53.6) is unstable in the intersection of the spherical annulus and the sector $\theta_0 < \theta < \pi - \theta_0$. Show that $\theta_0 = \pi/4$ when $B = 0$.

Appendix A

Elementary Properties of Almost Periodic Functions

Almost periodic functions are a generalization of the periodic functions *which leave intact the property of completeness of Fourier series.* An AP function on the line $-\infty<x<\infty$ can wiggle more or less arbitrarily but in such a way that any value of the function is very nearly repeated at least once in every sufficiently large but finite interval.

A (complex) continuous function $f(x)$ $(-\infty<x<\infty)$ is *almost periodic* (AP) if for each $\varepsilon>0$ there exists $l=l(\varepsilon)>0$ such that each real interval of length $l(\varepsilon)$ contains at least one number τ for which

$$|f(x+\tau)-f(x)|\leqslant\varepsilon.$$

Each such τ is called a translation number. If $\varepsilon=0$, $f(x)$ is a periodic function and τ is its period.

The definition given above implies that all of the values of $f(x)$ on the whole line can be produced nearly (within ε) on intervals of finite size $l(\varepsilon)$. For small values of ε, however, it may be necessary to make the interval very large.

The theory of AP functions of a single variable can be generalized to the case of functions of more than one variable. The generalization does not introduce any difficulties and the theory of AP functions, say of x, y is just a copy of what is true for AP functions of a single variable (see § 12 of Besicovitch, 1932 for a discussion of AP functions of x and y).

AP functions have some of the properties believed to characterize turbulent flow. The Fourier series property of AP functions, however, implies that each realization of "AP turbulence" has only a denumerable infinite collection of "eddy sizes". The approximation theorems for AP functions further guarantee that such "turbulence" could be approximated uniformly and in the mean by trigonometric polynomials.

AP representations of turbulence have been criticized by Bass (1962) and others. They note that the Fourier series imply that fluctuation correlations between distant points are not weaker than at nearby points (see property (10), below).

The simple properties of AP functions which are mentioned below are discussed in detail in the monographs of Bohr (1932) or Besicovitch. The function $f(x)$ mentioned below is assumed to be AP.

(1) An AP function is bounded and uniformly continuous on $-\infty < x < \infty$.
(2) The product and sum of AP functions are AP.
(3) The indefinite integral $F(x)$ of an AP function $f(x)$ is AP if $F(x)$ is bounded.
(4) The mean value

$$\mathbf{M}\{f(x)\} = \lim_{L\to\infty} \frac{1}{2L}\int_{-L}^{L} f(x)\,dx = \lim_{L\to\infty} \frac{1}{2L}\int_{-L+X}^{L+X} f(x)\,dx$$

has a definite finite value which is independent of X. (In the text we use the overbar notation of Reynolds rather than the $\mathbf{M}$ notation of Bohr.)

(5) The function $a(\lambda) = \mathbf{M}\{f(x)e^{-i\lambda x}\}$ is zero for all λ with the exception of a countable set. The $a_n = a(\lambda_n)$ for which $a(\lambda)$ is not zero are called the Fourier coefficients and the values λ_n are the Fourier exponents of f.

(6) $e^{i\lambda x}$, being periodic, is an AP function. Then the trigonometric polynomial

$$\sum_{-N}^{N} a_n e^{i\lambda_n x} \tag{A.1}$$

is an AP function. If the series $\sum |a_n|$ is convergent, the sum

$$\sum_{-\infty}^{\infty} a_n e^{i\lambda_n x}$$

is an AP function. If f is AP, so is $\bar{f}$. Then $f\bar{f} = |f|^2$ has a mean value, called the quadratic mean value of f. The polynomials (A.1) approximate $f(x)$ in the mean; that is,

$$\mathbf{M}\{|f(x) - \sum_{-N}^{N} a_n e^{i\lambda_n x}|^2\}$$

is smallest if the a_n are the Fourier coefficients and λ_n are the Fourier exponents for $f(x)$.

(7) $f(x)$ has a Fourier series[1]

$$f(x) \sim \sum_{-\infty}^{\infty} a_n e^{i\lambda_n x} \tag{A.2}$$

which is complete in the sense of Parseval's equality

$$\lim_{N\to\infty} \mathbf{M}\{|f(x) - \sum_{-N}^{N} a_n e^{i\lambda_n x}|^2\} = 0$$

that is

$$\sum_{-\infty}^{\infty} |a_n|^2 = \mathbf{M}\{|f(x)|^2\}.$$

[1] The $\sim$ sign in (A.2) denotes convergence in the Bohr mean norm. Conditions sufficiently general to guarantee uniform convergence of the Fourier series in applications are presently unknown. In contrast, the convergence of Fourier series for periodic $C^1(x)$ functions is uniform, and the convergence of the Fourier integral for any $C^1(x)$ function which is absolutely integrable is uniform. Convergence in the mean, rather than uniform convergence, is enough for the energy calculations which are developed in this book.

(8) If $\mathbf{M}\{|f(x)|^2\}=0$ then $f(x)=0$. There is no AP function, which does not identically vanish, for which $a(\lambda)=0$ for all λ.

(9) Given an AP function (A.2) and a positive number ε, there exists a trigonometric polynomial $p(x)$, whose exponents are Fourier exponents λ_n of $f(x)$, satisfying the inequality $|f(x)-p(x)|<\varepsilon$ for all values of x.

(10) The auto-correlation function of $f(x)$ is defined by

$$c(y)=\mathbf{M}\{f(x)\bar{f}(x+y)\},$$

where $\bar{f}$ is the complex-conjugate of f.

If $f(x)$ is AP then

$$c(y)=\sum_{-\infty}^{\infty}|a_n|^2 e^{-i\lambda_n y}$$

is an AP function of y and does not decay as $y^2\to\infty$.

(11) Let df/dt (and f) be AP in x and continuous on a closed t interval. On this interval differentiation with respect to a parameter and formation of the mean value in x are commutative. This implies the Reynolds transport theorem

$$\frac{d}{dt}\mathbf{M}\{f(x,t)\}=\mathbf{M}\left\{\frac{df}{dt}\right\}$$

with respect to AP functions of x.

In stability theory it is usual to restrict one's attention to periodic disturbances (of arbitrary period). It is relatively easy to achieve rigorous mathematical results for periodic disturbances because *periodic disturbances define bounded domains* in the sense that well defined conditions are imposed on the boundary of an enclosed region of space. In contrast, AP disturbances are more general but also more complicated. Though each AP function has a Fourier series, one needs a continuum of orthonormal vectors $e^{i\lambda x}$ $(-\infty<x<\infty)$ to represent the class. For periodic functions of period L, the complete set of orthonormal vectors $e^{i(n\pi/L)x}$ is denumerable. The conditional stability and instability theorems (§ 8) of linear theory have not yet been proved for AP disturbances though they do hold for periodic disturbances. The existence of spatially AP solutions of the Navier-Stokes equations is also an open mathematical question.

Appendix B

Variational Problems for the Decay Constants and the Stability Limit

This appendix will introduce beginners to variational problems defined over solenoidal vector fields. We want to reformulate the extremum problems for the decay constant (4.5) and the energy stability limit (4.6) as eigenvalue problems.

Looking forward to our later work on convection where the boundary conditions are more complicated, we shall consider functionals of a slightly more general form ((B1.1) and (B1.2) below) than those mentioned in (4.5) and (4.6). The conditions of admissibility for competitors in these more general variational problems are specified in Section B1; in Sections B2 and B8, we show how to convert the extremum problems for the functionals into eigenvalue problems. The relations among these problems are discussed in Section B9. Throughout this book we must consider the effect of varying parameters on the eigenvalues. Some of the more important mathematical methods used in such studies are explained in the study of the dependence of eigenvalues on the boundary and domain parameters (Section B4). One result is that when $\mathbf{u}|_S = 0$, then $\hat{\Lambda}(\mathscr{V})$ is a monotone decreasing function of the domain so that $\hat{\Lambda}(\mathscr{V}_1) < \hat{\Lambda}(\mathscr{V})$ when $\mathscr{V}$ can be entirely contained in $\mathscr{V}_1$. One can, therefore, estimate $\hat{\Lambda}(\mathscr{V})$ with values $\hat{\Lambda}(\mathscr{V}_1)$ obtained in larger but simpler domains.

As we learned from (4.12), an estimate of $\hat{\Lambda}(\mathscr{V})$ from below gives an estimate of the energy stability limit $R_{\mathscr{E}}(\mathscr{V})$. The simplest inequality is for a strip (see inequality (4.1)). Better estimates of $\hat{\Lambda}(\mathscr{V})$ can be achieved when $\mathscr{V}$ can be contained in a tube (Section B5) or a sphere or spherical annulus (Section B7). The methods for obtaining the values of $\hat{\Lambda}$ or estimates of these values are themselves of interest and they teach techniques for handling the kinematic constraint $\operatorname{div}\mathbf{u} = 0$ for incompressible flow. For this purpose the representation theorems for poloidal and totoidal vector fields discussed in B6 are particularly useful.

B1. Decay Constants and Minimum Problems

This section concerns the two functionals

$$\hat{\Lambda}[\mathbf{u}; \mathscr{V}, \hbar] = \{2\langle \mathbf{D}[\mathbf{u}] : \mathbf{D}[\mathbf{u}] \rangle + \hbar \langle h(\mathbf{x})\, |\mathbf{u}|^2 \rangle_S\} / \langle |\mathbf{u}|^2 \rangle \qquad \text{(B1.1)}$$

and

$$\Lambda[\theta; \mathscr{V}, \mathscr{h}] = \{\langle|\nabla\theta|^2\rangle + \mathscr{h}\langle h(\mathbf{x})\theta^2\rangle_S\}/\langle\theta^2\rangle. \tag{B 1.2}$$

Here $\mathscr{V}$ is a bounded domain or a cylinder (or channel) of arbitrary cross section A, and the angle brackets are volume-averaged integrals. The boundary function $h(\mathbf{x}) \geqslant 0, \mathbf{x} \in S$ is a piecewise continuous function which is defined on the part of the boundary $S - S_1$ of $\mathscr{V}$. On S_1 we require that $\mathbf{u} = 0$ and $\theta = 0$. The condition $\mathscr{h} \to \infty$, implies that $u = \theta = 0$ at every point of the boundary S of $\mathscr{V}$. It will be convenient to use the same function $h(\mathbf{x})$ in both (B 1.1) and (B 1.2). On infinite cylinders and limiting planes, we shall require $h = \text{constant}$; but if more than one cylinder is present we can have several different constants.

Our understanding of the conditions of kinematic admissibility is set out in § 58, and we use notation H and $\mathbf{H}$ to designate the kinematically admissible functions and vectors. It will be convenient in this chapter to consider only bounded domains. Hence, when considering infinite cylinders, we shall require periodic disturbances; then the region of integration is a period cell.

The decay constants of the energy estimates for monotonic stability are given by minimum values for (B 1.1, 2). These are

$$\hat{\Lambda}_1(\mathscr{V}, \mathscr{h}) = \min_{\mathbf{u} \in \mathbf{H}} \hat{\Lambda}[\mathbf{u}; \mathscr{V}, \mathscr{h}] = \hat{\Lambda}[\mathbf{u}_1; \mathscr{V}, \mathscr{h}] \tag{B 1.3}$$

and

$$\Lambda_1(\mathscr{V}, \mathscr{h}) = \min_{\theta \in H} \Lambda[\theta; \mathscr{V}, \mathscr{h}] = \Lambda[\theta_1; \mathscr{V}, \mathscr{h}]. \tag{B 1.4}$$

These values appear also as eigenvalues of differential equations. There are a countably infinite set of such eigenvalues, and they may all be characterized as minimum problems. For example,

$$\Lambda_2 = \min_{\substack{\theta \in H \\ \langle\theta\theta_1\rangle = 0}} \Lambda[\theta; \mathscr{V}, \mathscr{h}] = \Lambda[\theta_2; \mathscr{V}, \mathscr{h}]. \tag{B 1.5}$$

The principal eigenvalues Λ_1 and $\hat{\Lambda}_1$ will be called decay constants. The higher eigenvalues are not central to the stability problem. In two dimensions these eigenvalues (Λ_i and $\hat{\Lambda}_i$) give the frequencies of the small amplitude vibrations of an elastic membrane and an incompressible elastic plate, respectively.

The functions which are to be admitted into the competition for the minimum values of (B 1.3, 4, 5) have been called "kinematically admissible". The conditions of admissibility are discussed in § 58. We recall that the set of vector fields $\mathbf{u} \in \mathbf{H}$ over which $\hat{\Lambda}$ is to range is as follows:

$$\mathbf{H} = [\mathbf{u}\colon \operatorname{div}\mathbf{u} = 0, \mathbf{u}\cdot\mathbf{n}|_S = 0, \mathbf{u}|_{S_1} = 0, \langle 2\mathbf{D}[\mathbf{u}] : \mathbf{D}[\mathbf{u}]\rangle + \mathscr{h}\langle h|\mathbf{u}|^2\rangle < \infty]. \tag{B 1.6}$$

In the same way,

$$H = [\theta\colon \theta|_{S_1} = 0, \langle|\nabla\theta|^2\rangle + \mathscr{h}\langle h\theta^2\rangle < \infty]. \tag{B 1.7}$$

B 2. Fundamental Lemmas of the Calculus of Variations

The minimum problems (B 1.3, 4, 5) are most easily treated as eigenvalue problems. To convert the minimum problems into eigenvalue problems defined by differential (Euler) equations, we use two lemmas[1]:

Fundamental lemma (1) *for scalar fields: If* $\Gamma(\mathbf{x}) \in C^0(\mathscr{V})$ *is a fixed function and*

$$\langle \Gamma \phi \rangle = 0 \tag{B 2.1}$$

for every function $\phi \in C^2(\mathscr{V})$ *which vanishes on the boundary S of* $\mathscr{V}$, *then* $\Gamma(\mathbf{x}) \equiv 0$ *in* $\mathscr{V}$.

Fundamental lemma (2) *for solenoidal vector fields: If* $\boldsymbol{\Gamma}(\mathbf{x}) \in C^1(\mathbf{x})$ *and if*

$$\langle \boldsymbol{\Gamma} \cdot \boldsymbol{\phi} \rangle = 0 \tag{B 2.2}$$

for every solenoidal vector field $\boldsymbol{\phi} \in C^3(\mathscr{V})$ *such that* $\boldsymbol{\phi} \cdot \mathbf{n} = 0$ *on S, then there exists a single-valued potential p(x) such that*

$$\boldsymbol{\Gamma} = -\nabla p\,. \tag{B 2.3}$$

The first lemma is proved in nearly every standard work on variational calculus.

To prove the second lemma, choose a $C^1(\mathscr{V})$ vector field $\boldsymbol{\Xi}$ which vanishes on S. Then, it must be that $\mathbf{n} \cdot \operatorname{curl} \boldsymbol{\Xi} = 0$ on S. To see this, draw an arbitrary circuit on S and form the line integral on the circuit $\oint \boldsymbol{\Xi} \cdot d\mathbf{x} = \int \mathbf{n} \cdot \operatorname{curl} \boldsymbol{\Xi} = 0$ in any neighborhood of any point $\mathbf{x} \in S$. Now choose $\boldsymbol{\phi} \equiv \operatorname{curl} \boldsymbol{\Xi}$. Clearly, $\operatorname{div} \boldsymbol{\phi} \equiv 0$ and $\boldsymbol{\phi} \cdot \mathbf{n}|_S = 0$. Let $\mathscr{V}$ be simply connected. Then, since $\boldsymbol{\Xi}|_S = 0$

$$0 = \langle \boldsymbol{\Gamma} \cdot \boldsymbol{\phi} \rangle = \langle \boldsymbol{\Gamma} \cdot \operatorname{curl} \boldsymbol{\Xi} \rangle = \langle \boldsymbol{\Xi} \cdot \operatorname{curl} \boldsymbol{\Gamma} \rangle \tag{B 2.4}$$

and since $\boldsymbol{\Xi}$ is arbitrary (and not necessarily solenoidal), Lemma 1 applies to each of the three scalar integrals of (B 2.4). Thus, $\operatorname{curl} \boldsymbol{\Gamma}|_{\mathscr{V}} = 0$ and

$$\boldsymbol{\Gamma} = -\nabla p\,. \tag{B 2.5}$$

Eq. (B 2.5) holds for every simply-connected part of $\mathscr{V}$. Suppose $\mathscr{V}$ is a doubly-connected region with a closed boundary S, and let $\mathscr{V}''$ be the simply-connected region with closed boundary $S - S''$ obtained from $\mathscr{V}$ by insertion of the barrier S'' (see Fig. B. 1). Using the divergence theorem in the simply-connected region, we find that

$$0 = \int_{\mathscr{V}''} \boldsymbol{\Gamma} \cdot \boldsymbol{\phi} = -\int_{\mathscr{V}''} \nabla p \cdot \boldsymbol{\phi} = -\int_{S-S''} \mathbf{n} \cdot \boldsymbol{\phi} p$$

---

[1] The regularity requirements stated in these lemmas are sufficient to guarantee that stationary points of functionals are smooth functions which solve Euler's differential equation. It is possible to develop theory for weak solutions of Euler's functional equation (cf. Eq. B 8.6). In the weak solution formulation the assumptions about regularity of $\Gamma(x)$ and $\phi(x)$ may be relaxed (see Morrey, 1966).

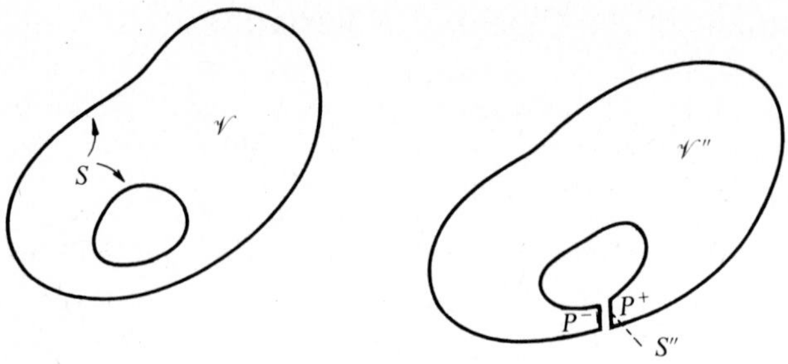

Fig. B1: The boundary of $\mathscr{V}$ is S. The boundary of $\mathscr{V}''$ is $S-S''$. The pressure is allowed to be discontinuous across S''; it is then shown to be continuous

for every solenoidal $\boldsymbol{\phi}$. Since $\boldsymbol{\phi}\cdot\mathbf{n}=0$ on S the last integral reduces to an integral over the barrier S'', which, for continuous $\boldsymbol{\phi}$, has the form

$$0=\int_{S-S''}\mathbf{n}\cdot\boldsymbol{\phi}p=\int_{S''}\mathbf{n}\cdot\boldsymbol{\phi}(p^+-p^-).$$

Since S'' is an arbitrary surface and the integrand is continuous, we find that $p^+=p^-$ and p is single-valued.

B3. The Eigenvalue Problem for the Decay Constants $\hat{\Lambda}_1(\mathscr{V},\hbar)$

First we consider the minimum problem for $\hat{\Lambda}_1$. Let $\mathbf{u}_1\in\mathbf{H}$ solve (B 1.3) and consider an arbitrary admissible field $\mathbf{u}=\mathbf{u}_1+\varepsilon\boldsymbol{\phi}\in\mathbf{H}$. For any fixed $\boldsymbol{\phi}$, $\hat{\Lambda}[\mathbf{u};\mathscr{V},\hbar]=\hat{\Lambda}[\mathbf{u}_1+\varepsilon\boldsymbol{\phi};\mathscr{V},\hbar]$ is a function of ε which has a minimum when $\varepsilon=0$. At the minimum point ($\varepsilon=0$), it must be true that

$$\frac{\partial}{\partial\varepsilon}\hat{\Lambda}[\mathbf{u}_1+\varepsilon\boldsymbol{\phi};\mathscr{V},\hbar]=0.$$

We work out the differentiation, and after integrating by parts find that

$$-\langle\boldsymbol{\phi}\cdot(\nabla^2\mathbf{u}_1+\hat{\Lambda}_1\mathbf{u}_1)\rangle+\langle\boldsymbol{\phi}\cdot(2\mathbf{n}\cdot\mathbf{D}[\mathbf{u}_1]+\hbar h\mathbf{u}_1)\rangle_{S-S_1}=0. \tag{B 3.1}$$

Consider the special choice of $\boldsymbol{\phi}\in\mathbf{H}$ such that $\boldsymbol{\phi}|_S=0$ (not just on S_1). For these $\boldsymbol{\phi}$, the second integral of (B 3.1) is zero, and fundamental lemma 2 applies to the vanishing volume integral. Hence,

$$\nabla^2\mathbf{u}_1+\hat{\Lambda}_1\mathbf{u}_1=-\nabla p. \tag{B 3.2}$$

Returning once again to (B 3.1) and using (B 3.2), we find that

$$\langle\boldsymbol{\phi}\cdot(2\mathbf{n}\cdot\mathbf{D}[\mathbf{u}_1]+\hbar h\mathbf{u}_1)\rangle_S=0 \tag{B 3.3}$$

for all $\boldsymbol{\phi}\in\mathbf{H}$ and not just those which vanish on S.

To deduce boundary conditions from (B 3.3), it is easiest to replace the tangential vector $\boldsymbol{\phi}$ with a vector $\boldsymbol{\phi}'$,

$$\boldsymbol{\phi}=\boldsymbol{\phi}'-\mathbf{n}(\boldsymbol{\phi}'\cdot\mathbf{n}),$$

which is like $\boldsymbol{\phi}$ except $\boldsymbol{\phi}'\cdot\mathbf{n}$ is arbitrary on S (recall that $\boldsymbol{\phi}\cdot\mathbf{n}|_S=0$). Then,

$$\langle\phi_l'\{2D_{lj}n_j-2n_l(\mathbf{n}\cdot\mathbf{D}\cdot\mathbf{n})+\hbar h(\mathbf{u}_1)_l\}\rangle_S=0,$$

and since $\boldsymbol{\phi}'$ is now sufficiently arbitrary

$$2\mathbf{D}[\mathbf{u}_1]\cdot\mathbf{n}-2\mathbf{n}(\mathbf{n}\cdot\mathbf{D}[\mathbf{u}_1]\cdot\mathbf{n})+\hbar h\mathbf{u}_1|_{S-S_1}=0 \tag{B 3.4}$$

where we have put $\mathbf{u}_1\cdot\mathbf{n}|_S=0$ and $\boldsymbol{\phi}'=\boldsymbol{\phi}=\mathbf{u}_1=0$ on S_1. Eq. (B 3.4) expresses natural boundary conditions; natural in the sense that the kinematically admissible competitors $\mathbf{u}\in\mathbf{H}$ for the minimum (B 1.3) need not satisfy (B 3.4), but the winner $\mathbf{u}=\mathbf{u}_1$ does need to satisfy (B 3.4).

In sum, the eigenvalue problem which is equivalent to (B 1.3) is

$$\nabla^2\mathbf{u}+\hat{\Lambda}\mathbf{u}+\nabla p=0, \tag{B 3.5}$$

$$\operatorname{div}\mathbf{u}=0, \quad \mathbf{u}\cdot\mathbf{n}|_S=0, \quad \mathbf{u}|_{S_1}=0 \tag{B 3.6a}$$

and

$$2\mathbf{D}[\mathbf{u}]\cdot\mathbf{n}-2\mathbf{n}(\mathbf{n}\cdot\mathbf{D}[\mathbf{u}]\cdot\mathbf{n})+\hbar h\mathbf{u}|_{S-S_1}=0. \tag{B 3.6b}$$

In the same way but more easily, one finds that

$$\nabla^2\theta+\Lambda\theta=0, \tag{B 3.7}$$

$$\theta|_{S_1}=0 \tag{B 3.8a}$$

and

$$\frac{\partial\theta}{\partial n}+\hbar h\theta|_{S-S_1}=0, \tag{B 3.8b}$$

is equivalent to the problem (B 1.4).

The eigenvalue problem (B 3.5, 6) can also be derived using fundamental lemma 1 instead of lemma 2. But then the components of $\mathbf{u}=\mathbf{u}_1+\varepsilon\boldsymbol{\phi}$ must be in $\mathbf{H}$, and it may be true that $\operatorname{div}\mathbf{u}=\operatorname{div}\boldsymbol{\phi}\neq0$ (or equivalently, $\langle p\nabla\cdot\mathbf{u}\rangle=-\langle\boldsymbol{\phi}\cdot\nabla p\rangle\neq0$ for arbitrary $p\in C^1(\mathscr{V})$). It is then natural to seek the stationary value of

$$\hat{\Lambda}[\mathbf{u},p;\mathscr{V},\hbar]=\frac{2\langle\mathbf{D}\cdot\mathbf{D}\rangle+\langle\hbar hu^2\rangle_S-2\langle\mathbf{u}\cdot\nabla p\rangle}{\langle u^2\rangle} \tag{B 3.9}$$

for arbitrary preassigned $p \in C^1(\mathcal{V})$ and $\tilde{u}_i = u_i + \varepsilon \phi_i \in \mathbf{H}$. The vector $\tilde{\mathbf{u}}(p)$ which makes

$$\frac{\partial \tilde{A}}{\partial \varepsilon}[\mathbf{u} + \varepsilon \phi, p; \mathcal{V}, \mathscr{h}] = 0 \tag{B 3.10}$$

when $\varepsilon = 0$ will not, in general, be solenoidal, but we can select $p = \tilde{p}$ such that $\nabla \cdot \tilde{\mathbf{u}}(\tilde{p}) = 0$ and $\tilde{A}[\tilde{\mathbf{u}}, \tilde{p}; \mathcal{V}, \mathscr{h}] = \hat{A}_1[\mathbf{u}, \mathcal{V}, \mathscr{h}]$. In seeking a solution $\mathbf{u}$ for (B 3.10) and a p such that $\nabla \cdot \tilde{\mathbf{u}} = 0$, we come back to the original problem.

In the problem (B 1.5), it is necessary that the competitors θ for the minimum satisfy an orthogonality condition $\langle \theta \theta_1 \rangle = 0$. Attempting to proceed as before, we set $\theta = \theta_2 + \varepsilon \phi \in H$, where $\theta_2 \in H$ solves (B 1.5) and $\phi \in H$ is arbitrary. But ϕ cannot be arbitrary because $\langle \theta \theta_1 \rangle = \langle \phi \theta_1 \rangle = 0$. Then, the differentiation of $\Lambda_2(\theta_2 + \varepsilon \phi; \mathcal{V}, \mathscr{h})$ with respect to ε at $\varepsilon = 0$ leads to the expression

$$-\langle \phi(\nabla^2 \theta_2 + \Lambda_2 \theta_2) \rangle + \left\langle \phi \left(\frac{\partial \theta_2}{\partial n} + \mathscr{h} h \theta_2 \right) \right\rangle_s = 0;$$

but since ϕ is not arbitrary, we cannot use lemma 1.

To resolve the problem, we let $\theta = \theta_2 + \varepsilon \phi \in H$ and do not require $\langle \theta \theta_1 \rangle = 0$ but only that $\langle \theta_2 \theta_1 \rangle = 0$. Then ϕ is arbitrary but, of course, the condition $\langle \theta \theta_1 \rangle = 0$ must be dropped and (B 1.5) must be replaced with a different but equivalent problem. In the same spirit that the arbitrary function p was introduced above, one here introduces the real (Lagrange) multiplier C_1 and seeks the minimum of the functional

$$\{ \langle |\nabla \theta|^2 \rangle + \mathscr{h} \langle h \theta^2 \rangle_s + C_1 \langle \theta \theta_1 \rangle \} / \langle \theta^2 \rangle$$

among arbitrary $\theta = \theta_2 + \varepsilon \phi$. When $\varepsilon = 0$, this gives the required minimum but does not restrict the class of admissible ϕ. Proceeding as before, we find that

$$\nabla^2 \theta_2 + \Lambda_2 \theta_2 = C_1 \theta_1, \qquad \theta_2|_{S_1} = 0$$

and

$$\frac{\partial \theta_2}{\partial n} + \mathscr{h} h \theta_2|_{S - S_1} = 0.$$

To find C_1, we note that

$$C_1 \langle \theta_1^2 \rangle = \langle \theta_1 (\nabla^2 \theta_2 + \Lambda_2 \theta_2) \rangle = \langle \theta_2 \nabla^2 \theta_1 \rangle = -\Lambda_1 \langle \theta_2 \theta_1 \rangle = 0$$

and

$$C_1 = 0.$$

In the same way, one can generate the membrane equation again and again by adding additional orthogonality constraints. Moreover, $\Lambda_i \geqslant \Lambda_j$ when $i > j$

since the addition of a constraint could only raise (not lower) the minimum. In bounded domains and for periodic functions in cylinders and strips, it is known that this process of raising the eigenvalue does not get "stuck", so that the eigenvalues tend to infinity with the addition of more and more orthogonality conditions (there is no accumulation point at any finite point of the spectrum). This implies that the eigenvalues are discrete and denumerable, and that there are at most a finite number of eigenvectors for each eigenvalue. The eigenvectors can all be made orthonormal, and the resulting set of orthonormal vectors is complete.

Exercise B 3.1 (Galdi, 1975): Show that the minimum value of the ratio

$$\Lambda[\theta, l] = \int_0^l \theta'^2 \, dz / \int_0^l \theta^2 \, dz$$

among functions for which $\theta(l) = \theta(0) = 0$ is π^2/l^2. Show that $\inf \Lambda[\theta, \infty] = 0$ among functions for which $\theta(0) = \theta(\infty) = 0$.

Hint: Consider the function $e^{-2\beta z} - e^{-\beta z}$ with $\beta > 0$.

B4. Domain and Transfer Constant ($\hbar$) Dependence of the Decay Constants

First we shall show that the membrane eigenvalues increase with $\hbar$ at a decreasing rate (see Fig. B. 2). Consider the principal eigenvalue $\Lambda_1 = \Lambda$ expressed variationally

$$\Lambda(\hbar) = \min_H \{\langle|\nabla\theta|^2\rangle + \hbar\langle h\theta^2\rangle_S\}/\langle\theta^2\rangle. \tag{B 4.1}$$

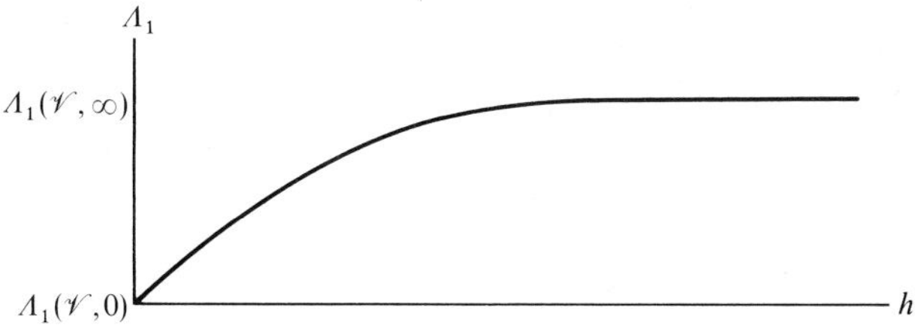

Fig. B2: Dependence of the membrane eigenvalue on the transfer coefficient h. On a perfectly conducting boundary $\hbar = \infty$; on a perfectly insulating boundary $\hbar = 0$

Let θ be the minimizing function for (B 4.1); it is also the principal eigenfunction of (B 3.7, 8). Designate derivatives with respect to $\hbar$ as $\theta^{(n)} = d^n\theta/d\hbar^n$, etc. Then $\Lambda(\hbar)$ has the following properties:

(a) *$\Lambda(\hbar)$ is an increasing function of $\hbar$*

$$\Lambda^{(1)} = \langle h\theta^2\rangle_S/\langle\theta^2\rangle, \tag{B 4.2}$$

$\Lambda(\hbar)$ *tends to a fixed value* $\Lambda(\infty)$ *as* $\hbar \to \infty$ *and*

$$\Lambda(\infty) = \langle |\nabla\theta|^2 \rangle / \langle \theta^2 \rangle, \qquad \theta|_S = 0$$

corresponding to zero boundary values on the whole boundary $(S = S_1)$.

(b) *If* $S_1 = 0$, *then on the "insulated boundary"*

$$\Lambda(0) = 0.$$

(c) $\Lambda^{(1)}(\hbar)$ *is a decreasing function of* $\hbar$ *and*

$$\Lambda^{(2)}(\hbar) = \frac{2\langle |\theta^{(1)}|^2 \rangle}{\langle \theta^2 \rangle} \left\{ \Lambda - \frac{\langle |\nabla\theta^{(1)}|^2 \rangle + \hbar \langle h |\theta^{(1)}|^2 \rangle_S}{\langle |\theta^{(1)}|^2 \rangle} \right\} < 0. \tag{B 4.3}$$

If both $\hbar$ and S_1 are zero, then $\partial\theta/\partial n = 0$ on the whole boundary and the unique solution of (B 3.7, 8) is $\theta = \text{const.} \neq 0$ when $\Lambda = 0$. To prove (B 4.2) and (B 4.3), consider the boundary value problems generated from (B 3.7, 8) by differentiation with respect to $\hbar$. At the boundary on S_1

$$\theta^{(1)} = \theta^{(2)} = 0 \tag{B 4.4a}$$

and on $S - S_1$

$$\frac{\partial\theta^{(1)}}{\partial n} + \hbar h \theta^{(1)} + h\theta = \frac{\partial\theta^{(2)}}{\partial n} + \hbar h \theta^{(2)} + 2h\theta^{(1)} = 0. \tag{B 4.4b}$$

Form scalar products from the equations for $\theta^{(1)}$ and $\theta^{(2)}$ and integrate by parts using (B 3.7, 8) and (B 4.4) to find

$$\begin{aligned} 0 &= \langle \theta(\nabla^2\theta^{(1)} + \Lambda\theta^{(1)} + \Lambda^{(1)}\theta) \rangle = \left\langle \frac{\partial\theta^{(1)}}{\partial n}\theta - \frac{\partial\theta}{\partial n}\theta^{(1)} \right\rangle_S \\ &\quad + \Lambda^{(1)}\langle \theta^2 \rangle = -\langle h\theta^2 \rangle_S + \Lambda^{(1)}\langle \theta^2 \rangle, \end{aligned} \tag{B 4.5}$$

$$\begin{aligned} 0 &= \langle \theta^{(1)}(\nabla^2\theta^{(1)} + \Lambda\theta^{(1)} + \Lambda^{(1)}\theta) \rangle \\ &= \langle |\theta^{(1)}|^2 \rangle \left\{ -\frac{\langle |\nabla\theta^{(1)}|^2 \rangle}{\langle |\theta^{(1)}|^2 \rangle} + \Lambda \right\} + \Lambda^{(1)}\langle \theta\theta^{(1)} \rangle + \left\langle \theta^{(1)} \frac{\partial\theta^{(1)}}{\partial n} \right\rangle_S \\ &= \Lambda^{(1)}\langle \theta\theta^{(1)} \rangle - \hbar\langle h\theta\theta^{(1)} \rangle_S + \langle |\theta^{(1)}|^2 \rangle \left\{ \Lambda - \frac{\langle |\nabla\theta^{(1)}|^2 \rangle + \hbar\langle h|\theta^{(1)}|^2 \rangle_S}{\langle |\theta^{(1)}|^2 \rangle} \right\}, \end{aligned} \tag{B 4.6}$$

$$\begin{aligned} 0 &= \langle \theta(\nabla^2\theta^{(2)} + \Lambda\theta^{(2)} + 2\Lambda^{(1)}\theta^{(1)} + \Lambda^{(2)}\theta) \rangle \\ &= \left\langle \frac{\partial\theta^{(2)}}{\partial n}\theta - \frac{\partial\theta}{\partial n}\theta^{(2)} \right\rangle_S + 2\Lambda^{(1)}\langle \theta^{(1)}\theta \rangle + \Lambda^{(2)}\langle \theta^2 \rangle \\ &= 2\{\Lambda^{(1)}\langle \theta\theta^{(1)} \rangle - \hbar\langle h\theta\theta^{(1)} \rangle_S\} + \Lambda^{(2)}\langle \theta^2 \rangle. \end{aligned} \tag{B 4.7}$$

Eq. (B 4.5) implies (B 4.2). Eq. (B 4.3) is obtained by combining (B 4.6) and (B 4.7). The inequality follows from the fact that $\theta^{(1)}$ is admissible in H for competition for the minimum in B 4.1. Hence, the ratio in the squiggly bracket of (B 4.3) is not smaller than Λ. It is actually larger because $\theta^{(1)}$ cannot be an eigenfunction of Λ.

We shall now study the variation of the eigenvalue $\hat{\Lambda}$ with the domain. Let the plate eigenvalue problem

$$\{\nabla^2\mathbf{u}+\hat{\Lambda}\mathbf{u}-\nabla p=0,\ \nabla\cdot\mathbf{u}=0,\ \mathbf{u}|_S=0\} \tag{B 4.8}$$

hold identically in a one-parameter (τ) family of domains $\mathscr{V}(\tau)$. We may think of τ as the time and form a substantial derivative

$$\frac{d\mathbf{u}}{d\tau}=\frac{\partial\mathbf{u}}{\partial\tau}+\mathbf{v}\cdot\nabla\mathbf{u}=\mathbf{u}_\tau+\mathbf{v}\cdot\nabla\mathbf{u} \tag{B 4.9}$$

where $\mathbf{v}=d\mathbf{x}/d\tau$ is the "velocity" of the deformation.

Since (B 4.8) is an identity in τ, so too is $d^n/d\tau^n\{\quad\}=0$ an identity. Observing that

$$\mathbf{v}\cdot\nabla\{\nabla^2\mathbf{u}+\hat{\Lambda}\mathbf{u}-\nabla p\}=0 \quad \text{and} \quad (\mathbf{v}\cdot\nabla)\nabla\cdot\mathbf{u}=0,$$

and with $n=1$, we find that

$$\nabla^2\mathbf{u}_\tau+\hat{\Lambda}\mathbf{u}_\tau-\nabla p_\tau+\mathring{\hat{\Lambda}}\mathbf{u}=0, \qquad \nabla\cdot\mathbf{u}_\tau=0, \qquad \mathbf{u}_\tau=-\mathbf{v}\cdot\nabla\mathbf{u}|_S \tag{B 4.10}$$

where $\mathring{\hat{\Lambda}}=d\hat{\Lambda}/d\tau$.

We find, using (B 4.9), (B 4.10) and (B 4.15) that

$$0=\langle\mathbf{u}\cdot(\nabla^2\mathbf{u}_\tau+\hat{\Lambda}\mathbf{u}_\tau-\nabla p_\tau+\mathring{\hat{\Lambda}}\mathbf{u})\rangle=-\left\langle\mathbf{u}_\tau\cdot\frac{\partial\mathbf{u}}{\partial n}\right\rangle_S+\mathring{\hat{\Lambda}}\langle|\mathbf{u}|^2\rangle.$$

Since $\mathbf{u}|_S\equiv 0$, $\mathbf{u}_\tau=-\mathbf{v}\cdot\nabla\mathbf{u}=-v_n\,\partial\mathbf{u}/\partial n|_S$, where $v_n=\mathbf{n}\cdot\mathbf{v}$, we have

$$\mathring{\hat{\Lambda}}=-\left\langle v_n\left|\frac{\partial\mathbf{u}}{\partial n}\right|^2\right\rangle_S\Big/\langle|\mathbf{u}|^2\rangle. \tag{B 4.11}$$

Eq. (B 4.11) is Hadamard's (1908) formula, but here it is proved for plate eigenvalue rather than membrane eigenvalue. It shows that $\hat{\Lambda}(\mathscr{V})$ *decreases as the domain is enlarged* ($v_n>0$). Another way to state this is:

$$\hat{\Lambda}(\mathscr{V}_2)<\hat{\Lambda}(\mathscr{V}_1) \tag{B 4.12}$$

whenever $\mathscr{V}_1$ *can be entirely contained in* $\mathscr{V}_2$. This means we may estimate $\hat{\Lambda}$ from below when the domain is complicated by calculating $\hat{\Lambda}$ in a larger domain of simple shape.

Exercise B4.1: (a) Show that the higher membrane eigenvalues are monotone and convex in $\mathscr{h}$. (b) Show that the plate eigenvalues $\hat{\Lambda}$ are monotone and convex in $\mathscr{h}$.

Exercise B4.2: Consider the membrane problem (B3.7,8). Prove the generalized Hadamard formula

$$\dot{\Lambda}\langle\theta^2\rangle=\langle v_n\{(\partial_i\theta)^2-\Lambda\theta^2\}\rangle_S-2\left\langle(\mathbf{v}\cdot\nabla\theta)\frac{\partial\theta}{\partial n}\right\rangle_S$$
$$+\mathscr{h}\langle\{(\nabla_2\cdot\mathbf{v})+\mathbf{v}\cdot\nabla\mathscr{h}\}\breve{\theta}^2\rangle_S \tag{B4.13}$$

where

$$\nabla_2\cdot\mathbf{v}=(\delta_{ij}-n_in_j)\partial_jv_i$$

is the surface dilatation. When $\theta_{|S}=0$ show that (B4.13) reduces to Hadamard's formula. If $\mathscr{h}=0$, show that there are domain enlarging transformations $v_n>0$ for which Λ may increase.

Exercise B4.3: Let $\mathbf{u}=0$ and $\theta=0$ on S. Note that $2\langle\mathbf{d}:\mathbf{d}\rangle=\langle|\nabla\mathbf{u}|^2\rangle$ and compare Λ and $\hat{\Lambda}$. Prove that

$$\hat{\Lambda}_1\geqslant\Lambda_1\,. \tag{B4.14}$$

Exercise B4.4 (Velte, 1962): Consider the variational problem for the plate eigenvalue $\hat{\Lambda}_1$ when A is a simply-connected region of the plane. Show that this eigenvalue is also an eigenvalue of the problem

$$\nabla^4\psi+\Lambda\nabla^2\psi=0\,,\qquad \psi=\partial\psi/\partial n=0|_{\partial A}$$

which governs the buckling of a thin elastic plate which is fixed and clamped on $S(A)$.

Exercise B4.5: Show that at each point $\mathbf{x}$ on the (smooth) boundary S of $\mathscr{V}$ every solenoidal field $\mathbf{u}$ which vanishes on S has

$$\mathbf{n}\cdot\nabla(\mathbf{u}\cdot\mathbf{n})\equiv0\quad\text{for}\quad\mathbf{x}\in S\,. \tag{B4.15}$$

Hint: Consider local coordinates in a neighborhood of $\mathbf{x}\in S$ and decompose vectors into normal and tangential components.

Exercise B4.6: Suppose that $\mathscr{G}[\boldsymbol{\theta}(\mathbf{x}),\alpha]$ is a (sufficiently regular) functional whose domain is $\boldsymbol{\theta}\in\mathbf{H}$ where $\mathbf{H}$ is a suitably defined class of functions and α is a parameter. Suppose further that $\mathscr{G}[\tilde{\boldsymbol{\theta}}(\mathbf{x},\alpha),\alpha]$ is stationary in the sense that

$$\frac{\partial}{\partial\varepsilon}\mathscr{G}[\tilde{\boldsymbol{\theta}}+\varepsilon\boldsymbol{\theta}',\alpha]=0$$

when $\varepsilon=0$, $\boldsymbol{\theta}'$ is any admissible vector field and that $\partial\tilde{\boldsymbol{\theta}}/\partial\alpha\in\mathbf{H}$. Show that

$$\frac{d}{d\alpha}\mathscr{G}[\tilde{\boldsymbol{\theta}},\alpha]=\frac{\partial}{\partial\alpha}\mathscr{G}[\tilde{\boldsymbol{\theta}},\alpha] \tag{B4.16}$$

where the partial derivative means that the dependence of $\tilde{\boldsymbol{\theta}}$ on α is to be ignored in the differentiation. Use this to prove (B4.2).

Exercise B4.7: Suppose $f(\mathbf{x},\tau)=0$ identically for $\mathbf{x}\in\bar{\mathscr{V}}(\tau)$ when τ is in some open interval $\mathscr{I}$. Show that $\partial^nf/\partial\tau^n=0$ when $\mathbf{x}\in\bar{\mathscr{V}}$ and $\tau\in\mathscr{I}$. Suppose that $\nabla^2\phi=0$ in $\mathscr{V}(\tau)$ for $\tau\in\mathscr{I}$ and show that $\nabla^2(\partial^n\phi/\partial\tau^n)=0$. Suppose $f(\mathbf{x},\tau)=0$ when $\mathbf{x}\in S(\tau)$ for $\tau\in\mathscr{I}$. Show that

$$\left(\frac{\partial}{\partial\tau}+\frac{d\mathbf{x}}{d\tau}\cdot\nabla\right)^n f=0,\ \mathbf{x}\in S(\tau).$$

Develop an algorithm for computing the solution of (B4.8) as a power series in τ. Show that the series is independent of the interior values of the position vector $\mathbf{x}(\mathbf{x}_0,\tau)$ of a point in $\mathscr{V}(\tau)$ relative to a reference domain $\mathscr{V}_0$, $\mathbf{x}_0\in\mathscr{V}_0$, $\mathbf{x}_0=\mathbf{x}(x_0,0)$.

B 5. The Decay Constant $\hat{\Lambda}_1$ in a Box

It is frequently possible to simplify variational problems by relaxing side constraints. The price paid for this simplicity is dear. Without the original side constraint the exact extreme value may not be attained; instead one gets only an estimate of the extreme value. The solenoidality constraint is interesting to consider in this regard; if we drop $\operatorname{div}\mathbf{u}=0$ althogether from the definition of $\mathbf{H}$, then the minimum value of $\hat{\Lambda}$ among $\mathbf{u}$ with $\mathbf{u}|_{S_1}=0$ is just $\hat{\Lambda}_1$. In complicated situations one can find a value in this trade off of precision for simplicity.

Here we are going to exhibit a nice procedure (Velte, 1962) in which the constraint $\operatorname{div}\mathbf{u}=0$ is relaxed, but only slightly. The problem is (B 1.3) when $\mathscr{V}=B$ is a box of length l_1, depth l_2 and height l_3. Velte treated the cube $l_1=l_2=l_3$. In the general case we shall show that

$$\frac{4}{l_1^2}+\frac{1}{l_2^2}+\frac{1}{l_3^2}\leqslant\frac{\hat{\Lambda}_1(B)}{\pi^2}\leqslant\frac{\{\frac{16}{3}+4(l_2^2+l_3^2)/l_1^2+(5l_1^2+l_3^2)/l_2^2+(5l_1^2+l_2^2)/l_3^2\}}{(2l_1^2+l_2^2+l_3^2)} \tag{B 5.1}$$

where

$$\hat{\Lambda}_1(B)=\min\{\langle|\nabla\mathbf{u}|^2\rangle/\langle|\mathbf{u}|^2\rangle,\ \nabla\cdot\mathbf{u}=0,\ \mathbf{u}|_S=0\}. \tag{B 5.2}$$

Proof: If we allow more functions to compete for the minimum, we may lower it. Hence, we shall have $\Lambda^*\leqslant\hat{\Lambda}_1$ where

$$\Lambda^*=\min\{\langle|\nabla\mathbf{u}|^2\rangle/\langle|\mathbf{u}|^2\rangle,\langle\mathbf{u}\rangle=0,\ \mathbf{u}|_S=0\}$$

provided that we can show that all functions satisfying $\operatorname{div}\mathbf{u}=0$ also satisfy $\langle\mathbf{u}\rangle=0$. This is clearly true since $0=\langle x_j\operatorname{div}\mathbf{u}\rangle=-\langle u_j\rangle$. Define

$$\Gamma_1=\min\{\langle|\nabla\phi|^2\rangle/\langle\phi^2\rangle,\ \langle\phi\rangle=0,\ \phi|_S=0\} \tag{B 5.3}$$

and note that if $\tilde{\mathbf{u}}$ is the minimizing vector for Λ^*, then $\langle|\nabla u_i|^2\rangle\geqslant\Gamma_1\langle u_i^2\rangle$ (for each i) so that by adding, we find that $\Lambda^*\geqslant\Gamma_1$. Then $\Gamma_1\leqslant\hat{\Lambda}_1(B)$. It needs to be shown that Γ_1 is the value on the left of (B 5.1).

To find Euler's equation for (B 5.3), one considers the minimum value of the functional

$$I(\varepsilon)=\langle(|\nabla\phi|^2+a\phi)\rangle/\langle\phi^2\rangle \tag{B 5.4}$$

over functions $\phi=\phi_0+\varepsilon\eta$, where ε and the Lagrange multiplier "a" are real numbers, ϕ vanishes at the boundary and ϕ_0 solves the minimum problem (B 5.3). Of course $\langle\phi_0\rangle=0$, but this property is not necessarily shared by η, which is now sufficiently arbitrary for the fundamental lemma 1 to be applied to the relation

$$\langle\eta(\nabla^2\phi_0+\Gamma\phi_0-a)\rangle=0.$$

This arises from the minimum condition $\partial I(0)/\partial\varepsilon = 0$ for each fixed (but arbitrary) η.

In this way, one arrives at the eigenvalue problem

$$\nabla^2\phi_0 + \Gamma\phi_0 = a\,,$$
$$\phi_0|_{S(B)} = 0\,, \qquad \text{(B 5.5)}$$

and

$$\langle\phi_0\rangle = 0\,. \qquad \text{(B 5.6)}$$

There are two cases to consider:

(1) $a = 0$. Then Γ is an eigenvalue of (B 5.5). The eigenfunctions have the form

$$\sin\frac{l\pi x_1}{l_1}\sin\frac{m\pi x_2}{l_2}\sin\frac{n\pi x_3}{l_3} \qquad \text{(B 5.7)}$$

with corresponding eigenvalues $\pi^2\{l^2/l_1^2 + m^2/l_2^2 + n^2/l_3^2\} \equiv \Gamma_{lmn}$. Without loss of generality, we take $l_1 \geqslant l_2 \geqslant l_3$. The smallest of the values $\Gamma_{lmn} = \Gamma_{111}$ is not an eigenvalue of (B 5.5) and (B 5.6) because its eigenfunctions are of one sign on B. The second of the values Γ_{lmn},

$$\Gamma_{211} = \pi^2\left\{\frac{4}{l_1^2} + \frac{1}{l_2^2} + \frac{1}{l_3^2}\right\},$$

is an eigenvalue of (B 5.5) and (B 5.6).

(2) $a \neq 0$. The minimizing function belongs to a value of Γ which is not an eigenvalue. We show that any such value is larger than Γ_{211}. Any solution of (B 5.5) can be represented by the Fourier series

$$\phi_0 = \sum_{l,m,n=1}^{\infty} A_{lmn}\sin\frac{l\pi x_1}{l_1}\sin\frac{m\pi x_2}{l_2}\sin\frac{n\pi x_3}{l_3}$$

where

$$A_{lmn} = a\left(\frac{4}{\pi}\right)^3\left[\Gamma - \left(\frac{l^2}{l_1^2} + \frac{m^2}{l_2^2} + \frac{n^2}{l_3^2}\right)\pi^2\right]^{-1}[lmn]^{-1}$$

when l, m and n are odd and $A_{lmn} = 0$ in all other cases. The side condition $\langle\phi_0\rangle = 0$ then leads to a relation for $\Gamma(l,m,n$ odd):

$$F(\Gamma) = \sum_{l,m,n=1}\left[\Gamma - \left(\frac{l^2}{l_1^2} + \frac{m^2}{l_2^2} + \frac{n^2}{l_3^2}\right)\pi^2\right]^{-1}(lmn)^{-2} = 0\,. \qquad \text{(B 5.8)}$$

Obviously, this relation does not hold when $\Gamma \leqslant \Gamma_{111}$. It cannot hold when $\Gamma_{111} \leqslant \Gamma \leqslant \Gamma_{211}$ for it is easily verified that $F(\Gamma_{211}) > 0$. When $\Gamma = \Gamma_{211}$ the first term of (B 5.8) is positive, and the rest of terms (all negative) have a sum less that the first term. In the interval $\Gamma_{111}<\Gamma<\Gamma_{211}$, the first term of (B 5.8) is even larger, and the other negative terms are smaller in absolute value than when $\Gamma=\Gamma_{211}$. This proves the left inequality of (B 5.1).

An upper bound for $\hat{\Lambda}_1$ can be found by inserting any vector of the admissible class into the functional (B 5.2). The vector

$$
\begin{aligned}
-u_1 &= \left(\sin\frac{\pi x_1}{l_1}\right)^2 \left(\sin\frac{\pi x_3}{l_3}\sin\frac{2\pi x_2}{l_2}+\sin\frac{\pi x_2}{l_2}\sin\frac{2\pi x_3}{l_3}\right),\\
u_2 &= \frac{l_2}{l_1}\left(\sin\frac{\pi x_2}{l_2}\right)^2 \sin\frac{\pi x_3}{l_3}\sin\frac{2\pi x_1}{l_1},\\
u_3 &= \frac{l_3}{l_1}\left(\sin\frac{\pi x_3}{l_3}\right)^2 \sin\frac{2\pi x_1}{l_1}\sin\frac{\pi x_2}{l_2}
\end{aligned}
$$

is solenoidal and vanishes at the boundary of the box. This field is inserted in (B 5.2) to obtain the inequality on the right of (B 5.1).

Exercise B5.1: Each of the six sides of a free surface box is a free surface on which the normal velocity and shear stress vanish.

(a) Verify that (B3.6b) gives the vanishing shear stress condition. Show that on a face whose normal is $x_I (I=1,2,3)$ (B3.6b) and $\operatorname{div}\mathbf{u}=0$ together imply that

$$u_I = \frac{\partial^2 u_I}{\partial x_I^2} = 0. \tag{B5.9}$$

(b) Show that the decay constant $\hat{\Lambda}_1(B,0)$ for the free surface box has

$$\hat{\Lambda}_1(B,0)=\pi^2\left(\frac{1}{l_1^2}+\frac{1}{l^2}\right). \tag{B5.10}$$

where l_1 is the length of the side along x_1 and $l_2=l_3=l$.

Hint: Show first that $\mathbf{n}\cdot\nabla^2\mathbf{u}=-\partial p/\partial n$ on the boundary. Deduce that p is constant in B. Using these results, solve (B3.5, 6).

Any domain which can be contained in a box can be contained in some sphere. The value of $\hat{\Lambda}_1$ can be obtained exactly in the sphere without relaxing the solenoidality constraint. We treat the sphere problem in (Section B 7) using a general representation theorem for solenoidal fields described in Section B 6.

B6. Representation Theorem for Solenoidal Fields

A solenoidal vector field is composed of three scalar fields which satisfy a first-order differential equation, $\operatorname{div}\mathbf{u}=0$. It is sometimes possible to "integrate" this equation and to reduce the number of defining scalar fields to two: the "toroidal" and "poloidal" potential fields. Arbitrary solenoidal vector fields can be obtained from these two "potential" fields by differentiation.

Below, we have given the form of the decomposition into toroidal and poloidal fields in forms appropriate for application.

Toroidal and poloidal fields in the spherical annulus: Let $\mathbf{u}\in C^1(\mathscr{V}(\eta))$ *where* $\mathscr{V}(\eta)$ *is a spherical annulus and* $\int_{S(r)}\mathbf{r}\cdot\mathbf{u}=0$ *on each sphere* $S(r)$, $r^2=x_1^2+x_2^2+x_3^2=\text{const.}$ *Then* $\mathbf{u}$ *has the unique decomposition*

$$\mathbf{u}=\mathbf{u}_1+\mathbf{u}_2, \tag{B 6.1}$$

where

$$\operatorname{div}\mathbf{u}_1=\mathbf{r}\cdot\operatorname{curl}\mathbf{u}_1=0, \tag{B 6.2}$$

and

$$\operatorname{div}\mathbf{u}_2=\mathbf{r}\cdot\mathbf{u}_2=0. \tag{B 6.3}$$

Moreover, there exist poloidal and toroidal "potentials" χ *and* Ψ *such that*

$$\mathbf{u}_1=\boldsymbol{\delta}\chi=\operatorname{grad}\left(r\frac{\partial\chi}{\partial r}+\chi\right)-\mathbf{r}\nabla^2\chi\equiv\operatorname{curl}^2(\mathbf{r}\chi) \tag{B 6.4}$$

and

$$\mathbf{u}_2=\mathbf{r}\wedge\operatorname{grad}\Psi\equiv-\operatorname{curl}(\mathbf{r}\Psi). \tag{B 6.5}$$

The potentials are unique to within an additive function of r alone.

The decomposition of solenoidal fields in spheres is discussed in detail in the book of Chandraskhar (1961). The proof which is given below is due to Payne and Weinberger (private communication).

We note that $\mathbf{u}$ is not affected if Ψ and χ are made unique by requiring that

$$0=\langle f(r)\chi\rangle=\langle f(r)\Psi\rangle \tag{B 6.6}$$

for arbitrary functions $f(r)$ of r alone.

Proof of the decomposition in spheres: The second of the conditions (B 6.2) means that the tangential part of $\mathbf{u}_1$ is a gradient on each sphere $x_1^2+x_2^2+x_3^2=\text{const.}$

Hence, $\mathbf{u}_1$ can be written as

$$\mathbf{u}_1=\operatorname{grad}\phi+\Psi\mathbf{r}. \tag{B 6.7}$$

Since $\operatorname{div}\mathbf{u}_1=0$, we must have

$$\operatorname{div}\mathbf{u}_1=\nabla^2\phi+\mathbf{r}\cdot\operatorname{grad}\Psi+3\Psi=0. \tag{B 6.8}$$

Define a function χ by the relation $\phi=\dfrac{\partial}{\partial r}(r\chi)$. The relation

$$\nabla^2\phi\equiv\mathbf{r}\cdot\operatorname{grad}(\nabla^2\chi)+3\nabla^2\chi \tag{B 6.9}$$

is then an identity and, by comparison of (B 6.8) and (B 6.9), we note that we can have solenoidal $\mathbf{u}_1$ of the form (B 6.7) if we take $\Psi = -\nabla^2\chi$. Thus,

$$\mathbf{u}_1 = \boldsymbol{\delta}\chi \equiv \operatorname{grad}\left(r\frac{\partial\chi}{\partial r} + \chi\right) - \mathbf{r}\nabla^2\chi. \tag{B 6.10}$$

It must be shown that χ can be determined uniquely if arbitrary solenoidal $\mathbf{u}$ is given. Since $\mathbf{u}_2 \cdot \mathbf{r} = 0$, we have $\mathbf{r}\cdot\mathbf{u}_1 = \mathbf{r}\cdot\mathbf{u}$ which, using (B 6.10), may be written in polar spherical coordinates (r, θ, ϕ)

$$\nabla_S^2\chi \equiv r^2\left[\nabla^2\chi - \frac{\partial^2\chi}{\partial r^2} - \frac{2}{r}\frac{\partial\chi}{\partial r}\right] = \left[\frac{1}{\sin\theta}\frac{\partial}{\partial\theta}\sin\theta\frac{\partial\chi}{\partial\theta} + \frac{1}{\sin^2\theta}\frac{\partial^2\chi}{\partial\phi^2}\right] = -\mathbf{r}\cdot\mathbf{u}. \tag{B 6.11}$$

This equation is integrable provided that the right side is orthogonal to solutions of the homogeneous problem (functions $f(r)$ of r alone) over each sphere, $r = \text{const}$. If there is no source of "mass", then (B 6.11) is integrable. The function χ is determined to within an additive function of r alone, and this does not contribute to the form (B 6.10) of $\mathbf{u}_1$.

To determine $\mathbf{u}_2$, note that one can always find $\mathbf{q}$ such that

$$\mathbf{u}_2 = \mathbf{r} \wedge \mathbf{q},$$

and since

$$\operatorname{div}\mathbf{u}_2 = 0,$$

$$\mathbf{r}\cdot\operatorname{curl}\mathbf{q} = 0.$$

As in the earlier argument, $\mathbf{q}$ can be represented as a gradient on spheres, and $\mathbf{q} = \operatorname{grad}\Psi + \mathbf{r}\hat{\psi}$, so that

$$\mathbf{u}_2 = \mathbf{r} \wedge \operatorname{grad}\Psi. \tag{B 6.12}$$

It must be shown that given arbitrary solenoidal $\mathbf{u}$, Ψ can be uniquely determined. Since $\mathbf{r}\cdot\operatorname{curl}\mathbf{u}_1 = 0$, we must have $\mathbf{r}\cdot\operatorname{curl}\mathbf{u}_2 = \mathbf{r}\cdot\operatorname{curl}\mathbf{u}$. Using (B 6.12), one finds that

$$\nabla_S^2\Psi = \mathbf{r}\cdot\operatorname{curl}\mathbf{u}. \tag{B 6.13}$$

As before, the nature of the LHS of (B 6.13) is such that the RHS must satisfy the condition $\int_S \mathbf{r}\cdot\operatorname{curl}\mathbf{u} = 0$ on each sphere, $r = \text{const}$. This relation expresses Stokes' theorem on closed spheres and holds for any $\mathbf{u}$. It is clear that Ψ is determined by (B 6.13) only up to an additive function of r alone, but that $\mathbf{u}_2$ is determined uniquely by Ψ.

The representation theorem for the spherical annulus has a corollary for plane layers. Let $P(\alpha, \beta)$ be a period cell

$$0 \leqslant z \leqslant d, \quad x_0 \leqslant x \leqslant x_0 + 2\pi/\alpha, \quad y_0 \leqslant y \leqslant y_0 + 2\pi/\beta$$

where x_0, y_0, α, β are arbitrary. $\mathscr{C}(\alpha,\beta)$ is the intersection of $P(\alpha,\beta)$ and plane $z=$const. The boundary $\partial\mathscr{C}(\alpha,\beta)$ is a closed curve of length l with unit tangent vector $\boldsymbol{\tau}$.

Toroidal and poloidal fields in a period cell: Let $\mathbf{u}$ *be a solenoidal, doubly-periodic vector with periods* $2\pi/\alpha$ *and* $2\pi/\beta$ *and such that*

$$0=\int\int_{\mathscr{C}(\alpha,\beta)} \mathbf{e}_z\cdot\mathbf{u}(x,y,0)\,dx\,dy\,. \tag{B 6.14}$$

Then $\mathbf{u}=\mathbf{u}_1+\mathbf{u}_2$ *is the sum of a poloidal and a toroidal field such that*

$$\mathbf{e}_z\cdot\operatorname{curl}\mathbf{u}_1=\mathbf{e}_z\cdot\mathbf{u}_2=\operatorname{div}\mathbf{u}_1=\operatorname{div}\mathbf{u}_2=0, \tag{B 6.15}$$

$$\mathbf{u}_1=\boldsymbol{\delta}\chi\equiv\operatorname{grad}\left(\frac{\partial\chi}{\partial z}\right)-\mathbf{e}_z\nabla^2\chi\equiv\operatorname{curl}^2\mathbf{e}_z\chi\,, \tag{B 6.16}$$

and

$$\mathbf{u}_2=\mathbf{e}_z\wedge\operatorname{grad}\Psi\equiv-\operatorname{curl}\mathbf{e}_z\Psi \tag{B 6.17}$$

where χ *and* Ψ *are doubly periodic and satisfy the equations*

$$\nabla_2^2\chi\equiv\frac{\partial^2\chi}{\partial x^2}+\frac{\partial^2\chi}{\partial y^2}=-\mathbf{e}_z\cdot\mathbf{u}\,, \tag{B 6.18}$$

and

$$\nabla_2^2\Psi=\mathbf{e}_z\cdot\operatorname{curl}\mathbf{u}\,. \tag{B 6.19}$$

The periodic cell decomposition can be formally obtained from the spherical annulus decomposition in the limit of large radius in which the gap size is fixed. Otherwise, one can start in the strip and prove the theorem anew. We must show that (B 6.18) and (B 6.19) are solvable for arbitrary given fields $\mathbf{u}$ only when their right sides are orthogonal to (bounded) solutions of the homogeneous problem $\nabla_2^2\phi=0$, that is $\phi=\phi(z)$. The right side of (B 6.19) has the required form since the orthogonality relation

$$0=\int\int_{\mathscr{C}} \mathbf{e}_z\cdot\operatorname{curl}\mathbf{u}\,dx\,dy=\oint_{\partial\mathscr{C}}\mathbf{u}\cdot\tau\,dl \tag{B 6.20}$$

is an identity for periodic $\mathbf{u}$. If the condition (B 6.14) holds, then by the periodicity and solenoidality of $\mathbf{u}$, so does the required orthogonality condition

$$0=\int\int_{\mathscr{C}(\alpha,\beta)} \mathbf{e}_z\cdot\mathbf{u}(x,y,z)\,dx\,dy \tag{B 6.21}$$

for the right of (B 6.18).

Exercise B6.1: (Toroidal and poloidal fields in a bounded domain $\mathscr{V}$):

Let $\mathbf{u}=0$ on the boundary $\mathscr{A}$ of a closed right cylinder. Show that $\mathbf{u}$ may be decomposed into poloidal and toroidal fields $\mathbf{u}_1$ and $\mathbf{u}_2$ given by (B6.16) and (B6.17) with

$$\chi(x,y,z)=\frac{1}{2\pi}\int\int_{\mathscr{A}}(-\mathbf{e}_z\cdot\mathbf{u}(x_0,y_0,z))\log((x-x_0)^2+(y-y_0)^2)^{-1/2}\,dx_0\,dy_0$$

and

$$\psi(x,y,z) = \frac{1}{2\pi} \int\int_{\mathscr{A}} (\mathbf{e}_z \cdot \operatorname{curl} \mathbf{u}(x_0,y_0,z)) \log((x-x_0)^2+(y-y_0)^2)^{-1/2} \, dx_0 \, dy_0 \, .$$

The corresponding result for a general bounded domain is given by W. Warner (1972).

Excercise B6.2: Suppose that in the spherical annulus there is a uniform suction at the outer wall, a uniform injection at the inner wall and that the mass flux $C_1 = \int_S \mathbf{u}\cdot\mathbf{r}/r$ is constant on all spheres $S(r)$. Show that the vector $\mathbf{v}=\mathbf{u}-C_1\mathbf{r}/4\pi r^3$ can be resolved into toroidal and poloidal fields.

Exercise B6.3: On a rigid surface $\mathbf{u}=0$. On a free surface the normal component of $\mathbf{u}$ and the shear stresses vanish. Show that

$$\chi = \frac{\partial \chi}{\partial r} = \Psi = 0 \qquad \text{(on rigid spherical surfaces)}$$

$$\chi = \frac{\partial \chi}{\partial z} = \Psi = 0 \qquad \text{(on rigid planar surfaces } z=\text{constant)}$$

$$\chi = \frac{\partial^2 \chi}{\partial r^2} = \frac{\partial \Psi/r}{\partial r} = 0 \qquad \text{(on free spherical surfaces)}$$

and

$$\chi = \frac{\partial^2 \chi}{\partial z^2} = \frac{\partial \Psi}{\partial z} = 0 \qquad \text{(on free planar surfaces } z=\text{constant).}$$

What are the appropriate boundary conditions for χ and Ψ on rigid surfaces of arbitrary shape?

Exercise B6.4: Write the Navier-Stokes equations for the fluid motion which is induced by a prescribed deformation of the walls in terms of the functions χ and Ψ. Formulate the boundary conditions in terms of these two functions.

B7. The Decay Constant $\hat{\Lambda}_1$ in the Spherical Annulus with $\mathbf{u}=0|_S$

The minimum problem (B1.3) can be separated into independent problems for the solenoidal fields $\mathbf{u}_1$ and $\mathbf{u}_2$. The eigenvalue problem may be split into two simpler problems because the decomposition $\mathbf{u}=\mathbf{u}_1+\mathbf{u}_2$ generates two fields which are orthogonal in the sense that

$$\int_{\mathscr{V}(\eta)} \mathbf{u}_1\cdot\mathbf{u}_2 = \int_{\mathscr{V}(\eta)} \left(\nabla\left(\frac{\partial r\chi}{\partial r}\right)\cdot\mathbf{u}_2 - \mathbf{r}\cdot\mathbf{u}_2\nabla^2\chi\right)$$

$$= \int_{S(\eta)\cup S(1)} \frac{\mathbf{r}}{r}\cdot\mathbf{u}_2\frac{\partial r\chi}{\partial r} - \int_{\mathscr{V}(\eta)}\left(\frac{\partial(r\chi)}{\partial r}\nabla\cdot\mathbf{u}_2 + \nabla^2\chi\mathbf{r}\cdot\mathbf{u}_2\right) = 0 \qquad \text{(B 7.1)}$$

and

$$\int_{\mathscr{V}(\eta)} \mathbf{D}[\mathbf{u}_1]:\mathbf{D}[\mathbf{u}_2] = -\int_{\mathscr{V}(\eta)} \mathbf{u}_1\cdot\nabla^2\mathbf{u}_2 = -\int_{\mathscr{V}(\eta)}\left[\nabla^2\mathbf{u}_2\cdot\operatorname{grad}\frac{\partial r\chi}{\partial r} - \nabla^2(\mathbf{r}\cdot\mathbf{u}_2)\nabla^2\chi\right] \tag{B 7.2}$$

$$= -\int_{S(\eta)\cup S(1)} \nabla^2\mathbf{u}_2\cdot\frac{\mathbf{r}}{r}\frac{\partial r\chi}{\partial r} + \int_{\mathscr{V}(\eta)}\left[\frac{\partial r\chi}{\partial r}\nabla^2\operatorname{div}\mathbf{u}_2 - \nabla^2(\mathbf{r}\cdot\mathbf{u}_2)\nabla^2\chi\right] = 0\,.$$

The problems generated by these two fields are formulated as separate boundary value problems for the functions χ and Ψ in the following way: For $\hat{\Lambda}$ we have the eigenvalue problem

$$\nabla^2\mathbf{u} + \hat{\Lambda}\mathbf{u} = -\nabla P\,. \tag{B 7.3}$$

The radial component of vorticity is then given by

$$\nabla^2(\mathbf{r}\cdot\operatorname{curl}\mathbf{u}) + \hat{\Lambda}\mathbf{r}\cdot\operatorname{curl}\mathbf{u} = 0\,.$$

This is equivalent, by (B 6.13), to $\nabla_S^2(\nabla^2+\hat{\Lambda})\Psi=0$. Now, form the radial component of the curl^2 of (B 7.3). Since $\mathbf{u}$ is solenoidal, $\operatorname{curl}^2\mathbf{u} = -\nabla^2\mathbf{u}$, $\mathbf{r}\cdot\nabla^2\mathbf{u} = \nabla^2(\mathbf{u}\cdot\mathbf{r})$ and $\nabla^4\mathbf{u}\cdot\mathbf{r} + \hat{\Lambda}\nabla^2\mathbf{u}\cdot\mathbf{r} = 0$.

On taking account of (B 6.11), this is the same as

$$\nabla_S^2(\nabla^4\chi + \hat{\Lambda}\nabla^2\chi) = 0\,.$$

The equations

$$(\nabla^2 + \hat{\Lambda})\,\Psi = 0 \tag{B 7.4}$$

and

$$(\nabla^4 + \hat{\Lambda}\nabla^2)\,\chi = 0 \tag{B 7.5}$$

are implied by the conditions (B 6.6).

The minimum eigenvalue of (B 7.3) *when* $\mathscr{V}(\eta)$ *is the spherical annulus of outer radius one and inner radius* η *and* $\mathbf{u}=0$ *at* $r=\eta, 1$ *is* $\hat{\Lambda}_1=\gamma^2$ *where* γ *is the smallest zero* $\gamma=\sqrt{\hat{\Lambda}_1}$ *of*

$$J_{3/2}(\gamma\eta)\,Y_{3/2}(\gamma) - J_{3/2}(\gamma)\,Y_{3/2}(\gamma\eta)\,. \tag{B 7.6}$$

The proof follows by comparing eigenvalues of (B 7.4) and (B 7.5). This smallest zero γ^2 is the second to smallest eigenvalue $\hat{\Lambda}$ of (B 7.4) with $\Psi=0$ at $r=\eta,1$. This second eigenvalue is smaller than the first eigenvalue of (B 7.5) with $\chi=\dfrac{\partial\chi}{\partial r}=0$ at $r=\eta,1$. To show this we note first that all solutions of (B 7.4) can be represented by series of spherical harmonics with terms

$$\Psi_{lm}(r)\,Y_l^m(\theta,\phi) = \Psi_{lm}(r)\,P_l^{(m)}(\cos\theta)\,e^{im\phi}$$

$(l \geq m)$ and

$$\Psi_{lm}(r) = r^{-1/2}\{J_{l+1/2}(r\sqrt{\Lambda})\, Y_{l+1/2}(\sqrt{\Lambda}) - J_{l+1/2}(\sqrt{\Lambda})\, Y_{l+1/2}(r\sqrt{\Lambda})\}. \tag{B 7.7}$$

Eigenvalues are found as roots of the equation (B 7.7) when $r=\eta$ and the smallest root is associated with value $l=0$. This eigenvalue cannot give $\hat{\Lambda}_1$, because the eigenfunction $\Psi_{00}(r)$ is a function of r alone, does not give a field $\mathbf{u} \neq 0$ and could not satisfy (B 6.6). The second smallest eigenvalue is associated with eigenfunctions for which $l=1$. There are acceptable solutions which satisfy (B 6.6).

The possibility that $\hat{\Lambda}_1$ could appear as the smallest eigenvalue of (B 7.5) for functions χ for which (B 6.6) holds and such that $\chi = \frac{\partial \chi}{\partial r} = 0$ at $r=\eta, 1$ can be excluded. This smallest eigenvalue appears as a solution of the minimum problem

$$\left[\Lambda_1 = \min \frac{\langle |\nabla^2 \chi|^2 \rangle}{\langle |\nabla \chi|^2 \rangle};\ \chi = \frac{\partial \chi}{\partial r} = 0 \text{ at } r=\eta, 1;\ \langle f(r)\chi \rangle = 0\right]. \tag{B 7.8}$$

Now, clearly, $\Lambda_1 > \lambda_1$ where

$$\left[\lambda_1 = \min \frac{\langle |\nabla^2 \phi|^2 \rangle}{\langle |\nabla \phi|^2 \rangle};\ \phi = 0 \text{ at } r=\eta, 1;\ \langle f(r)\phi \rangle = 0\right]. \tag{B 7.9}$$

The Euler problem for λ_1 is $\nabla^4\phi + \lambda_1 \nabla^2 \phi = 0$; $\phi = \nabla^2\phi = 0$ at $r=\eta, 1$ where the boundary condition $\nabla^2\phi = 0$ arises as a natural boundary condition and, of course, $\langle f(r)\phi \rangle = 0$. Now $\nabla^2\phi + \lambda_1\phi$ is harmonic and vanishes at the boundary; by potential theory, $\nabla^2\phi + \lambda_1\phi = 0$ at each point in the annulus. Conversely, all solutions of $\nabla^2\phi + \lambda_1\phi = 0$ which have $\phi = 0$ at $r=\eta, 1$ and $\langle f(r)\phi \rangle = 0$ satisfy B 7.9. It follows that the smallest eigenvalue of (B 7.4), (B 7.5) and the side conditions is taken by an eigenfunction of (B 7.4), that is by (B 7.6).

The values $\eta = 0$ and $\eta \to 1$ are of special interest. When $\eta = 0$ we have no inner sphere; then (B 7.6) reduces to $J_{3/2}(\gamma) = 0$ and we find the principal value of γ as the smallest solution of $\gamma = \tan\gamma$ so that $\gamma = 4.4934$. This eigenvalue has multiplicity three. The eigenfunctions are all obtained by rotating the eigenfunction

$$\Psi = r^{-3} x_3 (\sin\gamma r - \gamma r \cos\gamma r).$$

The corresponding eigenfunction $\mathbf{u}$ is given by

$$u_1 = x_2 x_3 r^{-3}(\sin\gamma r - \gamma r \cos\gamma r),$$

$$u_2 = -x_1 x_3 r^{-3}(\sin\gamma r - \gamma r \cos\gamma r)$$

and

$$u_3 = 0.$$

In the limit $\eta\to 1$ one can find from (B 7.6) that

$$(1-\eta)^2\,\hat{\Lambda}_1(\eta)=\pi^2\,.$$

Here $\mathscr{V}(\eta)$ is the strip of height $1-\eta$ $(\eta\to 1)$. Alternately, one can prove this same result directly in the strip of unit height.

Here, an acceptable periodic solution of $\nabla^2\Psi+\Lambda\Psi=0$, $\Psi=0$ at $z=0,1$ is $\Psi=\sin\pi z\cos\alpha x\cos\beta y$, $\Lambda=\pi^2+\alpha^2+\beta^2$. Moreover, this solution is orthogonal to every function of z alone on the period cell $P(\alpha,\beta)$.

In the case when the boundaries of the strip are free, one cannot even guarantee the existence of a nonzero decay constant. Here,

$$\lim_{\eta\to 1}(1-\eta)^2\,\hat{\Lambda}_1(\eta)=0\,.$$

In the unit strip one finds that $\Psi=\cos\alpha x\cos\beta y$ solves $\nabla^2\Psi+(\alpha^2+\beta^2)\Psi=0$ and $\partial\Psi/\partial z=0$ at $z=0,1$. Hence, both $\pi^2(\Psi=0$ at $z=0,1)$ and $0(\partial\Psi/\partial z=0$ at $z=0,1)$ are limiting eigenvalues of acceptable eigensolutions when $\alpha^2+\beta^2\to 0$.

The eigensolutions which we have found are such that $\chi=0$ $(\mathbf{u}_1=0)$, so that the radial velocity component is identically zero. However, one can find velocity fields having small nonzero values of $\mathbf{u}_1$, i. e., $\int_{P(\alpha,\beta)}\mathbf{u}_1^2=\varepsilon\to 0$, where $P(\alpha,\beta)$ is the period cell of the last paragraph, and such that the ratio

$$2\int_{P(\alpha,\beta)}\mathbf{D}[\mathbf{u}]:\mathbf{D}[\mathbf{u}]/\int_{P(\alpha,\beta)}|\mathbf{u}|^2$$

differs by terms of order ε from the minimum value of this ratio which is attained when $\varepsilon=0$. This means that we can get stronger values for $\hat{\Lambda}_1$ only among fields with no vertical vorticity $(\mathbf{u}_2=0)$. Such fields sometimes occur in stability problems (see Chapters IX and X).

B8. The Energy Eigenvalue Problem

We shall now leave the problem of the decay constant and turn to the problem of the energy stability limit. This limit is defined by (4.6) as

$$\nu_{\mathscr{E}}=\max_H\frac{-\langle\mathbf{u}\cdot\mathbf{D}[\mathbf{U}]\cdot\mathbf{u}\rangle}{\langle|\nabla\mathbf{u}|^2\rangle}\quad(\text{here } \mathbf{u}|_S=0)\,. \tag{B 8.1}$$

When written in variables made dimensionless by introducing a length scale l and a velocity scale U_m, (B 8.1) becomes

$$\frac{1}{R_{\mathscr{E}}}=\max_H(-\langle\mathbf{u}\cdot\mathbf{D}[\mathbf{U}]\cdot\mathbf{u}\rangle/\langle|\nabla\mathbf{u}|^2\rangle) \tag{B 8.2}$$

where the same notation has been used for dimensionless and dimensional variables.

We want to convert (B 8.2) into an eigenvalue problem, to characterize the set of eigenvalues with respect to completeness and to show that $R_{\mathscr{E}}$ defined by (B 8.2) can also be found as the principal eigenvalue of a partial differential equation.

Again keeping in mind problems of convection where the boundary conditions may be more complicated, we shall consider a slightly more general problem than (B 8.2), i. e.

$$\frac{1}{\rho} = \max_H (\mathscr{I}/\mathscr{D}) \tag{B 8.3}$$

where

$$\mathscr{I} = -\langle \mathbf{u}\cdot \mathbf{D}[\mathbf{U}]\cdot \mathbf{u}\rangle, \qquad \mathscr{D} = \langle 2\,\mathbf{D}[\mathbf{u}] : \mathbf{D}[\mathbf{u}]\rangle + \hbar\langle h|\mathbf{u}|^2\rangle_{S-S_1}$$

and $\mathbf{H}$ is the linear vector space defined by (B 1.6).

Suppose that the maximum of (B 8.3) is attained when $\mathbf{u}=\tilde{\mathbf{u}}$. Consider the value of $\mathscr{I}/\mathscr{D}$ when $u_i = \tilde{u}_i + \varepsilon\eta_i$ where $\eta_i = \left.\frac{\partial u_i}{\partial\varepsilon}\right|_{\varepsilon=0}$ is an arbitrary admissible vector (it satisfies Eq. B 8.1). For each fixed η_i we have

$$\frac{1}{\rho(\varepsilon)} = \frac{\mathscr{I}(\varepsilon)}{\mathscr{D}(\varepsilon)}. \tag{B 8.4}$$

Clearly $1/\rho(\varepsilon)$ is a maximum when $\varepsilon=0$ and

$$\left.\frac{d}{d\varepsilon}\right|_{\varepsilon=0} (\rho(\varepsilon)\mathscr{I} - \mathscr{D}) = \rho(0)\frac{d\mathscr{I}}{d\varepsilon} - \frac{d\mathscr{D}}{d\varepsilon} = 0. \tag{B 8.5}$$

Using (B 8.3) we may write (B 8.5) as

$$\rho\left\langle \tilde{\mathbf{u}}\cdot \mathbf{D}[\mathbf{U}]\cdot \frac{\partial \mathbf{u}}{\partial\varepsilon}\right\rangle + 2\left\langle \mathbf{D}[\tilde{\mathbf{u}}] : \frac{\partial \mathbf{D}[\mathbf{u}]}{\partial\varepsilon}\right\rangle + \hbar\left\langle h\tilde{\mathbf{u}}\cdot\frac{\partial \mathbf{u}}{\partial\varepsilon}\right\rangle_S = 0. \tag{B 8.6}$$

Here all quantities are evaluated at $\varepsilon=0$ and we have used the symmetry of $\mathbf{D}$ to write

$$\left\langle \frac{\partial \mathbf{u}}{\partial\varepsilon}\cdot \mathbf{D}[\mathbf{U}]\cdot\tilde{\mathbf{u}}\right\rangle = \left\langle \tilde{\mathbf{u}}\cdot \mathbf{D}[\mathbf{U}]\cdot\frac{\partial \mathbf{u}}{\partial\varepsilon}\right\rangle.$$

Eq. (B 8.6) may be regarded as Euler's functional equation. It holds for every kinematically admissible vector $\partial\mathbf{u}/\partial\varepsilon$. To convert this equation into an eigenvalue problem for a system of differential equations, we note that

$$2\left\langle \mathbf{D}[\tilde{\mathbf{u}}] : \frac{\partial \mathbf{D}[\mathbf{u}]}{\partial\varepsilon}\right\rangle = 2\left\langle \nabla\cdot\left(\mathbf{D}[\tilde{\mathbf{u}}]\cdot\frac{\partial \mathbf{u}}{\partial\varepsilon}\right)\right\rangle - \left\langle \frac{\partial \mathbf{u}}{\partial\varepsilon}\cdot\nabla^2\tilde{\mathbf{u}}\right\rangle,$$

and apply the divergence theorem and the fundamental lemma 2 of the calculus of variations (Section B 2). This leads one to the equation

$$\mathbf{u}\cdot\mathbf{D}[\mathbf{U}]-\frac{1}{\rho}\nabla^2\mathbf{u}=-\nabla p \tag{B 8.7a}$$

and the natural boundary conditions

$$(\mathbf{u}\cdot\mathbf{n})\mathbf{n}+2\,\mathbf{D}[\mathbf{u}]\cdot\mathbf{n}-2\mathbf{n}(\mathbf{n}\cdot\mathbf{D}[\mathbf{u}]\cdot\mathbf{n})+\hbar h\mathbf{u}|_{S-S_1}=0. \tag{B 8.7b}$$

The Euler eigenvalue problem is defined by (B 8.7a, b) and the admissibility conditions

$$\mathbf{u}\cdot\mathbf{n}|_S=0,\quad \mathbf{u}|_{S_1}=0,\quad \operatorname{div}\mathbf{u}=0. \tag{B 8.7c, d, e}$$

B9. The Eigenvalue Problem and the Maximum Problem

Let there be two different eigensolutions $\mathbf{u}_I$ and $\mathbf{u}_J$ belonging to different eigenvalues $\rho_I\neq\rho_J$ of the same Euler eigenvalue problem (B 8.7). By comparing the problem satisfied by the two different solutions it is easy to show that

$$(\rho_I-\rho_J)\langle\mathbf{u}_I\cdot\mathbf{D}[\mathbf{U}]\cdot\mathbf{u}_J\rangle=0 \tag{B 9.1}$$

and

$$2\langle\mathbf{D}[\mathbf{u}_I]:\mathbf{D}[\mathbf{u}_J]\rangle+\hbar\langle\mathbf{u}_I\cdot\mathbf{u}_J h\rangle_S=0. \tag{B 9.2}$$

The Eqs. (B 9.1, 2) are orthogonality relations. If it should happen that for one eigenvalue ρ there are $n>1$ eigenfunctions (ρ is an eigenvalue of multiplicity n), then these eigenfunctions can be formed into an orthogonal set. Hence, eigenfunctions of finite multiplicity may be assumed to satisfy orthogonality conditions.

It is easily shown that if $\mathbf{u}_I$ is an eigensolution with eigenvalue ρ_I, then (no summation over I)

$$\frac{1}{\rho_I}=\frac{-\langle\mathbf{u}_I\cdot\mathbf{D}[\mathbf{U}]\cdot\mathbf{u}_I\rangle}{2\langle\mathbf{D}[\mathbf{u}_I]:\mathbf{D}[\mathbf{u}_I]\rangle+\hbar\langle h\mathbf{u}_I\cdot\mathbf{u}_I\rangle}. \tag{B 9.3}$$

Conversely, it is known that all the eigenvalues can be obtained from maximum problems

$$\frac{1}{\rho_J}=\max_{\mathbf{u}\in\mathbf{H}}\frac{-\langle\mathbf{u}\cdot\mathbf{D}[\mathbf{U}]\cdot\mathbf{u}\rangle}{2\langle\mathbf{D}[\mathbf{u}]:\mathbf{D}[\mathbf{u}]\rangle+\hbar\langle h|\mathbf{u}|^2\rangle_S} \tag{B 9.4}$$

and side (orthogonality) conditions $\langle\mathbf{u}_J\cdot\mathbf{D}[\mathbf{U}]\cdot\mathbf{u}_I\rangle=0$ $(I<J)$. The variational characterization guarantees that there is a decreasing sequence (to zero) of eigen-

values ($\rho_J^{-1} < \rho_I^{-1}$) of finite multiplicity and a complete set of eigenfunctions. The largest of the eigenvalues ρ_I^{-1} is clearly the one for which no side orthogonality conditions are imposed. This eigenvalue is just the maximum required by (B 9.3).

Noting now that with $\mathbf{u}|_{S_1}=0$ and letting $S_1=S$ we may reduce (B 9.4) to (B 8.1), set $\rho=R_{\mathscr{E}}$ and replace (B 8.7) with

$$\mathbf{u}\cdot\mathbf{D}[\mathbf{U}]-\frac{1}{R_{\mathscr{E}}}\nabla^2\mathbf{u}=-\nabla p, \qquad \operatorname{div}\mathbf{u}=0 \quad\text{and}\quad \mathbf{u}|_S=0. \tag{B 9.5}$$

The equivalence of the maximum problem and the eigenvalue problem allows one to find criteria for monotonic global stability from the latter.

Suppose $R<R_{\mathscr{E}}$ *where* $1/R_{\mathscr{E}}$ *is the largest eigenvalue of the Euler eigenvalue problem* (B 9.5). *Then the basic motion is monotonically and globally stable.*

Notes for Appendix B

The material in this Appendix is at the mathematical level of the well-known text "Methods of Mathematical Physics" by Courant-Hilbert. Some of the topics treated here (in B.3 and B.4) are identical to topics in Vol. 1 of Courant-Hilbert but the methods used here to establish the dependence of functionals on parameters are different and more explicit. These explicit methods are given as a theorem in Exercise B4.6 and can be used to determine how eigenvalues depend on parameters (see Exercises (89.3,4) for applications to non-selfadjoint problems). Methods and results which are germane to problems of variational calculus on solenoidal vector fields are stressed in this Appendix. The condition of solenoidality is a constraint on the functions which are allowed to compete in variational problems. The fundamental lemma (2) of B.2 is one consequence of this constraint. Solenoidal vector fields are conveniently treated by the poloidal-toroidal decompositions of the type given in B.6. The proof of the decomposition theorem in spheres is due to Payne and Weinberger and I am indebted to them for letting me publish it here.

The monotonicity of $\Lambda(h)$ is sufficient to establish the existence of $\Lambda(\infty)>0$ (Poincaré's inequality) but without an explicits estimate from below. The monotonicity result may have been given first in an expository paper by Poincaré (1890). In that paper Poincaré also develops procedures of the Ritz-Rayleigh type for obtaining upper and lower bounds for the eigenvalues.

The mapping method for domain peturbations which is used in B.4 to treat Hadamard's formula is an extension and improvement of methods introduced first by Joseph (1967). This method is useful in problems like Hadamard's where the deformation of the domain is prescribed as well as in problems, like free surface problems (Joseph, 1973; Joseph and Fosdick; 1973; Joseph, 1974C; Joseph and Sturges, 1974) in which the shape of the domain is unknown and must be obtained as part of the solution. This method is applied in § 94 to obtain expressions for the shape of the free surface above a convecting pool of liquid.

Eq. (B7.6) was proved by Sorger (1966). The equivalent result when there is no inner sphere was given by Payne and Weinberger (1963). The proof given here is due to the last mentioned authors and appears here for the first time.

The equivalence of the Euler problem and the maximum problem was first proved by Serrin (1959B).

Many of the most useful computational precedures in the calculus of variations are discussed by Finlayson (1972). This book gives many applications and also treats problems of approximation in a more general context. The book by Morrey (1966) gives a complete mathematical treatment of the variational calculus.

Appendix C

Some Inequalities

Certain inequalities arise in the energy theory of stability and the variational theory of turbulence. Five such inequalities are proved in this appendix; two others are stated in Exercise C.1. The inequalities (C.1), (C.6) and (C.11) are due to V. Gupta and D. Joseph and appear here for the first time. The inequality (C.9) and (C.10) those stated in Exercise (C.1) are due to L. N. Howard (1972). Other interesting inequalities along with applications to turbulence theory are given in Howard's paper.

If $w(z)$ is continuous, $w(0)=w(1)=0$ and dw/dz is in $L_2(0,1)$ then

$$\int_0^1 w^2\,dz \leqslant \tfrac{1}{8}\int_0^1 \left(\frac{dw}{dz}\right)^2 dz\,. \tag{C.1}$$

Actually,

$$\int_0^1 w^2\,dz \leqslant \frac{1}{\pi^2}\int_0^1 \left(\frac{dw}{dz}\right)^2 dz\,.$$

This last inequality can be proved by forming the Euler equation for the maximum value of the ratio

$$\int_0^1 w^2\,dz \Big/ \int_0^1 \left(\frac{dw}{dz}\right)^2 dz\,.$$

The technique of proof of (C.1) given here avoids Euler equations and can be used in more complicated situations, as in (C.5), where exact solutions of the Euler equations are unknown.

To prove (C.1) we first note that

$$w(z)=\int_0^z \frac{dw}{dz'}\,dz' \leqslant \sqrt{z}\left(\int_0^z \left|\frac{dw}{dz'}\right|^2 dz'\right)^{1/2},$$

and when $z\leqslant\frac{1}{2}$

$$w^2(z)\leqslant z\alpha\int_0^1 \left|\frac{dw}{dz}\right|^2 dz \tag{C.2}$$

where

$$\alpha = \int_0^{1/2} \left|\frac{dw}{dz}\right|^2 dz \Big/ \int_0^1 \left|\frac{dw}{dz}\right|^2 dz .$$

We next observe that by a nearly identical argument applied at the wall $z=1$ that when $z \geqslant \frac{1}{2}$ we have

$$w^2(z) \leqslant (1-z)(1-\alpha) \int_0^1 \left|\frac{dw}{dz}\right|^2 dz . \tag{C.3}$$

The inequalities (C.2) and (C.3) are now integrated over their domain of validity:

$$\int_0^{1/2} w^2\, dz \leqslant \frac{\alpha}{8} \int_0^1 \left|\frac{dw}{dz}\right|^2 dz \tag{C.4}$$

and

$$\int_{1/2}^1 w^2\, dz \leqslant \frac{1-\alpha}{8} \int_0^1 \left|\frac{dw}{dz}\right|^2 dz \tag{C.5}$$

adding (C.4) and (C.5) we find (C.1).

If $w(z)$ *and* $w'(z)$ *are continuous,* $w(z)=w'(z)=0$ *at* $z=0,1$ *and* d^2w/dz^2 *is in* $L_2(0,1)$ *then*

$$\int_0^1 w^2\, dz \leqslant \frac{1}{192} \int_0^1 \left(\frac{d^2 w}{dz^2}\right)^2 dz . \tag{C.6}$$

Actually, the number 192 can be replaced with a large value $(4.73)^4$ (Rayleigh, 1878). The larger value follows from analysis of the fourth order eigenvalue problem for the maximum value of the ratio $\int_0^1 w^2\, dz \big/ \int_0^1 \left|\frac{d^2 w}{dz^2}\right|^2 dz$.

To prove (C.6) we first note that

$$\frac{d}{dz}\left[(z-z_0)\frac{dw}{dz}\right] = \frac{dw}{dz} + (z-z_0)\frac{d^2 w}{dz^2},$$

then, we have

$$w(z_0) = \int_0^{z_0} (z_0 - z)\frac{d^2 w}{dz^2}\, dz \leqslant \sqrt{\frac{z_0^3}{3}} \left(\int_0^{z_0} \left|\frac{d^2 w}{dz^2}\right|^2 dz\right)^{1/2}$$

and, when $z_0 \leqslant \frac{1}{2}$

$$w^2(z_0) \leqslant \frac{z_0^3 \beta}{3} \int_0^1 \left|\frac{d^2 w}{dz^2}\right|^2 dz \tag{C.7}$$

where

$$\beta = \int_0^{1/2} \left|\frac{d^2 w}{dz^2}\right|^2 dz \Big/ \int_0^1 \left|\frac{d^2 w}{dz}\right|^2 dz .$$

By similar reasoning in the interval $\frac{1}{2} \leqslant z \leqslant 1$ one finds that

$$w^2(z_0) \leqslant \frac{(1-z_0)^3}{3}(1-\beta)\int_0^1 \left|\frac{d^2 w}{dz^2}\right|^2 dz . \tag{C.8}$$

Integration of the inequalities (C.7) and (C.8) leads one to

$$\int_0^{1/2} w^2\, dz \leqslant \frac{1}{192}\beta \int_0^1 \left|\frac{d^2 w}{dz^2}\right|^2 dz$$

and

$$\int_{1/2}^{1} w^2\, dz \leqslant \frac{1}{192}(1-\beta)\int_0^1 \left|\frac{d^2 w}{dz^2}\right|^2 dz .$$

The inequality (C.6) now folows by addition.

We turn next to a lemma of L. N. Howard (1972). First we note that $\langle\cdot\rangle = \frac{1}{2}\int_{-1}^{1} \cdot\, dt$ is the volume averaged integral.

If $f(z)$ is continuous, $f(\pm 1)=0$, f' is in $L_2(-1,1)$ and $\langle f^2\rangle = 1$, then

$$\langle f'^2\rangle \langle (1-f^2)^2\rangle \geqslant \tfrac{4}{9} \tag{C.9}$$

Let $x=(1-z)/2$; then, relative to the functions $g(x)=f(z)$ we have

$$\langle g'^2\rangle \langle (1-g^2)^2\rangle \geqslant \tfrac{16}{9} . \tag{C.10}$$

This estimate is needed for the bound (94.28) on turbulent heat transport in a porous layer.

Proof: For any constant A and any t in $(-1,1)$ we have

$$\begin{aligned} 0 &\leqslant \tfrac{1}{2}\int_{-1}^{t} (f' - A(1-f^2))^2\, dz + \tfrac{1}{2}\int_t^1 (f' + A(1-f^2))^2\, dz \\ &= \langle f'^2\rangle + A^2 \langle (1-f^2)^2\rangle - 2A(f(t) - f^3(t)/3) . \end{aligned}$$

Choose A so as to minimize this expression, namely set

$$A = (f(t) - f^3(t)/3)\, \langle (1-f^2)^2\rangle^{-1},$$

which then gives

$$0 \leqslant \langle f'^2\rangle - f^2(t)(1 - f^2(t)/3)^2 \langle (1-f^2)^2\rangle^{-1} .$$

Since $\langle f^2\rangle=1$ and f is continuous, there is a value of t in $(-1,1)$ such that $f^2(t)=\langle f^2\rangle=1$; choosing this t gives the result asserted in the lemma.

The following lemma is needed for the bound (32.2) on the pressure gradient needed to drive an assigned mass flux deficit.

If $\mathbf{u}(x,y,z)$ *is an almost periodic vector of* (x,y) *and a continuous vector field on the infinite layer* $-\frac{1}{2}<z<\frac{1}{2}$, $\mathbf{u}(x,y,\pm\frac{1}{2})=0$ *and* $\nabla\mathbf{u}$ *is in* $L_2(\mathscr{V})$ *where* L_2 *is defined with the volume-averaged scalar product, then*

$$\frac{\langle[\overline{wu}-\langle wu\rangle-12z\langle zwu\rangle]^2\rangle}{\langle zwu\rangle^2}\geqslant\frac{576}{D+48} \tag{C.11}$$

where

$$D=\langle|\nabla\mathbf{u}|^2\rangle/\langle zwu\rangle.$$

To prove (C.11) we first note that the equation

$$\langle\overline{wu}-\langle wu\rangle-12z\langle zwu\rangle\rangle=0$$

implies that there is a value $\hat{z}$, $-\frac{1}{2}<\hat{z}<\frac{1}{2}$ such that

$$\overline{wu}-\langle wu\rangle-12\hat{z}\langle zwu\rangle=0.$$

It is convenient to use the coordinate $\zeta=z+\frac{1}{2}$ and to define

$$\not{f}(\zeta)=\overline{wu}-\langle wu\rangle-12(\zeta-\tfrac{1}{2})\langle zwu\rangle.$$

At $\hat{\zeta}=\hat{z}+\frac{1}{2}$ we have $\not{f}(\hat{\zeta})=0$.

Define

$$D_{0\zeta}=\int_0^\zeta\overline{|\nabla\mathbf{u}|^2}\,d\zeta;$$

clearly

$$D_{0\hat{\zeta}}=\frac{D_{0\hat{\zeta}}}{D_{01}}D_{01}=\alpha D_{01}\quad\text{and}\quad D_{\hat{\zeta}1}=D_{01}\left[1-\frac{D_{0\hat{\zeta}}}{D_{01}}\right]=(1-\alpha)D_{01}$$

where $\alpha=D_{0\hat{\zeta}}/D_{01}$.

Now since the inequalities

$$w^2\leqslant\zeta\int_0^\zeta\left(\frac{\partial w}{\partial\zeta'}\right)^2d\zeta',\quad u^2\leqslant\zeta\int_0^\zeta\left(\frac{\partial u}{\partial\zeta'}\right)^2d\zeta'$$

hold at each (x,y) point and also when averaged over the x,y plane we have, using Schwarz's inequality and $2ab \leqslant a^2+b^2$ that

$$|\overline{wu}| \leqslant [\overline{w^2 u^2}]^{1/2} = \zeta \left[\int_0^\zeta \overline{\left(\frac{\partial w}{\partial \zeta'}\right)^2} d\zeta'\right]^{1/2} \left[\int_0^\zeta \overline{\left(\frac{\partial u}{\partial \zeta'}\right)^2} d\zeta'\right]^{1/2} \leqslant \frac{\zeta}{2} D_{0\zeta}\,.$$

When $\hat{\zeta} > \zeta$ we have

$$|\overline{wu}| \leqslant \frac{\zeta}{2} \frac{D_{0\zeta}}{D_{0\hat{\zeta}}} \frac{D_{0\hat{\zeta}}}{D_{01}} D_{01} \leqslant \frac{\alpha\zeta}{2} D_{01}\,,$$

and with $\beta = \langle wu \rangle / \langle zwu \rangle$ and $D = D_{01}/\langle zwu \rangle$,

$$\frac{|f(\zeta)|}{\langle zwu \rangle} = \left| \frac{\overline{wu}}{\langle zwu \rangle} - \beta - 12(\zeta - \tfrac{1}{2}) \right| \geqslant |6-\beta| - \left| 12\zeta - \frac{\overline{wu}}{\langle zwu \rangle} \right|$$

$$\geqslant |6-\beta| - \zeta\left(12 + \frac{\alpha}{2} D\right) \equiv g(\zeta)/\langle zwu \rangle\,. \tag{C.12}$$

At $\zeta = 0$ we have $|f(0)| = g(0) = |6-\beta| \langle zwu \rangle$. At $\zeta = \tilde{\zeta} = |6-\beta| \Big/ \left(12 + \frac{\alpha}{2} D\right)$ we have $g(\tilde{\zeta}) = 0$. Since $|f(\zeta)| \geqslant g(\zeta)$ when $0 < \zeta < \hat{\zeta}$ where $\hat{\zeta}$ is the first zero of $f(\zeta)$ we must have

$$\hat{\zeta} > \tilde{\zeta}\,. \tag{C.13}$$

When $\zeta < \tilde{\zeta}$, (C.12) holds and $|f(\zeta)| \geqslant |g(\zeta)|$.

Using (C.13) and (C.12) we find that

$$\int_0^{\hat{\zeta}} f^2 \, d\zeta \geqslant \int_0^{\tilde{\zeta}} |f|^2 \, d\zeta \geqslant \int_0^{\tilde{\zeta}} |g|^2 \, d\zeta$$

$$= \langle zwu \rangle^2 \int_0^{\tilde{\zeta}} \left\{ |6-\beta| - \zeta\left(12 + \frac{\alpha}{2} D\right)\right\}^2 d\zeta$$

$$= \langle zwu \rangle^2 |6-\beta|^2 \int_0^{\tilde{\zeta}} \left(1 - \frac{\zeta}{\tilde{\zeta}}\right)^2 d\zeta$$

$$= \langle zwu \rangle^2 |6-\beta|^2 \tilde{\zeta}/3 = \langle zwu \rangle^2 |6-\beta|^3 / 3 \left(12 + \frac{\alpha}{2} D\right). \tag{C.14}$$

In the interval $\hat{\zeta} \leqslant \zeta \leqslant 1$ we have

$$\overline{wu} \leqslant \frac{1-\zeta}{2} \frac{D_{\zeta 1}}{D_{\hat{\zeta} 1}} \frac{D_{\hat{\zeta} 1}}{D_{01}} D_{01} \leqslant \frac{1-\zeta}{2} \frac{D_{01} - D_{0\hat{\zeta}}}{D_{01}} D_{01} = \frac{1-\zeta}{2} (1-\alpha) D_{01}$$

and

$$\frac{|\mathscr{f}(\zeta)|}{\langle zwu\rangle} = \left|\frac{\overline{wu}}{\langle zwu\rangle} - \beta + 12(1-\zeta) - 6\right|$$

$$\geqslant |6+\beta| - (1-\zeta)\left(12 + \frac{1-\alpha}{2}D\right).$$

Reasoning as before we find that

$$\int_\zeta^1 \mathscr{f}^2 d\zeta \geqslant \frac{|6+\beta|^3 \langle zwu\rangle^2}{3\left(12 + \frac{1-\alpha}{2}D\right)} \tag{C.15}$$

and addition of (C.14) and (C.15) gives

$$\frac{\langle[\overline{wu} - \langle wu\rangle - 12z\langle zwu\rangle]^2\rangle}{\langle zwu\rangle^2} = \frac{1}{\langle zwu\rangle^2} \int_{-1/2}^{1/2} \mathscr{f}^2(z)\,dz$$

$$= \frac{1}{\langle zwu\rangle^2} \int_0^1 \mathscr{f}^2(\zeta)\,d\zeta \geqslant \tfrac{1}{3}\left\{\frac{|6+\beta|^3}{12 + \frac{1-\alpha}{2}D} + \frac{|6+\beta|^3}{12 + \frac{\alpha}{2}D}\right\}$$

$$\geqslant \tfrac{1}{3}\min_{\alpha,\beta}\left\{\frac{|6+\beta|^3}{\left(12 + \frac{1-\alpha}{2}D\right)} + \frac{|6-\beta|^3}{\left(12 + \frac{\alpha}{2}D\right)}\right\} \tag{C.16}$$

where $0 \leqslant \alpha \leqslant 1$, $-\infty < \beta < \infty$. The minimum of the RHS. of (C.16) is found at $\beta = 0$ and $\alpha = \frac{1}{2}$. Hence

$$\frac{\langle[\overline{wu} - \langle wu\rangle - 12z\langle zwu\rangle]^2\rangle}{\langle zwu\rangle^2} \geqslant \frac{2}{3}\frac{6^3}{12 + \frac{D}{4}} = \frac{576}{D+48}.$$

Proof of the Estimate (40.11)[1].

Since $u(a,x) = u(b,x) = 0$, we have

$$u^2(r,x) = 2\int_a^r u(\xi,x)\frac{\partial u}{\partial \xi}(\xi,x)\,d\xi = -2\int_r^b u(\xi,x)\frac{\partial u}{\partial \xi}(\xi,x)\,d\xi$$

and

$$u^2(r,x) \leqq \int_a^b |u(\xi,x)|\left|\frac{\partial u}{\partial \xi}(\xi,x)\right| d\xi\,.$$

[1] The inequality (40.11) was derived by Joseph and Hung (1972). This inequality extends to periodic functions a well known Sobolev imbedding inequality. The imbedding inequalities are defined over functions of compact support (convenient references are Ladyzhenskaya (1969) and Serrin (1963).

In the same way, since $u(r,x_1)=u(r,x_2)$ where $x_2=x_1+2\pi/\alpha$, we find

$$u^2(r,x)\leqq u^2(r,x_1)+\int_{x_1}^{x_2}|u(r,\eta)|\left|\frac{\partial u}{\partial \eta}(r,\eta)\right|d\eta\,.$$

Thus,

$$\int_{x_1}^{x_2}\int_a^b u^4(r,x)\,r\,dr\,dx\leqslant \int_{x_1}^{x_2}\int_a^b\left[u^2(r,x_1)\int_a^b|u(\xi,x)|\left|\frac{\partial u}{\partial \xi}(\xi,x)\right|d\xi\right]dx\,r\,dr$$
$$+\int_{x_1}^{x_2}\int_a^b\left[\int_a^b|u(\xi,x)|\left|\frac{\partial u}{\partial \xi}(\xi,x)\right|d\xi\cdot\int_{x_1}^{x_2}|u(r,\eta)|\left|\frac{\partial u}{\partial \xi}(r,\eta)\right|d\eta\right]dx\,r\,dr\,,$$

we find that and, after noting that $r\geqq a$ and using Schwarz's inequality, the right hand side

$$\leqq\frac{1}{a}\langle u^2\rangle^{1/2}\left\langle\left|\frac{\partial u}{\partial r}\right|^2\right\rangle\int_a^b u^2(r,x_1)\,r\,dr+\frac{1}{a}\langle u^2\rangle\left\langle\left|\frac{\partial u}{\partial r}\right|^2\right\rangle^{1/2}\left\langle\left|\frac{\partial u}{\partial x}\right|^2\right\rangle^{1/2}.$$

Here the angle brackets denote integrals

$$\int_{x_1}^{x_1+2\pi/\alpha}\int_a^b f(x,r)\,r\,dr\,dx=\langle f\rangle$$

where f is periodic in x. This integral is independent of x_1. Hence, we may integrate the whole inequality from $x_1=0$ to $x_1=2\pi/\alpha$ to find

$$\frac{a\langle u^4\rangle}{\langle u^2\rangle}\leqq\frac{\alpha}{2\pi}\langle u^2\rangle^{1/2}\left\langle\left|\frac{\partial u}{\partial r}\right|^2\right\rangle^{1/2}+\left\langle\left|\frac{\partial u}{\partial r}\right|^2\right\rangle^{1/2}\left\langle\left|\frac{\partial u}{\partial x}\right|^2\right\rangle^{1/2}.$$

Exercise C.1 (Howard, 1972): (a) If $f(z)$ is continuous, $f(\pm 1)=0$ and f' is in $L_2(-1,1)$ then $2f^2(z)\leqslant(1-z^2)\int_{-1}^{1}f'^2(t)\,dt$. (b) If $f(z)$ and $f'(z)$ are continuous, and both vanish at $z=\pm 1$, and if f'' is in $L_2(-1,1)$, then

$$24f^2(z)\leqslant(1-z^2)^3\int_{-1}^{1}f''^2(t)\,dt\,.$$

Appendix D

Oscillation Kernels

It is possible to reformulate many stability problems which arise in stability theory as integral equations with oscillatory kernels. The eigenvalues and eigenfunctions of such integral equations possess many of the special characteristics of eigenfunctions and eigenvalues of the Sturm-Liouville problems (positivity and simplicity for the eigenvalues and positivity for the principal eigenfunction and the interlacing zeros property for the higher eigenfunctions). A beautiful feature of oscillatory kernel theory is that it applies to nonself-adjoint differential equation eigenvalue problems of higher order. This makes the oscillation kernel theory a very substantial generalization of Sturm-Liouville theory. Moreover, a simple criterion (D.3 below) due to M. G. Krein (1939) allows one to determine when the Green function used in converting the system of differential equations to an integral equation is an oscillation kernel. Krein's criterion helps to make the whole theory useful and easy to apply.

Oscillation kernel theory then allows one to determine the structure of the critical disturbances of stability theory; the simplicity of the eigenvalues is an important fact both for the uniqueness of the critical disturbance and for the possibility of constructing nonlinear bifurcation solutions from analytic perturbation theory.

We shall first define an oscillation kernel; then we describe the spectral properties of integral equations with oscillatory kernels; this is followed by Krein's criterion for oscillation Green functions; finally, we show how the theory applies to stability problems.

Oscillation kernel theory is a generalization of oscillation matrix theory (see Gantmacher Vol. 2, § 9, 1959). Eigenvalues of oscillation matrices are real, positive and simple. The components of the k-th eigenvector have exactly $k-1$ changes of sign. Eigenvalues and eigenfunctions of integral equations with oscillatory kernels have analagous properties.

The main theorems of oscillation kernel theory rely on entirely classical procedures of calculus and matrix theory. However, the proofs are lengthy and sometimes delicate. An account of the theory is found in the monographs of Gantmacher and Krein (1960) and S. Karlin (1968). We have followed the notation of the first-named authors (see § 2 of Chapter IV).

A function of two variables $K(x,s)$ is called an oscillation kernel if

(a) $K(x,s)>0, \quad a<x<b, \quad a<s<b,$

(b) $\det K(x_i,s_j) \geqslant 0, \quad a < \begin{Bmatrix} x_1 < x_2 < \cdots < x_n \\ s_1 < s_2 < \cdots < s_n \end{Bmatrix} < b$

and

(c) $\det K(x_i,x_j) > 0, \quad a < x_1 < x_2 < \cdots < x_n,$

where n is an arbitrary positive integer.

Consider an integral equation of the form

$$\phi(x) = \lambda \int_a^b K(x,s)\,\phi(s)\,d\sigma(s) \tag{D.1}$$

where $K(x,s)$ is an oscillation kernel and $d\sigma/ds > 0$: Then

(a) The eigenvalues of (D.1) are all positive and simple: $0 < \lambda_0 < \lambda_1 < \lambda_2 < \cdots$.

(b) The eigenfunction $\phi_0(x)$ belonging to λ_0 has no zero in $a \leqslant x \leqslant b$.

(c) The eigenfunction $\phi_k(x)$ belonging to λ_k has exactly k nodes ($\phi_k(x)$ changes sign at a node) and no other zeros.

(d) For arbitrary numbers k and m $(0 \leqslant k \leqslant m)$ and real numbers $c_k, c_{k+1}, \ldots, c_m$ $(\sum_{i=k}^m c_i^2 > 0)$ the linear combination $\phi(x) = \sum_{i=k}^m c_i \phi_i(x)$ has at least k nodes and at most m zeros.

(e) The nodes of neighboring eigenfunctions interlace one another.

This theorem was proved by Kellog (1916, 1918) for the case in which $K(x,s) = K(s,x)$ and $d\sigma = ds$. Kellog's results were extended to nonsymmetric kernels by Gantmacher (1936). The simplicity of the smallest eigenvalue and the positivity of the first eigenfunction also follow from the results of Jentzch (1912) on integral equations with positive kernels.

In order to study the spectrum of eigenvalue problems for differential equations by the theory of oscillation kernels, it is necessary to convert these problems to integral equations by means of Green functions. Then it comes to a question of whether the Green functions are oscillatory.

Krein (1939) has given a simple criterion under which the operator

$$L(u) = \sum_{l=0}^n \gamma_l(x)\, d^l u/dx^l, \quad n \geqslant 2 \tag{D.2a}$$

defined on the closed interval $[a,b]$ with

$$\gamma_n(x) > 0$$

and subject to the boundary conditions

$$\begin{aligned} &u(a) = u'(a) = \cdots = u^{(p-1)}(a) = 0, \\ &u(b) = u'(b) = \cdots = u^{(q-1)}(b) = 0 \end{aligned} \tag{D.2b}$$

where $p+q=n$, has an oscillatory Green function. If the Green function for (D.2a) and (D.2b) exists, then $(-1)^q G(x,s)$ will be an oscillation kernel if $L(u)$

may be represented in (a,b) as an iterated operator

$$L(u)=\mu_0(x)\frac{d}{dx}\mu_1(x)\frac{d}{dx}\mu_2(x)\cdots\frac{d}{dx}\mu_n(x)u \tag{D.3}$$

with strictly positive weights $\mu_k(x)$ possessing k continuous derivatives in (a,b).

Proofs for Krein's results are only sketched in his (1939) paper. Proof of some of his results is found in the paper of J. Karon (1969). S. Karlin (1971) has extended Krein's theorem to a general class of boundary value problems

Now we are ready to consider the conversion of some of the stability problems considered in the text to integral equations with oscillatory kernels. The procedure is just like the one which led to (23.2). We ordinarily encounter a coupled set of equations; one of 2nd-order and one of 4th-order. The 4th-order operators which arise are of the type $\mathcal{N}_4=\mathcal{N}_2\mathcal{N}_2$ where $\mathcal{N}_2$ is a 2nd-order operator which can be written as an iterated operator. For example, the Fourier decomposition of ∇^2 would lead to

$$\text{(a)}\quad \mathcal{N}_2 u=(D^2-\alpha^2)u=\frac{1}{\cosh\alpha x}D\left[\cosh^2\alpha x\, D\frac{u}{\cosh\alpha x}\right] \tag{D.4}$$

for periodic functions in a strip,

$$\text{(b)}\quad \mathcal{N}_2 u=\left(D^2+\frac{2}{x}D-\frac{l(l+1)}{x^2}\right)u=x^{-l-2}D[x^{2l+2}Dx^{-l}u] \tag{D.5}$$

for functions defined on a sphere or spherical annulus and

$$\text{(c)}\quad \mathcal{N}_2 u=\mathcal{L}_n u=\frac{1}{xI_n(\alpha x)}D\left[xI_n^2(\alpha x)D\frac{u}{I_n(\alpha x)}\right] \tag{D.6}$$

for functions on a cylinder or a cylindrical annulus where

$$\mathcal{L}_n=\left(D^2+\frac{1}{x}D-n^2/x^2-\alpha^2\right)\quad\text{and}\quad I_n(\alpha x)=J_n(i\alpha x)$$

is the modified Bessel function of order n. For boundary conditions of the appropriate type (all those encountered in the stability problems treated in this book)

$\mathcal{N}_2$ and $\mathcal{N}_2\mathcal{N}_2=\mathcal{N}_4$ have oscillation Green functions which we shall call $G_2(x,x_0)$ and $G_4(x,x_0)$.

The operators defined under (a), (b) and (c) may be obtained as special cases of

$$\mathcal{N}_2 u=\{(D[f(x)D]-g(x)\}u$$

where $f(x)$ and $g(x)$ are strictly positive when $a\leqslant x\leqslant b$. *$\mathcal{N}_2 u$ may be written as an iterated operator with positive weights:*

$$\mathcal{N}_2 u = \frac{1}{y} D\left[f y^2 D\left(\frac{u}{y}\right)\right] \tag{D.7}$$

where y is a strictly positive solution of

$$\mathcal{N}_2 y = f D^2 y + f' D y - g y = 0\,. \tag{D.8}$$

The existence of a strictly positive solution of (D.8) is an elementary consequence of the maximum principle. A solution $y(x)$ with $y(a) > 0$, $y'(a) > 0$ cannot have a positive maximum; it follows that $y(x) > 0$ for all $x > a$.

Typically, the stability problems are of the form

$$\mathcal{N}_2 \theta + \lambda \mathscr{f}(x) w = 0\,, \tag{D.9 a}$$

$$\mathcal{N}_2^2 w - \lambda \mathscr{g}(x) \theta = 0\,, \tag{D.9 b}$$

$$\theta = w = Dw = 0|_{x=a,b} \tag{D.9 c}$$

This problem is equivalent to the pair of integral equations

$$\theta = -\lambda \int_a^b G_2(x,x_0)\, \mathscr{f}(x_0)\, w(x_0)\, dx_0$$

and

$$w = \lambda \int_a^b G_4(x,x_0) \mathscr{g}(x_0) \theta(x_0) dx_0\,.$$

These equations are equivalent to either one of the pair of equations

$$\theta = -\lambda^2 \int_a^b K_1(x,x_1) \mathscr{g}(x_1) \theta(x_1) dx_1$$

or

$$w = -\lambda^2 \int_a^b K_2(x,x_1)\, \mathscr{f}(x_1)\, w(x_1) dx_1$$

where

$$K_1(x,x_1) = \int_a^b \mathscr{f}(x_0)\, G_2(x,x_0)\, G_4(x_0,x_1)\, dx_0 \tag{D.10a}$$

and

$$K_2(x,x_1) = \int_a^b \mathscr{g}(x_0) G_4(x,x_0) G_2(x_0,x_1) dx_0 \tag{D.10b}$$

We would like next to apply the theorem concerning (D.1) to the problems (D.9a, b). To do this we need to show that K_1 and K_2 are oscillation kernels and, of course, we require that $\mathscr{f}(x)$ and $\mathscr{g}(x)$ are nonnegative on (a,b). That K_1 and K_2 are oscillation kernels follows (essentially from the fact that the product of oscillation matrices is again an oscillation matrix (Gantmacher and Krein, 1960, pg. 86). The composition K_1 or K_2 of oscillatory Green functions is again oscillatory (see Ivanilov and Iakovlev, 1966, or Karlin, 1968). Hence, the nice spectral properties apply for λ, w and θ of Eq. (D.9) if $\mathscr{f}$ and $\mathscr{g}$ are nonnegative.

Among the problems which may be reduced to integral equations with oscillation kernels are:

Poiseuille flow problems:

u, f satisfying (21.5 a, b) and

$$u = f = Df = 0 \quad \text{at} \quad r = \eta(1-\eta)^{-1} \quad \text{and} \quad (1-\eta)^{-1} \tag{D.11}$$

u, f satisfying (22.4 a, b) and (D.11) when DU is negative.

(b) Couette flow problems:

f, v satisfying (39.2) when $\kappa = n = 0$ and $B > 0$, $A < 0$.

$\hat{w}, \hat{\phi}$ arising from the Fourier reduction of (41.3) when $\dfrac{B}{r^2\sqrt{\lambda}} - \dfrac{(\lambda-1)}{\sqrt{\lambda}} A > 0$.

(c) Rotating Poiseuille flow problems:

f, u satisfying (46.3) when $\hat{\omega} = 0$.

(d) The generalized Bénard problem for convection in a porous container:

$\bar{\bar{\chi}}, \bar{\bar{\theta}}$ satisfying (73.8) when f and G' are positive.

$\bar{\bar{\chi}}^*, \bar{\bar{\theta}}^*$ satisfying (73.9) when f and G' are positive

(e) The Bénard problem in a spherical annulus:

$\bar{\bar{f}}_l, \bar{\bar{\theta}}_l$ satisfying (74.11) and (74.13) when $\mathscr{g}(r)$ and $\ell(r)$ are positive,

$\bar{\bar{f}}, \bar{\bar{\theta}}$ satisfying (74.12) and (74.13) when $\mathscr{g}\dfrac{(r)}{\sqrt{\lambda}} + \sqrt{\lambda}\,\ell(r)$ is positive.

(f) The generalized Bénard problem:

$\hat{\chi}, \hat{\theta}$ satisfying (75.13, 15) when $1/\sqrt{\lambda} + \sqrt{\lambda} + \sqrt{\lambda}\,\xi\hat{\mathscr{g}}(z) > 0$,

$\hat{\chi}, \hat{\theta}$ satisfaing (75.14, 15) when $1 + \xi\hat{\mathscr{g}}(z) > 0$;

(g) The Bénard problem in a spherical cap of porous material:

$\bar{\bar{\chi}}, \bar{\bar{\theta}}, \bar{\bar{\chi}}^*, \bar{\bar{\theta}}^*$ satisfying (74.31) when $\ell(r)$ and $\mathscr{g}(r)$ are positive.

Appendix E

Some Aspects of the Theory of Stability of Nearly Parallel Flow

There are strictly parallel flows (for example, Poiseuille-Couette flow down a sliding annulus) and "nearly" parallel flows. "Nearly" parallel flows are those for which the derivatives along the "streaming" axis are small relative to the transverse (r) derivatives. Flow in a boundary layer or jet is an example of nearly parallel flow.

It is customary to treat nearly parallel flows as if they were exactly parallel. There is no strict justification for this procedure and it is almost certainly not justified in all cases. However, the instability results which follow upon treating nearly parallel flow as strictly parallel are sometimes in good agreement with experiment. Since so many nearly parallel motions might occur, it is natural to study the stability problem for arbitrarily given "parallel" flows $U(r)$. This leads to the study of the Orr-Sommerfeld (OS) problem in the cylindrical annulus. The theory of the OS equation relies on asymptotic analysis and is very complicated (see Lin, 1955 for the older references; for more recent work see Reid, 1965); asymptotic methods were once used in computations but now computations are more easily, quickly and accurately done by numerical methods. It is possible however, to obtain delicate properties of solutions to the OS problem from extremely simple arguments. In § E1 we give these arguments and show how they are supported by numerical analysis of the exact problem.

The most important question to ask of the theory of arbitrarily given parallel flow is not how solutions behave but whether and in what sense these solutions are valid approximations to the problems they are claimed to represent. This question is discussed in § E2.

§ E1. Orr-Sommerfeld Theory in a Cylindrical Annulus

For some parallel motions the critical Reynolds number of linear theory is very high, even infinite. When R is large we expect the flow field to divide into inviscid and viscous regions. In the viscous regions, like those in the neighborhood of solid boundaries, the velocity must change rapidly and derivatives of velocity are so large they cannot be neglected even when they are divided by the large Reynolds number.

Linear stability theory leads to such a division into inviscid and viscous regions. The structure of the eigensolutions is so delicately balanced in this division that even slight changes in the problem can induce big changes in the linear stability results. In contrast, slight changes in the problem do not strongly effect the results of energy analysis (Carmi, 1969).

The hypersensitivity of the linear stability limit to slight changes in the velocity profile is related to the fact that the high critical Reynolds numbers are usually not observed. Naturally, if the limit for stability to small disturbances is large, it is more likely that the nonlinear terms in the disturbance equations will lead to instability under subcritical conditions.

To understand the delicate balance between viscous layers and the inviscid interior we shall study the Orr-Sommerfeld problem (34.6a). Though a complete analysis of this problem is beyond the scope of this work some of the more important properties of solutions can be determined by elementary analysis. This analysis starts from equation (34.6a) with $R=2\bar{R}$ and $w=-\frac{1}{r}\,\partial\Psi'/\partial x$,

$$i(U-c)\bar{w}\mathscr{L}w-i\hat{\Psi}|w|^2=(\alpha R)^{-1}\bar{w}\mathscr{L}^2 w \tag{E1.1}$$

where $\hat{\Psi}(r)$ is defined by (44.29a), $\mathscr{L}=\frac{1}{r}\frac{d}{dr}\left(r\frac{d}{dr}\right)-\frac{1}{r^2}-\alpha^2$, $c=c_r+ic_i=\sigma/i\alpha$ and w and Dw vanish at $\tilde{a}=\eta(1-\eta)^{-1}$ and $\tilde{b}=(1-\eta)^{-1}$. The basic variable of this analysis is the Reynolds stress distribution.

$$\tau\equiv -i\left[\bar{w}Dw-wD\bar{w}\right]. \tag{E1.2}$$

This variable arises naturally when the complex conjugate of (E1.1) is added to (E1.1). This addition leads to (44.26a, b). At the border of stability and instability $c_i=0$, and (44.26a) may be written as

$$\frac{1}{\alpha R}=\tfrac{1}{2}\langle\tau DU\rangle/\langle|\mathscr{L}w|^2\rangle. \tag{E1.3}$$

The local energy transfer between the basic and disturbed motion depends on the sign of τDU; when positive, energy is transferred from the basic motion to the disturbance motion. τ vanishes strongly at a rigid wall (see E1.1).

We wish to determine the distribution of $\tau(r)$ across the annulus. Since neutral disturbances are generally found only when αR is large it is reasonable to assume that a small disturbance of the basic shear flow $U(r)$ is strongly influenced by shearing forces only in boundary layers where steep gradients are required to bring the velocity to zero. One of the main mathematical difficulties in the theory of the Orr-Sommerfeld equation is that this assumption is not quite correct; there is a second type of layer in which the viscosity is important. The second type of viscous layer, which is called a critical layer, occurs at each root r_c, $\tilde{a}<r_c<\tilde{b}$ of the equation $U(r)-c_r=0$. If $\hat{\Psi}(r_c)=\hat{\Psi}_c\neq 0$ then any neutral solution of the inviscid ($\alpha R\to\infty$) Orr-Sommerfeld problem (the Rayleigh problem)

$$(U-c)\mathscr{L}w_I-\hat{\Psi}w_I=0\,,\qquad w_I=0\quad\text{at}\quad r=\tilde{a},\tilde{b} \tag{E1.4}$$

suffers a jump in τ_I (or Dw_I) as r crosses r_c. Viscosity must become important in any region where the derivatives of the velocity become large.

The existence of critical layers is fundamental to the Orr-Sommerfeld problem because the eigenfunctions of Rayleigh's equation have eigenvalues whose real parts c_r lie in the range of $U(r)$ [Exercise (E 1.1)]. Hence the existence of r_c is guaranteed.

The distribution of the Reynolds stress $\tau(r)$ for the actual flow is determined in a basic way by this jump in the inviscid Reynolds stress $\tau_I(r)$ and τ_I is governed by the equation (Foote and Lin, 1950)

$$D(r\tau_I)=2c_i r\hat{\Psi}|w_I|^2/|U-c|^2 \tag{E 1.5}$$

which may be obtained from (E 1.4). For amplified disturbances $c_i>0$, $|U-c|^2>0$ and, since $\tau_I=0$ at the walls,

$$\langle\hat{\Psi}|w_I|^2/|U-c|^2\rangle=0\,. \tag{E 1.6}$$

Eq. (E 1.6) shows that there are no amplified solutions of the Rayleigh problem for parallel flows for which $\hat{\Psi}$ is one-signed. This means that the weighted vorticity, ζ/r, which is conserved on fluid particles in axisymmetric, inviscid flow, is a relative extremum at $r=r_c$. In the usual circumstances this extremum gives a relative maximum of ζ/r (Exercise E 1.2).

Most of the remarks which we have made for amplified disturbances of Rayleigh's problem hold formally for damped disturbances, $c_i<0$. However, Lin (1945) argues that Rayleigh's problem is a proper $\alpha R\to\infty$ limit of the Orr-Sommerfeld problem only when $c_i>0$.

Suppose that $c_i\downarrow 0$; then (E 1.5) shows that $r\tau_I$ is constant on any interval in which $U-c_r\neq 0$. To compute the jump in τ as r crosses r_c we note that for sufficiently small values of ε

$$[r\tau_I]_{r_c-\varepsilon}^{r_c+\varepsilon}=\lim_{c_i\downarrow 0} 2c_i r_c\hat{\Psi}_c|w_1|_c^2\int_{r_c-\varepsilon}^{r_c+\varepsilon}dr/[(U-c_r)^2+c_1^2]\,.$$

Noting now that if $U'_c\neq 0$ we have $dr=d(U-c_r)/U'$ and

$$\lim_{c_i\downarrow 0}\int_{r_c-\varepsilon}^{r_c+\varepsilon}\frac{c_i dr}{[(U-c_r)^2+c_i^2]}=\lim_{c_i\downarrow 0}\frac{1}{U'_c}\int_{-U'_c\varepsilon}^{U'_c\varepsilon}\frac{c_i dx}{x^2+c_i^2}$$

$$=\lim_{c_i\downarrow 0}\frac{1}{|U'_c|}\int_{-\infty}^{\infty}\frac{c_i dx}{x^2+c_i^2}=\pi/|U'_c|\,.$$

It follows that

$$[r\tau_I]_{r_c-\varepsilon}^{r_c+\varepsilon}=r_c[\tau_I]_{r_c-\varepsilon}^{r_c+\varepsilon}=2\pi\hat{\Psi}_c r_c|w_I|_c^2/|U'|_c\,. \tag{E 1.7}$$

Since $\tau_I=0$ on the walls, Eqs. (E 1.5) and (E 1.7) show that eigensolutions of Rayleigh's equation with $c_i=0$ can exist only if the total sum of the jumps

$$\int_a^b r\tau_I dr=\sum[r\tau_I]=0. \tag{E 1.8}$$

Suppose that $U(r)$ is a monotone profile. Then there can be only one value of $r=r_c$ for which $U(r)-c_r=0$. The jump formula (E1.7) then shows that a neutral eigensolution of Rayleigh's problem can exist only if $\hat{\Psi}_c=0$ (the possibility that $[w_I]_c^2=0$ can be excluded by the argument given in Exercise (E1.1). For plane parallel flow the same conclusion about monotone velocity holds; in this case, however, the condition $\hat{\Psi}_c=0$ reduces to Rayleigh's inflection point criterion $U_c''=0$. Rayleigh's problem for symmetric profiles in a channel can be shown to be equivalent to the monotone case and the condition $U_c''=0$ is necessary and also sufficient (Tollmien, 1935) for the existence of neutral eigensolutions of Rayleigh's problem.

Flows which are unstable by the criteria of Rayleigh's inviscid theory are relatively unstable by the criteria of the full Orr-Sommerfeld theory. This relative instability is particularly evident in the stability properties of the flow as $R\to\infty$. When Rayleigh's instability criterion is satisfied there is a band of wave numbers α for which the basic flow is judged unstable by the Orr-Sommerfeld theory. On the other hand, if Rayleigh's criterion is not satisfied the basic flow is always judged stable by the Orr-Sommerfeld theory in the limit $\alpha R\to\infty$. The relative instability of flows satisfying Rayleigh's instability criterion is reflected in an appreciable lowering of the critical Reynolds numbers as is indicated in Fig. E1.1.

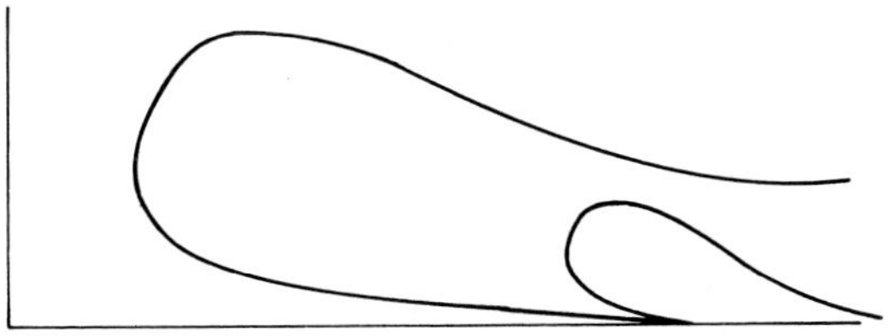

Fig. E1.1: Schematic sketch comparing the neutral curves of two flows; one flow satisfies the inviscid instability criterion of $\hat{\Psi}(r_c)=0$ of Rayleigh, the other low $\hat{\Psi}(r_c)\neq 0$ is judged stable by Rayleigh's inviscid theory

The lowering of the critical Reynolds number when a vorticity maximum is introduced in the basic flow can be seen in the study of retarded boundary layer flows by Obremski, Morkovin and Landahl (1969).

The critical Reynolds number of the Orr-Sommerfeld theory is very sensitive to changes in velocity profile. Small changes of $U(r)$ in round pipes (Gill, 1965) and in channels (Fu and Joseph, 1970) can lead to large changes in the critical Reynolds number. It must be remembered, however, that arbitrary profiles $U(r)$ are not steady solutions of the full Navier-Stokes equations. Though such profiles could arise locally as a perturbation of some basic shear flow due to, say, a long bump on the annulus wall, it would not be possible to maintain this perturbation further downstream; local instability produced in one part of the annulus would disappear further downstream where the basic flow is again of a stable type.

In the absence of a maximum of the weighted vorticity (E1.8) must hold. The vanishing of the sum of the jumps of the Reynolds stress at each critical layer is a more stringent condition than one would expect to hold generally. For example, the two jumps which would occur when $U(r_c)=c_r$ in plane Poiseuelle flow could not sum to zero. Thus, at first glance, the inviscid analysis leading to (E1.7) would appear to have only a limited applicability to the Orr-Sommerfeld problem.

The conclusion just reached goes too far; the reason is that the viscous boundary layers have been ignored; the wall layers imply that (E 1.7) could not hold very near the wall and (E 1.8) cannot then be deduced from (E 1.7). Considerable insight into the Orr-Sommerfeld problem at finite Reynolds numbers follows from (E 1.7) and an elementare analysis of the wall boundary layer given below.

Consider a velocity profile $U(r)>0$, $U(\tilde{a})=U(\tilde{b})=0$ (see Fig. E 1.3 a). Imagine that there is an inviscid neutral solution in the interior of the annulus and that the instability wave moves forward with velocity c_r where

$$0<c_r<\max U(r)\,, \qquad \tilde{a}<r<\tilde{b}\,.$$

This is compatible with the form (E 1.1) of the full Orr-Sommerfeld problem in the neighborhood of the wall where the neutral oscillation must come to rest; assuming a boundary layer of small size δ we may drastically simplify (E 1.1). In the boundary layer

$$\frac{d^2w}{dy^2}=\frac{i}{2}\frac{d^4w}{dy^4}\,, \qquad w=Dw=0 \quad \text{at} \quad y=0 \tag{E 1.9}$$

where $y=(r-a)/\delta$ and $\delta=(2/\alpha c_r R)^{1/2}$ is the boundary layer thickness. Eq. (E 1.9) follows from elementary scaling of variables when δ is small using the fact that U, w and Dw vanish at the wall and are small in the boundary layer.

There are four independent solutions of (E 1.9)

$$[A_1,\, A_2 y,\, A_3\exp(i-1)y,\, A_4\exp(1-i)y]\,. \tag{E 1.10}$$

The first two of these solutions are "inviscid", the fourth solution grows exponentially with y and is discarded $(A_4=0)$; application of the boundary conditions gives

$$w=A[1+(i-1)y-\exp(i-1)y]\sim iA[y^2+\tfrac{1}{3}(i-1)y^3] \tag{E 1.11}$$

where the last equation follows from a power series expansion of the exponential carried to lowest significant order. We compute

$$\tau\equiv -i(\overline{w}Dw-wD\overline{w})\sim\tfrac{2}{3}y^4|A|^2\,. \tag{E 1.12 a}$$

Hence τ increases with distance from the wall $r=\tilde{a}$. At $r=\tilde{b}$ we introduce $\hat{y}=(\tilde{b}-r)/\delta$

$$\tau\sim -\tfrac{2}{3}\hat{y}^4|A|^2\,. \tag{E 1.12 b}$$

Given the form of $U(r)$ (Fig. E 1.3 a) we may conclude that

$$\tau U'>0 \quad \text{near the inner and outer wall}\,.$$

It follows that the energy supply near the boundaries is positive. This important result was first found by C.C. Lin (1954) by a somewhat different analysis from

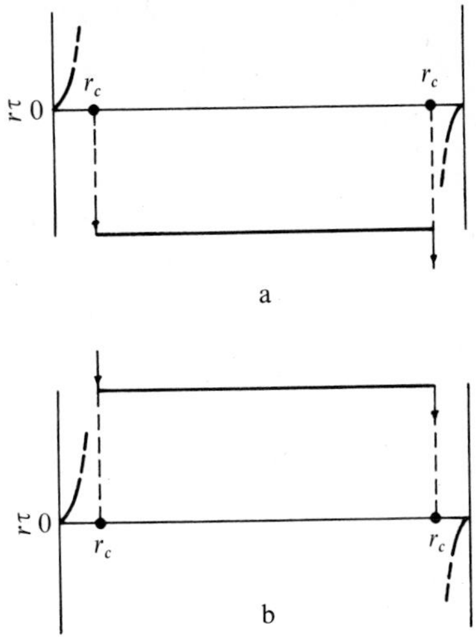

Fig. E1.2: The disturbance boundary-layer analysis and the jump condition for the Reynolds stress imply conditions (E1.13). These conditions do not determine the sign or magnitude of the Reynolds stress in the inviscid center. If the sign of the Reynolds stress is assumed to be positive, we have the situation sketched in (a) whereas a negative Reynolds stress implies the sketch in (b). Both distributions occur (see Figs. E1.4 and E1.5). A good qualitative picture of these distributions (which were calculated numerically by finite differences) also follows upon completion of the sketches of Fig. E1.2 as smooth curves in the most obvious way

the one just given. Lin's result was motivated by Prandtl's conjecture: by promoting a positive energy supply at the wall viscosity can have a destabilizing effect.

The delicately balanced structure of solutions to the Orr-Sommerfeld equation at finite Reynolds numbers, to which we alluded at the beginning of this section, is implied in the inviscid jump formula (E1.7) and the wall Reynolds stress results (E1.12). We have found that for profiles $U(r) \geq 0$ which increase from zero at the inner wall and decrease to zero at the outer wall

(i) $\tau > 0$ *near the inner wall and* $\tau < 0$ *at the outer wall*

(ii) $\tau U' > 0$ *near both walls*

(iii) $r\tau \simeq r\tau_I = const.$ *in the region between the critical layers*

(iv) *The sign in the jump* τ *of the Reynolds stress is the same as the sign of* $\hat{\Psi}(r_c)$. (E1.13)

Conditions (E1.13) are represented in the sketch of the Reynolds stress distribution shown in Fig. E1.2.

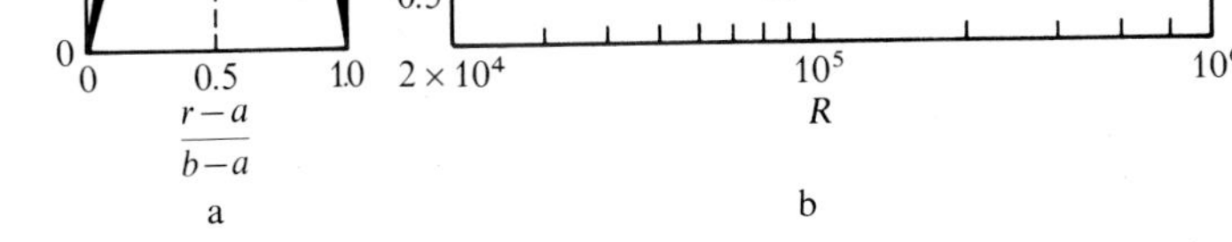

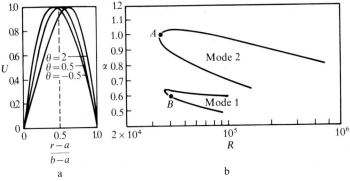

Fig. E1.3: (a) Velocity Profiles. (b) Neutral curves for the profile $\theta=2$ shown in Fig. E1.3a. The different profiles arise from changing the viscosity distribution by heating the cylinder walls (Mott and Joseph, 1968)

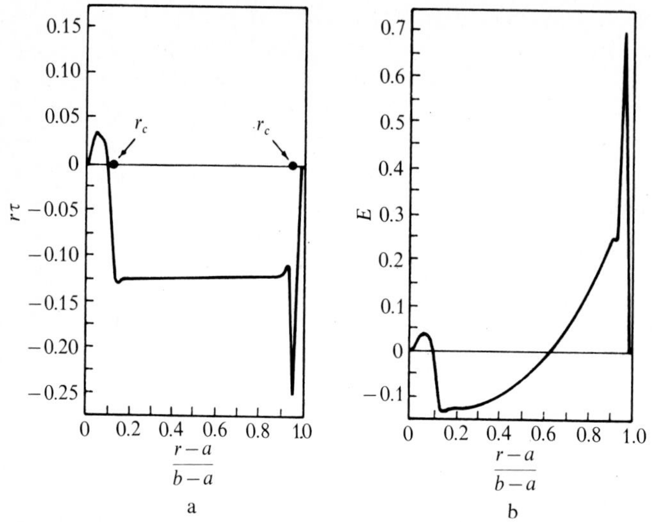

Fig. E1.4: Reynolds stress and energy supply for neutral disturbance A of Fig. E1.3b (Mott and Joseph, 1968)

An exact solution of the Orr-Sommerfeld problem was computed numerically by Mott and Joseph (1968). Relative to the velocity profile $U(r)$ corresponding to $\theta=2$ defined in Fig. E1.3a. Mott and Joseph have computed the two neutral curves shown in Fig. E1.3b and labeled Mode 1 and Mode 2.

The numerically computed Reynolds stress and energy supply distributions for points A and B of Fig. E1.3b are shown in Figs. E1.4 and E1.5.

The Orr-Sommerfeld theory is not in good agreement with experimental observations of the instability of laminar Poiseuille and Couette flow, to which it applies strictly. The most impressive success of the Orr-Sommerfeld theory

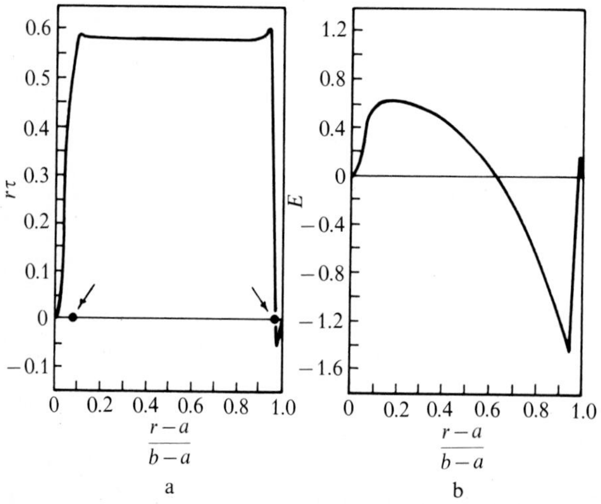

Fig. E1.5: Reynolds stress and energy supply for neutral disturbances B of Fig. E1.3b (Mott and Joseph, 1968)

is for the boundary layer flow along a flat plate with a zero pressure gradient. It is perhaps ironic that this is a nonparallel flow to which the theory does not strictly apply. The asymptotic theory for this flow was given first by Tollmien (1929) and was later calculated by Schlichting (1935), Lin (1945) and Shen (1954). When the fluctuations are kept small, these asymptotic results are in good qualitative agreement with experiments. However, more exact numerical results are in less good agreement with experiments (Fig. E1.6) than is the asymptotic theory, especially in the neighborhood of the critical Reynolds number ($R=U\delta/\nu \sim 500$, where U is the free stream velocity and δ is four times the boundary layer displacement thickness.)

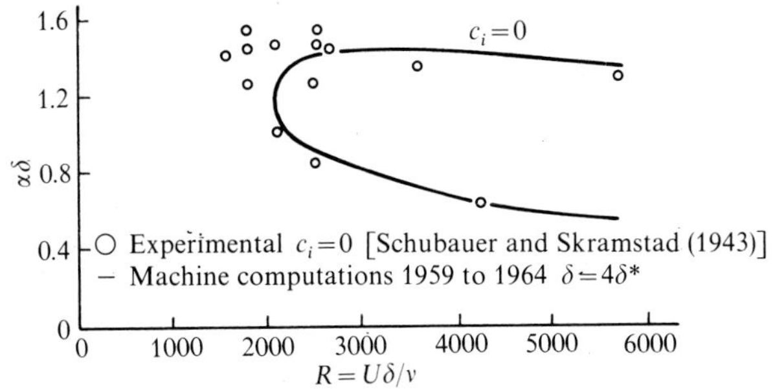

Fig. E1.6: Experimental and theoretical results for neutrally stable oscillations of the Blasius layer. The length δ is four times the boundary layer displacement thickness (Betchov and Criminale, 1967)

In the experiments of Schubauer and Skramstad (1943), (Fig. E1.6) and in other experiments, considerable care is taken to suppress large amplitude disturbances. When such disturbances are allowed the boundary layer instability can be triggered at even smaller subcritical values of the Reynolds number.

Exercise E1.1 (Rayleigh, 1913): Show that

$$\int_a^b (U-c)^2\left[\frac{1}{r}|(r\chi)'|^2+\alpha^2 r|\chi|^2\right]dr=0$$

where $c=c_r+ic_i$ and

$$\chi=w/(U-c)\,.$$

Prove that $U(r)-c_r=0$ has one root $r=r_c$ on the interval of integration. Suppose that $c_i=0$, $w(r_c)=0$, and χ is continuously differentiable on the interval of integration; show that $w(r_c)\equiv 0$.

Exercise E1.2 (Fjørtoft, 1950; Høiland, 1953): Show that $|\zeta/r|$ is a maximum at a point $r=r_c$ at which $\hat{\Psi}(r_c)$ changes sign whenever $U(r)$ is a monotone function in the annulus and $c_i>0$.

Exercise E1.3: Evaluate (E1.5) for Hagen-Poiseuille flow and plane Couette flow. Construct an argument to support the conclusion that these two flows are absolutely stable to infinitesimal disturbances.

§ E2. Stability and Bifurcation of Nearly Parallel Flows

It is very hard to decide about the validity of the assumption that the stability nearly parallel flow may be studied through the OS equation. Given the mathematical difficulty involved in studying stability problems governed by non-separable partial differential equations, it is perhaps natural to seek justification through comparison with experiments. In fact the comparisons are often indecisive and the question of validity is left unanswered. The difficulty in assessing validity through experiment is compounded by nonlinear effects; even in strictly parallel shear flows there is no apparent agreement with experiment because the bifurcating solution is unstable and the stable solutions are not "close" to eigenfunctions of the OS equation. Thus the problem of validity involves discussion of the parallel flow approximation in the linear and nonlinear theory.

Some persons believe that it is possible to take advantage of the basic directionality of nearly parallel flow by a perturbation method based on the notion of two space scales. Formal theories of this type have been given by Bouthier (1972, 1973) and Ling and Reynolds (1973) for the linear part of the problem and by Joseph (1974 B) for the linear and bifurcation part of the problem.

The existence of two space scales in nearly parallel flow is motivated by the boundary layer equations. In two-dimensions these equations are in the form:

$$U\partial_x U + V\partial_y U = \nu\partial^2_{yy} U\,,$$

$$\partial_x U + \partial_y V = 0\,.$$

Solutions of these equations are often in the form

$$U = \hat{U}(\nu x, y)\,,$$

$$V = \nu\hat{V}(\nu x, y)\,.$$

In the study of the stability of these solutions the dependence of the nearly parallel basic flow on the axial variable x is always in the form νx. In the two scale method $\chi = \nu x$ and x are treated as independent variables; χ is a "slow" variable if ν is small.

The choice of a proper zeroth order in a bifurcation theory for nearly parallel flow is very important. The earliest mathematical study of the effects of non-parallelism (Lanchon and Eckhaus, 1964) already indicated that though the use of the OS theory at zeroth order is valid for the Blasuis boundary layer, this same approximation could not be expected to give correctly the linear stability limit for flows like those in jets. Even earlier, Tatsumi and Kakutani (1958) noted that the OS theory might not apply to jets. If a flow is not well represented at the zeroth order, it cannot be approximated by perturbations.

The problem of the correct zeroth order is basic in developing a perturbation theory which will apply equally to flows in boundary layers and jets. Haaland (1972) has noted that the difference between flows of the boundary layer type and flows of the jet type can be characterized by the behaviour of the velocity component V at large distances from the axis $y=0$ of the basic flow. The boundary layer grows by the diffusion of vorticity and does not require inflow from infinity.

On the other hand, the conservation of the axial momentum of the jet together with the slowing of the jet with distance x downstream requires the entrainment of new fluid. The spreading of the jet implies a non-zero inflow ($V \neq 0$) at infinity.

In his study of the linear theory of stability of nearly parallel flows, Haaland modifies the Orr-Sommerfeld theory to include some of the effects of inflow. The effect of retaining these terms is to confine the vorticity of disturbances to the regions of the main flow where the flow is essentially irrotational. These inflow terms make a big difference in the critical Reynolds numbers especially when the wave numbers are small (see Fig. E2.1).

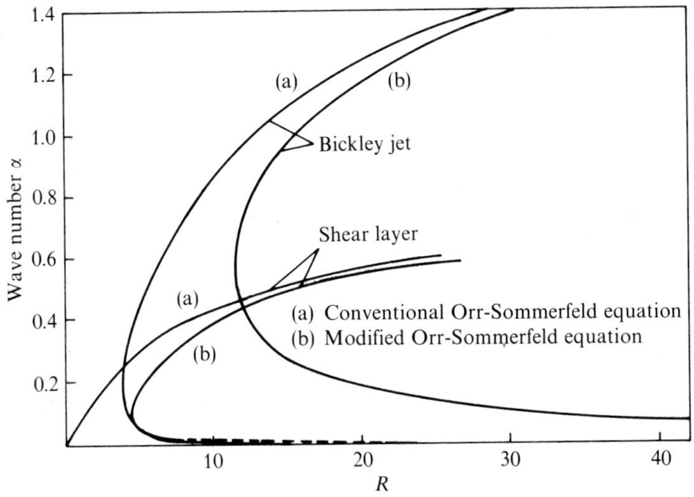

Fig. E2.1: Neutral curves for the Shear layer and the Bickley jet (after Haaland, 1972).

A bifurcation theory for nearly parallel flow should therefore allow for a certain flexibility in the choice of the zeroth order. The bifurcation theory of Joseph (1974 B) allows this flexibility of choice; the theory then corrects this zeroth order for effects of nonlinear terms and of linear terms which are neglected at the zeroth order (the extra terms). We shall not give this formal bifurcation theory in detail here. The basic mathematical procedures are as follows: first, one introduces two scales x and χ and in the equation

$$\frac{\partial}{\partial x} \to \frac{\partial}{\partial x} + v \frac{\partial}{\partial \chi};$$

at the zeroth order some terms containing v are retained and some are neglected. The second step then is to introduce a false parameter $\hat{v}$ replacing v wherever it is going to be set to zero at zeroth order. Having done this we may then introduce ω and ε as in the perturbation theory of Chapters II and IV. At this point we have a well-defined perturbation problem depending on parameters ε and $\hat{v}$. The bifurcation problem, so defined, is not a true problem because the coefficient $\hat{\nu}$ of the extra terms has been introduced artificially; some of the terms which like are neglected in the OS theory are kept at the zeroth order whilst the others appear in the theory at higher orders, as corrections. In the perturbation the parameters $\nu(\varepsilon, \hat{\nu})$ and $\omega(\varepsilon, \hat{\nu})$ are developed into a double power series along the lines set out in Chapter II. The solution of the false problem is supposed to solve the true problem when $\nu(\varepsilon, \hat{\nu}) = \hat{\nu}$. This method is called the method of false problems. The technical details are in the paper by Joseph (1974B).

References

(The section in the text where the given reference is cited appears as a cross-reference in the square bracket which follows the citations listed below).

Achenbach, E.: Experiments on the flow past spheres at very high Reynolds numbers. J. Fluid Mech. **54**, 565 (1972). [Notes for II]

Andreichikov, I. P., Yudovich, V. I.: Self-oscillating regimes branching from Poiseuille flow in a two-dimensional channel. Sov. Phys. Dokl. **17**, 120 (1972). [Notes for II, IV]

Baker, N. H., Moore, D. W., Spiegel, E. A.: Aperiodic behaviour of a non-linear oscillator. Quart. J. Mech. Appl. Math. **XXIV**, 391 (1971). [Notes for II]

Bass, J.: Les Fonctions Pseudo-Aléatoires (Mémorial Des Sciences Mathématiques). Paris: Gauthier-Villars 1962. [A]

Batchelor, G. K.: An Introduction to Fluid Dynamics. Cambridge University Press, 1967. [Notes for II]

Batchelor, G. K., Gill, A. E.: Analysis of the stability of axisymmetric jets. J. Fluid Mech. **14**, 529 (1962). [21]

Bateman, H., Dryden, H. L., Murnaghan, F. D.: Hydrodynamics. New York: Dover 1956 (Repub. from Bull. Nat. Res. Council. **84**, 1932). [35, Notes for IV]

Beavers, G. S., Sparrow, E. M., Magnuson, R. A.: Experiments on the breakdown of laminar flow in a parallel-plate channel. Int. J. Heat Mass Transfer. **13**, 809 (1970). [35]

Benjamin, T. B.: Applications of Leray-Schauder degree theory to problems of hydrodynamic stability. (forthcoming). [Notes for II]

Besicovitsch, A. S.: Almost Periodic Functions. Cambridge, England: The University Press 1932. [A]

Betchov, R., Criminale, W. O.: Stability of Parallel Flow. New York: Academic Press 1967. [Notes for II, 34, E.]

Block, R.: Ph. D. Thesis. Dept. of Mech. Eng., University of Minn., 1973. [36]

Bohr, H. A.: Almost Periodic Functions. Berlin: Julius Springer, 1932. In English translation: New York: Chelsea Publishing Co. 1947. [A]

Bouthier, M.: Stabilité linéaire des écoulements presque paralleles. J. Mecan. **11**, 599 (1972). [E]

Bouthier, M.: Stabilité linéaire des écoulements presque paralleles. Partie II. La couche limite de Blasius. J. Mécan. **12**, 75 (1973). [E]

Bratukhin, Iu. K.: On the evalution of the critical Reynolds number for the flow of fluid between two rotating spherical surfaces. J. Appl. Math. Mech. **25**, 1286 (1961). [53]

Browder, F.: On the eigenfunctions and eigenvalues of the general linear elliptic differential operator. Proc. U.S. Nat. Acad. Science **39**, 433 (1953). [Notes for II]

Busse, F. H.: Dissertation. Munich, 1962. See also: The stability of finite amplitude cellular convection and its relation to an extremum principle. J. Fluid Mech. **30**, 625 (1967). [Notes for II]

Busse, F. H.: Bounds on the transport of mass and momentum by turbulent flow between parallel plates. Z. Angew. Math. Phys. **20**, 1 (1969 A). [22, 30, 31, 32, Notes for IV]

Busse, F. H.: On Howard's upper bound for heat transport by turbulent convection. J. Fluid Mech. **37**, 457 (1969 B). [Notes for II]

Busse, F. H.: Bounds for turbulent shear flow. J. Fluid Mech. **41**, 219 (1970 A). [Notes for II, IV; 31]

Busse, F. H.: Über notwendige und hinreichende Kriterien für die Stabilität von Strömungen. Z. Angew. Math. Mech. **50**, 173 (1970 B). [49]

Busse, F. H.: The bounding theory of turbulence and its physical significance in the case of turbulent Couette flow: Springer Lecture Notes in Physics. Ed.: M. Rosenblatt and C. Van Atta **12.** Statistical Models and Turbulence 1971. [30, 37]

Busse, F. H.: The Bounding theory of turbulence and its physical significance in the case of pipe flow: Istituto Nazionale di Alta Matemica, Symposia Mathematica **IX,** (1972 A). [31]

Busse, F. H.: A property of the energy stability limit for plane parallel shear flow. Arch. Rational Mech. Anal. **47,** 28 (1972 B). [48]

Busse, F. H., Joseph, D. D.: Bounds for heat transport in a porous layer. J. Fluid Mech. **54,** 521 (1972). [Notes for II]

Carmi, S.: Linear stability of axial flow in an annular pipe. Phys. Fluids **13,** 829 (1970). [44]

Carmi, S.: Energy stability of channel flows. Z. Angew. Math. Phys. **20,** 487 (1969). [E]

Carothers, S. D.: Portland experiments on the flow of oil in tubes. Proc. Roy. Soc. A **87,** 154 (1912). [26]

Chandrasekhar, S.: Hydrodynamic and Hydromagnetic Stability. Oxford University Press 1961. [Notes for II; 37, 42, 43, 48, 53, A]

Chen, T. S., Joseph, D. D.: Subcritical bifurcation of plane Poiseuille flow. J. Fluid Mech. **58,** 337 (1973). [34]

Coddington, E. A., Levinson, N.: Theory of Ordinary Differential Equations. New York: McGraw Hill 1955 [7, 8]

Coles, D.: Transition in circular Couette flow. J. Fluid Mech. **21,** 385 (1965). [36, 37, 38, 39, 53]

Coles, D.: A note on Taylor instability in circular Couette flow. J. App. Mech. **34,** 529 (1967). [38]

Couette, M.: Études sur le frottement des liquides. Ann. Chim. Phys. **21,** 433 (1890). [37]

Courant, R., Hilbert, D.: Methods of Mathematical Physics, Vol. 1. New York: Interscience 1953. [B]

Crandall, M. G., Rabinowitz, P. H.: Bifurcation, perturbation of simple eigenvalues, and linearized stability. Arch. Rational Mech. Anal. **52,** 161 (1973). [Notes for II]

Datta, S. K.: Stability of spiral flow between concentric cylinders at low axial Reynolds numbers. J. Fluid Mech. **21,** 635 (1965). [43]

Davey, A.: The growth of Taylor vortices in flow between rotating cylinders. J. Fluid Mech. **14,** 336 (1962). [39]

Davey, A.: On the stability of plane Couette flow to infinitesimal disturbances. J. Fluid Mech. **57,** 369 (1973). [52]

Davey, A., DiPrima, R. C., Stuart, J. T.: On the instability of Taylor vortices. J. Fluid Mech. **31,** 17 (1968). [39]

Davey, A., Drazin, P. G.: The stability of Poiseuille flow in a pipe. J. Fluid Mech. **36,** 209 (1969). [35]

Davey, A., Nguyen, H.: Finite-amplitude stability of pipe flow. J. Fluid Mech. **45,** 701 (1971). [Notes for IV]

Davies, S. J., White, C. M.: An experimental study of the flow of water in pipes of rectangular section. Proc. Roy. Soc. A **119,** 92 (1928). [26, 35]

Davis, S. H.: On the principle of exchange of stabilities. Proc. Roy. Soc. A **310,** 341 (1969). [Notes for II]

Davis, S. H., von Kerczek, D.: A reformulation of energy stability theory. Arch. Rational Mech. Anal. **52,** 112 (1973). [4, Notes for I]

Deardorff, J. W.: On the stability of viscous plane Couette flow. J. Fluid Mech. **15,** 623 (1963). [52]

Denn, M.: Stability of Reaction and Transport Processes. Englewood Cliffs, New Jersey: Prentice-Hall 1975. [Notes for II]

Dikii, L. A.: The stability of plane-parallel flows of an ideal fluid (in Russian). Dokl. Akad. Nauk SSSR **135,** 1068 (1960). English transl.: Soviet Phys. Doklady **5,** 1179 (1961). [52]

DiPrima, R. C., Grannick, R. N.: A nonlinear investigation of the stability of flow between counter rotating cylinders. IUTAM-Symposium of Instability of Continuous System, Ed.: H. Leipholz, Berlin-Heidelberg-New York: Springer 1971. [36, 39]

DiPrima, R. C., Habetler, G. J.: A completeness theorem for non-selfadjoint eigenvalue problems in hydrodynamic stability. Arch. Rational Mech. Anal. **34,** 218 (1969). [Notes for II]

Drazin, P., Howard, L. N.: Hydrodynamic stability of parallel flow of inviscid fluid in Advances in Appl. Mech. **9,** 1, 1966. [Notes for II]

Eagles, P. M.: The stability of a family of Jeffery-Hamel solutions for divergent channel flow. J. Fluid Mech. **24,** 191 (1966). [44]

Eagles, P. M.: On the torque of wavy vortices. J. Fluid Mech. **62,** 1 (1974). [39]

Eckhaus, W.: Studies in Non-linear Stability Theory. Berlin-Heidelberg-New York: Springer 1965. [Notes for II; 39]

Ellingsen, T., Gjevik, B., Palm, E.: On the nonlinear stability of plane Couette flow. J. Fluid Mech. **40,** 93 (1970). [Notes for IV]

Fife, P.: The Bénard problem for general fluid dynamical equations and remarks on the Boussinesq approximations. Indiana Univ. Math. J. **20,** 303 (1979). [Notes for II]

Fife, P.: Branching phenomena in fluid dynamics and chemical reaction-diffusion theory: Proceedings of 1974 CIME symposium on eigenvalues of nonlinear problems, Varenna, June 1974 (to appear). [Notes for II]

Fife, P., Joseph, D. D.: Existence of convective solutions of the generalized Bénard problem which are analytic in their norm. Arch. Rational Mech. Anal. **33,** 116 (1969). [Notes for II]

Finlayson, B. A.: The Method of Weighted Residuals and Variational Principles. New York: Academic Press 1972. [B]

Fjørtoft, R.: Application of integral theorems in deriving criteria of stability for laminar flows and for the baroclinic circular vortex. Geofys. Publ. Oslo **17,** No. 6 (1950). [E]

Foà, E.: Sull'impiego dell'analisi dimensionale nello studio del moto turbolento. L'Industria (Milan) **43,** 426 (1929). [Notes for I]

Foote, J. R., Lin, C. C.: Some recent investigations in the theory of hydrodynamic stability. Quart. Appl. Math. **8,** 265 (1950). [E]

Fox, J. A., Lessen, M., Bhat, W. V.: Experimental investigations of the stability of Hagen-Poiseuille flow. Phys. Fluids **11,** 1 (1968). [35]

Fu, T. S., Joseph, D. D.: Linear instability of asymmetric flow in channels. Phys. Fluids **13,** 217 (1970). [E]

Galdi, G.: Alcuni casa di non esistenza del massimo funzionale associato ad una riformulazione del metado Dell'Energia. Arch. Rational Mech. Anal. **59,** 1 (1975). [Notes for I B]

Gallagher, A. P., Mercer, A. M.: On the behaviour of small disturbances in plane Couette flow. J. Fluid Mech. **13,** 91 (1962). [52]

Gantmacher, F. R.: On nonsymmetric Kellogg kernels. Dokl. Akad. Nauk SSSR **1,** 10: 3 (1936). [D]

Gantmacher, F. R.: The Theory of Matrices, Vol. 2. New York: Chelsea 1959. [D]

Gantmacher, F. R., Krein, M. G.: Oszillationsmatrizen, Oszillationskerne und Kleine Schwingungen Mechanischer Systeme. Berlin: Akademie Verlag 1960. [23, D]

Georgescu, A.: A note on Joseph's inequalities in stability theory. Z. Angew. Math. Phys. **21,** 258 (1970). [44]

Gill, A. E.: A mechanism for instability of plane Couette flow and of Poiseuille flow in a pipe. J. Fluid Mech. **21,** 503 (1965). [E]

Gollub, J. P., Swinney, H. L.: Onset of turbulence in a rotating fluid. Phy. Rev. Letters. **35,** 921 (1975). [Notes for II, 39]

Gor'kov, L.: Stationary convection in a plane liquid layer near the critical heat transfer point. Sov. Phys., JETP **6,** 311 (1958). [Notes for II]

Görtler, H., Velte, W.: Recent mathematical treatments of laminar flow and transition problems. Phys. Fluids **10,** 93 (1967). [Notes for II]

Greenspan, H. P.: The Theory of Rotating Fluids. Cambridge University Press 1969. [Notes for II]

Greenspan, D.: Numerical studies of steady, viscous incompressible flow between two rotating spheres. Computers and Fluids **3,** 69 (1975). [53]

Grindley, J. H., Gibson, A. H.: On the frictional resistance to the flow of air through a pipe. Proc. Roy. Soc. **A 80,** 114 (1907/1908). [26]

Grohne, R.: Über das Spektrum bei Eigenschwingungen ebener Laminarströmungen. Z. Angew. Math. Mech. **34,** 344 (1954). [52]

Guiraud, J. P., Iooss, G.: Sur la stabilité des écoulements laminaires. C. R. Acad. Sc., **266A,** 1283 (1968). [Notes for II]

Haaland, S.: Contributions to linear stability theory of nearly parallel flow. Ph. D. Thesis. Dept. of Aero. Eng., Univ. of Minn., 1972 [E]

Habetler, G., Matkowsky, B.: On the validity of a nonlinear dynamic stability theory. Arch. Rational Mech. Anal. **57,** 166 (1974). [Notes for II]

Haberman, W. L.: Secondary flow about a sphere rotating in a viscous liquid inside a coaxially rotating spherical container. Phys. Fluids **5,** 625 (1962). [53]

Hadamard, J.: Mémoire sur le problème d'analyse relatif à l'equilibre des plaques elastiques encastrées. Mémoires des Savants Etrangers **33,** (1908). [B]

Hale, J. K.: Integral manifolds of perturbed differential systems. Ann. Math. **73,** 496 (1971). [Notes for II]

Hanks, R. W.: The laminar-turbulent transition for flow in pipes, concentric annuli, and parallel plates. A. I. Ch. E. Jour. **9,** 45 (1963). [35]

Harrison, W.: On the stability of the steady motion of viscous liquid contained between two rotating coaxal circular cylinders. Proc. Camb. Phil. Soc. **20,** 455 (1921). [Notes for IV, 37, 42]

Havelock, T. H.: The stability of fluid motion. Proc. Roy. Soc. A. **98,** 428 (1921). [Notes for IV]

Hayakawa, N.: Bounds on mass transport of flow on an inclined plate. Z. Angew. Math. Phys. **21,** 685 (1970). [31]

Heisenberg, W.: Über Stabilität und Turbulenz von Flüssigkeitsströmen. Annalen der Physik **74,** 577 (1924). The English translation of this paper appears as NACA Tech. Memo. 1291, 1951. [Notes for IV]

Hills, R. N., Knops, R. J.: Continuous dependence for the compressible linearly viscous fluid. Arch. Rational Mech. Anal. **51,** 54 (1973). [Notes for I]

Høiland, E.: On two-dimensional perturbations of linear flow. Geofys. Publikasjoner, Norske Videnskaps – Akad. Oslo **10,** 9: 1 (1953). [E]

Hopf, E.: Ein allgemeiner Endlichkeitssatz der Hydrodynamik. Math. Annalen **117,** 764 (1941). [Notes for I, 2, 4]

Hopf, E.: Abzweigung einer periodischen Lösung eines Differentialsystems. Berichte der Mathematisch-Physikalischen Klasse der Sächsischen Akademie der Wissenschaften zu Leipzig **XCIV,** 1942. [Intro., Notes for II]

Hopf, E.: Über die Anfangswertaufgabe für die hydrodynamischen Grundgleichungen. Math. Nachrichten **4,** 213 (1951). [Notes for I]

Hopf, E.: A mathematical example displaying features of turbulence. Comm. Pure and Appl. Math. **1,** 303 (1948). See also: Hopf, E.: Repeated branching through loss of stability, an example. Proc. of the Conf. on Diff. Eqs. Univ. of Maryland, 49, 1956. [Introduction, Notes for II]

Hopf, L.: Der Verlauf kleiner Schwingungen auf einer Strömung reibender Flüssigkeit. Ann. Phys. Lpg. **44,** 1 (1914). [52]

Hoppensteadt, F., Gordon, N.: Nonlinear stability analysis of static states which arise through bifurcation. Comm. Pure Appl. Math. **28,** 355 (1975). [Notes for II]

Howard, L. N.: Heat transport by turbulent convection. J. Fluid Mech. **17,** 405 (1963). [Notes for II, 29]

Howard, L. N.: Bounds on flow quantities. Ann. Rev. of Fluid Mech. **4,** 473 (1972). [31, 32, Notes for IV, C]

Howard, L. N., Gupta, A. S.: On the hydrodynamic and hydromagnetic stability of swirling flows. J. Fluid Mech. **14,** 463 (1962). [43]

Howarth, L.: Note on the boundary layer on a rotating sphere. Phil. Mag. **42,** 1308 (1951). [53]

Hughes, T. H., Reid, W. H.: The stability of spiral flow between rotating cylinders. Phil. Trans. Roy. Soc. A **263,** 57 (1968). [43]

Hung, W. L.: Stability of Couette flow by the method of energy. M. S. Thesis. University of Minnesota 1968. [37]

Hung, W. L., Joseph, D. D., Munson, B. R.: Global stability of spiral flow. Part II. J. Fluid Mech. **51,** 593 (1972). [51, 52]

Iooss, G.: Theorie non linéaire de la stabilité des écoulements laminaires dans le cas de „l'Echange des Stabilités". Arch. Rational Mech. Anal. **40,** 166 (1971). [Notes for II]

Iooss, G.: Existence et stabilité de la solution périodique secondaire intervenant dans les problèmes d'évolution du type Navier-Stokes. Arch. Rational Mech. Anal. **47,** 301 (1972). [Intro., 7, 8, Notes for II]

Iooss, G.: Bifurcation et stabilité. Cours de 3ème cycle Orsay: 1973—1974. [16, Notes for II]

Iooss, G.: Bifurcation of a periodic solution of the Navier-Stokes equations into an invariant torus. Arch. Rational Mech. Anal. **58,** 35 (1975). [Notes for II]

Iooss, G.: Sur la deuxieme bifurcation d'une solution stationaire de systemes du type Navier-Stokes. Arch. Rational Mech. Anal. (forthcoming)

Ivanilov, I. P., Iakovlev, G. N.: The bifurcation of fluid flow between rotating cylinders. J. Appl. Math. Mech. (PMM) **30,** 910 (1966). [39, D]

Jeffreys, H.: The stability of a layer of fluid heated below. Phil. Mag. **2,** 833 (1926). [Notes for II]

Jentzch, R.: Über Integralgleichungen mit positivem Kern. J. Math. **141,** 235 (1912). [D]

Joseph, D. D.: Nonlinear stability of the Boussinesq equations by the method of energy. Arch. Rational Mech. Anal. **22,** 163 (1966). [Notes for I, 48]

Joseph, D. D.: Parameter and domain dependence of eigenvalues of elliptic partial differential equations. Arch. Rational Mech. Anal. **24,** 325 (1967). [B]

Joseph, D. D.: Eigenvalue bounds for the Orr-Sommerfeld equations. Part I. J. Fluid Mech. **33,** 617 (1968). [44]

Joseph, D. D.: Eigenvalue bounds for the Orr-Sommerfeld equation. Part II. J. Fluid Mech. **36,** 721 (1969). [44]

Joseph, D. D.: Stability of convection in containers of arbitrary shape. J. Fluid Mech. **47,** 257 (1971). [Notes for II, Notes for IV]

Joseph, D. D.: Remarks about bifurcation and stability of quasi-periodic solutions which bifurcate from periodic solutions of the Navier-Stokes equations: Nonlinear Problems in the Physical Sciences and Biology. (Lecture Notes in Mathematics, **322**). Eds.: I. Stakgold, D. D. Joseph and D. H. Sattinger. Berlin-Heidelberg-New York: Springer 1973. [Notes for II]

Joseph, D. D.: Domain perturbations: The higher order theory of infinitesimal water waves. Arch. Rational Mech. Anal. **51,** 295 (1973). [B]

Joseph, D. D.: Repeated supercritical branching of solutions arising in the variational theory of turbulence. Arch. Rational Mech. Anal. **53,** 101 (1974A). [Notes for II]

Joseph, D. D.: Response curves for plane Poiseuille flow in Advances in Applied Mechanics XIV, Ed.: C. S. Yih. New York: Academic Press 1974B. [27, 30, Notes for IV, E]

Joseph, D. D.: Slow motion and viscometric motion, stability and bifurcation of a simple fluid. Arch. Rational Mech. Anal. **56,** 99 (1974C). [B]

Joseph, D. D., Carmi, S.: Stability of Poiseuille flow in pipes, annuli, and channels. Quart. Appl. Math. **26,** 575 (1969). [21]

Joseph, D. D., Chen, T. S.: Friction factors in the theory of bifurcating flow through annular ducts. J. Fluid Mech. **66,** 189 (1974). [27, Notes for IV, 34, 35]

Joseph, D. D., Fosdick, R.: The free surface on a liquid between cylinders rotating at different speeds. Part I. Arch. Rational Mech. Anal. **49,** 321 (1973). [B]

Joseph, D. D., Hung, W.: Contributions to the nonlinear theory of stability of viscous flow in pipes and between rotating cylinders. Arch. Rational Mech. Anal. **44,** 1 (1971). [24, 37, 40, 42, C]

Joseph, D. D., Munson, B.: Global stability of spiral flow. J. Fluid Mech. **43,** 545 (1970). [22, 44, 45, 46, 49, 52]

Joseph, D. D., Nield, D. A.: Stability of bifurcating time-periodic and steady solutions of arbitrary amplitude. Arch. Rational Mech. Anal. **58,** 369 (1975). [15, Notes for II]

Joseph, D. D., Sattinger, D. H.: Bifurcating time periodic solutions and their stability. Arch. Rational Mech. Anal. **45,** 79 (1972). [12, Notes for II, IV]

Joseph, D. D., Sturges, L.: The free surface on a liquid filling a trench heated from its side. J. Fluid Mech. **69,** 565 (1975). [B]

Joseph, D. D., Tao, L. N.: Transverse velocity components in fully developed unsteady flows. J. Appl. Mech. **30,** 147 (1963). [24]

Kao, T. W., Park, C.: Experimental investigations of the stability of channel flow: Part 1. Flow of a single liquid in a rectangular channel. J. Fluid Mech. **43,** 145 (1970). [35]

Karlin, S.: Total Positivity, Vol. 1. Stanford Universtiy Press 1968. [D]

Karlin, S.: Total positivity, interpolation by splines and Green's functions of differential operators. J. Appx. Th. **4,** 91 (1971). [D]

Kármán, T. von: Über die Stabilität der Laminarströmung und die Theorie der Turbulenz. Proc. 1st Int. Cong. Appl. Mech., Delft, 97 (1924). [Notes for IV]

Kármán, T. von: Some aspects of the turbulence problem. Proc. 4th Intl. Cong. Appl. Mech., Cambridge, 54 (1934). [38]

Karon, J. M.: The sign-regularity properties of a class of Green's functions for ordinary differential equations. J. Diff. Eqs. **6,** 484 (1969). [D]

Kellogg, O. D.: The oscillation of functions of an orthogonal set. Amer. J. Math. **38,** 1 (1916). [D]

Kellogg, O. D.: Orthogonal function sets arising from integral equations. Amer. J. Math. **40,** 145 (1918). [D]

Khlebutin, G. N.: Stability of fluid motion between rotating and stationary concentric sphere. Izv. Akad. Nauk SSSR, Mekh. Zhid. Gaza, **3,** No. 1 (1968). (Fluid Dynamics, **3,** No. 6 (1968)). [53]

Kiessling, I.: Über das Taylorsche Stabilitätsproblem bei zusätzlicher axialer Durchströmung der Zylinder. Deutsche Versuchsanstalt für Luft- und Raumfahrt, Bericht 290, 1963. [49]

Kirchgässner, K., Kielhöfer, H.: Stability and bifurcation in fluid mechanics. Rocky Mountain J. Math. **3,** 275 (1972). [Intro., Notes for II]

Kirchgässner, K., Sorger, P.: Stability analysis of branching solutions of the Navier-Stokes equations. Proc. 12th Intl. Cong. Appl. Mech., Stanford (1968). [Notes for II, 39]

Kirchgässner, K., Sorger, P.: Branching analysis for the Taylor problem. Quart. J. Mech. Appl. Math. **22,** 183 (1969). [Notes for II, 39]

Kogleman, S., DiPrima, R. C., Stability of spatially periodic supercritical flows in hydrodynamics. Phys. Fluids **13,** 1 (1970). [39]

Kogleman, S., Keller, J. B.: Transient behavior of unstable nonlinear systems with applications to the Bénard and Taylor problems. SIAM Journal of Applied Math. **20,** 619 (1971). [Notes for II]

Krein, M. G.: Sur les fonctions de Green non-symétriques oscillatoires des opérateurs différentials ordinaires. Comptes Rendus (Dokl.) de l'Académie des Sciences de l'URSS, **26** (8) 643 (1939). [D]

Krein, S. G.: Sur les propriétés fonctionells des opérateurs de l'analyse vectorielle et de l'hydrodynamique. Dokl. Akad. Nauk SSSR **93,** 969 (1953). [Notes for II]

Krueger, E. R., DiPrima, R. C.: The stability of viscous fluid between rotating cylinders with an axial flow. J. Fluid Mech. **19,** 528 (1964). [43]

Krueger, E. R., Gross, A., DiPrima, R. C.: On the relative importance of Taylor-vortex and non-axisymmetric modes in flow between rotating cylinders. J. Fluid Mech. **24,** 521 (1966). [36, 39]

Ladyzhenskaya, O. A.: The Mathematical Theory of Viscous Incompressible Flow (2nd Edition, 1969). New York: Gordon and Breach 1963. [C]

Ladyzhenskaya, O. A.: Mathematical analysis of the Navier-Stokes equations for incompressible liquid. Ann. Rev. of Fluid Mech. **7,** 249 (1975) [Notes for I, II]

Lanchon, H., Eckhaus, W.: Sur l'analyse de la stabilité des écoulements faiblement divergents. J. Mécan. **3,** 445 (1964). [E]

Landau, L. D.: On the problem of turbulence. C. R. Acad. Sci., U.R.S.S. **44,** 311 (1944). See also L. D. Landau and E. M. Lifschitz. Fluid Mechanics. Oxford: Pergamon Press 1959. [Intro., Notes for II]

Lanford, O. E.: III, Bifurcation of periodic solutions into invariant tori: the work of Ruelle and Takens: Nonlinear Problems in the Physical Sciences and Biology (Lecture Notes in Mathematics, **322**). Eds.: I. Stakgold, D. D. Joseph, D. H. Sattinger. Berlin-Heidelberg-New York: Springer 1973. [Notes for II]

Lasalle, J., Lefschetz, S.: Stability by Liapunov's Direct Method. New York: Academic Press 1961. [8]

Leite, R. J.: An experimental investigation of the stability of Poiseuille flow. J. Fluid Mech. **5,** 81 (1959). [35]

Leray, J.: Etude de diverses équations intégrales non linéaires et de quelques problèmes que pose l'Hydrodynamique. Jour. de Math. Pures et Appl. **12,** 1 (1933). [2]

Lezius, D. K., Johnston, J. P.: Roll-cell instabilities in rotating laminar and turbulent channel flows. J. Fluid Mech. (forthcoming). [46]

Liang, S. F., Vidal, A., Acrivos, A./ Buoyancy-driven convection in cylindrical geometries. J. Fluid Mech. **36,** 239 (1969). [Notes for IV]

Liapounov, A. M.: Sur les figures d'équilibre peu différentes des ellipsoides d'une masse liquide homogène douée d'un mouvement de rotation. Zap. Akad. Nauk, St. Petersburg, **1,** 1 (1906). [Notes for II]

Lin, C. C.: On the stability of two-dimensional parallel flows. Part III. Quart. Appl. Math. **3,** 277 (1945). [E]

Lin, C. C.: Some physical aspects of the stability of parallel flows. Proc. Nat. Acad. Sci. **40,** 741 (1954). [E]

Lin, C. C.: The Theory of Hydrodynamic Stability. Cambridge University Press 1955. [Notes for II, 34, E]

Lindgren, E. R.: The transition process and other phenomena in viscous flow. Ark. Fysik **12,** 1 (1957). [26]

Lindgren, E. R.: Liquid flow in tubes I. The transition process under highly disturbed entrance flow conditions. Ark. Fysik **15,** 97 (1959). [26]

Lindstedt, A.: Beitrag zur Integration der Differentialgleichungen der Störungstheorie: Mémoires de l'Academie de Saint-Petersbourg 1882. [Intro., Notes for II]

Ling, C. H., Reynolds, W. C.: Non-parallel flow corrections for the stability of shear flows. J. Fluid Mech. **59,** 571 (1973). [E]

Lorenz, E. N.: Deterministic nonperiodic flow. J. Atmos. Sci. **20,** 130 (1963). [Notes for II]

Lorentz, H. A.: Über die Entstehung turbulenter Flüssigkeitsbewegungen und über den Einfluß dieser Bewegungen bei der Strömung durch Röhren. Abhandlungen über theoretische Physik **1,** Leipzig: Teubner 1907. [Notes for IV]

Lortz, D.: Dissertation. Munich, 1961. See also A stability criterion for steady finite amplitude convection with an external magnetic field. J. Fluid Mech. **23,** 113 (1965). [Notes for II]

Ludwieg, H.: Ergänzung der Arbeit: „Stabilität der Strömung in einem zylindrischen Ringraum". Z. Flugwiss. **9,** 359 (1961). [43, 46]

Ludwieg, H.: Experimentelle Nachprüfung der Stabilitäts-Theorien für reibungsfreie Strömungen mit schrauben-linienförmigen Stromlinien. 11th Intl. Cong. Appl. Mech., 1045 (1964). [43, 52]

Malkus, W. V. R.: The heat transport and spectrum of thermal turbulence. Proc. Roy. Soc. A **225,** 196 (1954). [Notes for IV]

Malkus, W. V. R., Veronis, G.: Finite amplitude cellular convection. J. Fluid Mech. **4,** 225 (1958). [Notes for II]

Mallock, H.R.A.: Determination of the viscosity of water. Proc. Roy. Soc. A **45**, 126 (1888). [37, 52]

Marsden, J., McCracken, M.: The Hopf Bifurcation and its Applications (Lecture notes in Applied Mathematical Sciences, **18**). Berlin-Heidelberg-New York: Springer 1976. [Notes for II]

Matkowsky, B. J.: A simple dynamic stability problem. Bull. Amer. Math. Soc. **76,** 620 (1970). [Notes for II]

McIntire, L. V., Lin, C. H.: Finite amplitude instability of second order fluids in plane Poiseuille flow. J. Fluid Mech. **52,** 273 (1972). [Notes for IV]

McLaughlin, J. B., Martin, P. C.: Transition to turbulence of a statically stressed fluid system. Phys. Rev. Lett. **33,** 1089 (1974); Phys. Rev. A **12,** 186 (1975). [Notes for II]

McLeod, J. B., Sattinger, D. H.: Loss of stability and bifurcation at a double eigenvalue. J. Functional Anal. **14,** (1973). [39, Notes for II]

Meksyn, D., Stuart, J. T.: Stability of viscous motion between parallel planes for finite disturbances. Proc. Roy. Soc. A **208,** 517 (1951). [Notes for IV]

Menguturk, M.: Ph. D. Thesis. Dept. of Mech. Eng., Duke University, 1974. [53]

Monin, A. S., Yaglom, A. M.: Statistical Fluid Mechanics. Volume I. Mechanics of Turbulence. Cambridge: The MIT Press 1971. [Notes for II]

Morales-Gomez, J. R.: Ph. D. Thesis. Dept. of Mech. Eng. (Thesis advisor: R. Urban), New Mexico State Univ., Las Cruces. New Mexico 1974. [53]

Morawetz, C. S.: The eigenvalues of some stability problems involving viscosity. J. Rational Mech. Anal. **1,** 579 (1952). [52]

Morrey, C. B.: Multiple Integrals in the Calculus of Variations. New York-Heidelberg-Berlin: Springer-Verlag, 1966. [B]

Mott, T., Joseph, D. D.: Stability of parallel flow between concentric cylinders. Phys. Fluids **11,** 2065 (1968). [35, E]

Munson, B. R.: Hydrodynamic stability of flow between rotating-sliding cylinders. Ph. D. Thesis. Dept. of Aero. Eng. and Mech., University of Minnesota 1970. [53]

Munson, B. R.: Viscous incompressible flow between eccentric coaxially rotating spheres. Phys. Fluids **17,** 528 (1974) [53]

Munson, B. R., Joseph, D. D.: Viscous incompressible flow between concentric rotating spheres. Part 1. Basic flow. J. Fluid Mech. **49,** 289 (1971A). [53]

Munson, B. R., Joseph, D. D.: Viscous incompressible flow between concentric rotating spheres. Part 2. Hydrodynamic stability. J. Fluid Mech. **49,** 305 (1971B). [53]

Munson, B. R., Menguturk, M.: Viscous incompressible flow between concentric rotating spheres. Part 3. Linear stability and experiment. J. Fluid Mech. **69,** 705 (1975). [53]

Munson, B. R., Menguturk, M.: Stability characteristics of eccentric spherical annulus flow. Proc. 8th South Eastern Conf. Th. Appl. Mechs. April 1976. [53]

Nagib, H. M.: On instabilities and secondary motions in swirling flows through annuli. Ph. D. dissertation. Illinois Institute of Technology 1972. [38, 46]

Nagib, H. M., Lavan, Z., Fejer, A. A., Wolf, L.: Stability of pipe flow with superposed solid body rotation. Phys. Fluids **14**, 766 (1971). [46]

Naumann, A.: Experimentelle Untersuchungen über die Entstehung der turbulenten Rohrströmung. Forschung auf dem Gebiete des Ingenieurwesens, Berlin, **2**, 85 (1931). [26]

Newell, A. C., Whitehead, J. A.: Finite bandwidth, finite amplitude convection. J. Fluid Mech. **36**, 309 (1969). [Notes for II]

Nickerson, E. C.: Upper bounds on the torque in cylindrical Couette flow. J. Fluid Mech. **38**, 807 (1969). [30, 37]

Nirenberg, L.: A strong maximum principle for parabolic equations. Comm. Pure and Appl. Math. **6**, 167 (1953). [40]

Obremski, H. J., Morkovin, M. V., Landahl, J.: Agardograph 134, NATO Paris, (1969). [E]

Orr, W. McF.: The stability or instability of the steady motions of a liquid. Part II: A viscous liquid. Proc. Roy. Irish Acad. A **27**, 69 (1907). [Notes for I, IV; 2, 4, 21, 47, 48, 51]

Ovseenko, I. G.: On the motion of a viscous liquid between two rotating spheres. Izv. Vyssh. Ucheb. Zaved. Matematika no **4**, 129 (1963). [53]

Palm, E.: On the tendency towards hexogonal cells in steady convection. J. Fluid Mech. **8**, 183 (1960), [Notes for II]

Palm, E.: Nonlinear thermal convection. Ann. Rev. Fluid Mech. **7**, 39 (1975). [Notes for II]

Payne, L., Weinberger, H.: An exact stability bound for Navier-Stokes flow in a sphere: Nonlinear Problems, Ed.: R. E. Langer. Univ. of Wisc. Press 1963. [B]

Pearson, C. E.: A numerical study of the time-dependent viscous flow between two rotating spheres. J. Fluid Mech. **28**, 323 (1967). [53]

Pedley, T. J.: On the instability of viscous flow in a rapidly rotating pipe. J. Fluid Mech. **35**, 97 (1969). [46, 50]

Pekeris, C. L., Shkoller, B.: Stability of plane Poiseuille flow to periodic disturbances of finite amplitude in the vicinity of the neutral curve. J. Fluid Mech. **29**, 31 (1967). [Notes for IV]

Pekeris, C. L., Shkoller, B.: The neutral curves for periodic perturbations of finite amplitude of plane Poiseuille flow. J. Fluid Mech. **39**, 629 (1969). [Notes for IV]

Poincaré, H.: Sur l'équilibre d'une masse fluide animée d'un mouvement de rotation, Acta Math. **7**, 259 (1885). [Notes for II]

Poincaré, H.: Sur les equations aux dérivées partielles de la physique mathématique. Amer. J. Math, **12**, 211 (1890)

Poincaré, H.: Les Méthodes Nouvelle de la Mécanique Céleste: III, Invariants integraux-Solutions périodique du deuxieme genre. Solutions doublement asymptotiques. New York: Dover 1957/1892. [Intro., Notes for II]

Prodi, G.: Teoremi di tipo locale per il sistema di Navier-Stokes l'stabilita delle soluzioni stazionarie. Rend. Sem. Univ. Padova **32**, 374 (1962). [Notes for II]

Protter, M. H., Weinberger, H. F.: Maximum Principles in Differential Equations. Englewood Cliffs, New Jersey: Prentice Hall 1967. [40]

Proudman, I.: The almost-rigid rotation of viscous fluid between concentric spheres. J. Fluid Mech. **1**, 505 (1956). [53]

Rabinowitz, P. H.: Existence and nonuniqueness of rectangular solutions of the Bénard problem. Arch. Rational Mech. Anal. **29**, 32 (1968). [Notes for II]

Rabinowitz, P. H.: Some global results for nonlinear eigenvalue problems. J. Func. Anal. **7**, 487 (1971). [Notes for II]

Rabinowitz, P. H.: A priori bounds for some bifurcation problems in fluid dynamics. Arch. Rational Mech. Anal. **49**, 270 (1973). [40]

Rayleigh, Lord: On the stability of the laminar motion of an inviscid fluid in Scientific Papers **6**, 197. London and New York: Cambridge Univ. Press, 1913. [E]

Rayleigh, Lord: Theory of Sound, 1878 (2nd ed. Vol. 1. paragraph 174). New York: Dover 1945. [44, C]

Rayleigh, Lord: On the dynamics of revolving fluids. Proc. Roy. Soc. A **93**, 148 (1916). Scientific Papers **6**, 447. [38]

Reichardt, H.: Über die Geschwindigkeitsverteilung in einer geradlinigen turbulenten Couette-strömung. Z. Angew. Math. Mech. **36**, S 26 (1956). [52]

Reid, W. H.: Inviscid modes of instability in Couette flow. J. Math. Anal. Appl. **1**, 411 (1960). [39]

Reid, W. H.: The stability of parallel flows: Basic Developments in Fluid Dynamics (Vol. 1, p. 249). Ed.: M. Holt. New York: Academic Press 1965. [Notes for II, E]

Reynolds, O.: An experimental investigation of the circumstances which determine whether the motion of water shall be direct or sinuous, and of the law of resistance in parallel channels: Papers on Mechanical and Physical Subjects. Cambridge University Press, Vol. II, p. 51, 1901, which reproduces the complete report of Phil. Trans. Roy. Soc. **174,** 935 (1883). [Notes for I, 26]

Reynolds, O.: On the dynamical theory of incompressible viscous fluids and the determination of the criterion. Phil. Trans. Roy. Soc. A **186,** 123 (1895). [Notes for I, IV; 2, 4]

Reynolds, W., Potter, M. C.: Finite-amplitude instability of parallel shear flows. J. Fluid Mech. **27,** 465 (1967A). [Notes for II, IV]

Reynolds, W., Potter, M. C.: A finite-amplitude state selection theory for Taylor-vortex flow. Unpublished Stanford University report 1967B. [39]

Riis, E.: The stability of Couette-flow in non-stratified and stratified viscous fluids. Geofys. Publikarjoner, Norske Videnskaps Akad. Oslo, **23,** No. 4 (1962). [52]

Robbins, K. A.: Disk dynamos and magnetic reversal. Ph. D. Dissertation. Dept. of Math. M.I.T. (1975). [Notes for II]

Rosenhead, L. (Ed.): Laminar Boundary Layers. Oxford University Press 1963. [53]

Ruelle, D., Takens, F.: On the nature of turbulence. Commun. Math. Phys. **20,** 167 (1971A). [Notes for II]

Ruelle, D., Takens, F.: Note concerning our paper on the nature of turbulence. Commun. Math. Phys. **23,** 343 (1971B). [Notes for II, 39]

Salwen, H., Grosch, E.: Stability of Poiseuille flow in a pipe of circular cross-section. J. Fluid Mech. **54,** 93 (1972). [35]

Sattinger, D. H.: The mathematical problem of hydrodynamic stability. Jour. Math. & Mech. **19,** 797 (1970). [8, Notes for II]

Sattinger, D. H.: Bifurcation of periodic solutions of the Navier-Stokes equations. Arch. Rational Mech. Anal. **41,** 66 (1971A). [Notes for II]

Sattinger, D. H.: Stability of bifurcating solutions by Leray-Schauder degree. Arch. Rational Mech. Anal. **43,** 154 (1971B). [Notes for II]

Sattinger, D. H.: Topics in Stability and Bifurcation Theory (Lecture Notes in Mathematics, **309**). Berlin-Heidelberg-New York: Springer 1973. [Notes for II]

Sawatzki, O., Zierep, J.: Flow between a fixed outer sphere and a concentric rotating inner sphere (In German). Acta Mechanica **9,** 13 (1970). [53]

Schensted, I.: Contributions to the theory of hydrodynamic stability. Ph. D. Thesis, Dept. of Phys., University of Michigan 1960. [Notes for II]

Schlichting, H.: Amplitudenverteilung und Energiebilanz der kleinen Störungen bei der Plattengrenzschicht. Nachr. Akad. Wiss. Göttingen Math.-Phys. Klasse I, 47 (1935). [E]

Schlichting, H.: Boundary Layer Theory, 4th ed. New York: McGraw Hill, 1960. [35]

Schlüter, A., Lortz, D., Busse, F.: On the stability of steady finite amplitude cellular convection. J. Fluid Mech. **23,** 129 (1965). [Notes for II]

Schmidt, E.: Zur Theorie der linearen und nicht linearen Integralgleichungen, 3. Teil. Math. Annalen **65,** 370 (1908). [Notes for II]

Schubauer, G. B., Skramstad, H. K.: Laminar boundary layer oscillations and transition on a flat plate. N.A.C.A. Tech. Rept. No. 909. Originally issued as N.A.C.A.A.C.R., April 1943. [E]

Schulz-Grunow, F.: Zur Stabilität der Couette Strömung. Z. Angew. Math. Mech. **38,** 323 (1958). [52]

Schwiderski, E. W. (with appendix by Jarnagin, M. P.): Bifurcation of convection in internally heated fluid layers. Phys. Fluids **15,** 1882 (1972). [Notes for II]

Segel, L.: Nonlinear hydrodynamic stability theory and its application to thermal convection and curved flow: Non-Equilibrium Thermodynamics: Variational Techniques and Stability. Eds.: R. J. Donnelly, I. Prigogine and R. Herman. University of Chicago Press 1966. [Notes for II]

Serrin, J.: Mathematical principles of classical fluid mechanics: Handbuch der Physik, Bd VIII/1. Eds.: S. Flügge and C. Truesdell. Berlin-Göttingen-Heidelberg: Springer 1959A. [Notes for I; 4, 30]

Serrin, J.: On the stability of viscous fluid motions. Arch. Rational Mech. Anal. **3,** 1 (1959B). [Notes for I, 2, 4, 37, 40, 42, B]

Serrin, J.: A note on the existence of periodic solutions of the Navier-Stokes equations. Arch. Rational Mech. Anal. **3,** 120 (1959C). [Notes for I]

Serrin, J.: On the uniqueness of compressible fluid motion. Arch. Rational Mech. Anal. **3**, 271 (1959D). [Notes for I]

Serrin, J.: The initial value problem for the Navier-Stokes equations: Nonlinear Problems, Ed.: R. E. Langer. Univ. of Wisc. Press 1963. [Notes for I, C]

Shapiro, A. H.: Shape and Flow; The Fluid Dynamics of Drag. Anchor Books. Garden City, New York: Doubleday and Co., 1961. [Notes for II]

Sharpe, F. R.: On the stability of the motion of a viscous liquid. Trans. Amer. Math. Soc. **6**, 496 (1905). [Notes for IV]

Shen, S. F.: Stability of laminar flows: Theory of Laminar Flows (High Speed Aerodynamics and Jet Propulsion, Vol. 4), Section G. Ed.: F. K. Moore Princeton, N. J.: Princeton Univ. Press 1964. [Notes for II, IV; E]

Snyder, H. A.: Stability of rotating Couette flow. II. Comparison with numerical results. Phys. Fluids **11**, 1599 (1968B). [39]

Snyder, H. A.: Stability of rotating Couette flow. I. Asymmetric waveforms. Phys. Fluids **11**, 728 (1968A). [36, 39]

Snyder, H. A.: Waveforms in rotating Couette flow. Int. J. Non-Linear Mechs. **5**, 659 (1970). [36]

Sorger, P. Über ein Variationsproblem aus der nichtlinearen Stabilitätstheorie zäher, inkompressibler Strömungen. Z. Angew. Math. Phys. **17**, 201 (1966). [20, B]

Sorokin, M. P., Khlebutin, C. N., Shaidurov, G. F.: Study of the motion of a liquid between two rotating spherical surfaces. J. Appl. Mech. Tech. Phys. no. **6**, 73 (1966). [53]

Sorokin, V. S.: Non linear phenomena in closed flows near critical Reynolds numbers. J. Appl. Math. Mech. **25**, 366 (1961). [Notes for II]

Squire, H. B.: On the stability for three-dimensional disturbances of viscous fluid flow between parallel walls. Proc. Roy. Soc. A **142**, 621 (1933). [25, 34]

Stakgold, I.: Branching of solutions of nonlinear equations. SIAM review **13**, 289 (1971). [Notes for II]

Stewartson, K., Stuart, J. T.: A non-linear instability theory for a wave system in plane Poiseuille flow. J. Fluid Mech. **48**, 529 (1971). [Notes for II]

Stuart, J. T.: On the nonlinear mechanics of hydrodynamic stability. J. Fluid Mech. **4**, 1 (1958). [Notes for II, 39]

Stuart, J. T.: On the nonlinear mechanics of wave disturbances in stable and unstable parallel flows. Part I: The basic behavior in plane Poiseuille flow. J. Fluid Mech. **9**, 353 (1960). [Notes for II]

Stuart, J. T.: Hydrodynamic stability: in Laminar Boundary Layers, Ed.: L. Rosenhead. London: Oxford 1963. [Notes for II]

Stuart, J. T.: Nonlinear stability theory. Ann. Rev. of Fluid Mech. **3**, 347 (1971). [Intro., Notes for II, 34, 39]

Synge, J. L.: Hydrodynamic stability in Semicentennial Publications of the American Mathematical Society, Vol. **2**, 227, Amer. Math. Soc. (1938A). [Notes for II, 38, 40, 44]

Synge, J. L.: On the stability of a viscous liquid between two rotating coaxial cylinders. Proc. Roy. Soc. **A, 167**, 250 (1938 B). [38, 40]

Tamaki, K., Harrison, W. J.: On the stability of steady motion of viscous liquid contained between two rotating coaxial circular cylinders. Trans. Camb. Phil. Soc. **22**, 425 (1920). [Notes for IV, 37]

Tatsumi, T., Kakutani, T.: The stability of a two-dimensional jet. J. Fluid Mech. **4**, 261 (1958). [E]

Taylor, G. L.: Stability of a viscous liquid contained between two rotating cylinders. Phil. Trans. Roy. Soc. A **223**, 289 (1923). [36, 38, 39, 52]

Thomas, T. Y.: Qualitative analysis of the flow of fluids in pipes. Amer. J. Math. **64**, 754 (1942). [Notes for I, 2, 4, 30]

Tollmien, W.: Über die Entstehung der Turbulenz. Nachr. Akad. Wiss. Göttingen Math.-Phys. Klasse, **21** (1929). [E]

Tollmien, W.: Ein allgemeines Kriterium der Instabilität laminarer Geschwindigkeitsverteilungen. Nachr. Akad. Wiss. Göttingen Math.-Phys. Klasse, **50**, 79 (1935). [E]

Ukhovskii, M. R., Yudovich, V. I.: On the equation of steady-state convection. J. Appl. Math. Mech. **27**, 432 (1963). [Notes for II]

Vainberg, M. M., Trenogin, V. A.: The methods of Lyapunov and Schmidt in the theory of non-linear equations and their further development. Russ. Math. Surveys **17**, no. 2, p. 1 (1962). [Notes for II]

Velte, W.: Über ein Stabilitätskriterium der Hydrodynamik. Arch. Rational Mech. Anal. **9,** 9 (1962). [B]

Velte, W.: Stabilitätsverhalten und Verzweigung stationärer Lösungen der Navier-Stokesschen Gleichungen. Arch. Rational Mech. Anal. **16,** 97 (1964). [Notes for II]

Velte, W.: Stabilität und Verzweigung stationärer Lösungen der Navier-Stokesschen Gleichungen beim Taylorproblem. Arch. Rational Mech. Anal. **22,** 1 (1966). [Notes for II, 39]

Walker, J. E., Whan, G. A., Rothfus, R. R.: Fluid friction in noncircular ducts. A. I. Ch. E. J. **3,** 484 (1957). [27, 35]

Warner, W. H.: Poloidal and toroidal potentials for solenoidal fields. Z. Angew. Math. Phys. **23,** 221 (1972). [B]

Wasow, W.: One small disturbance of plane Couette flow. J. Res. Nat. Bur. Stand. **51,** 195 (1953). [52]

Watson, J.: On the nonlinear mechanics of wave disturbance in stable and unstable parallel flows. Part II: The development of a solution for plane Poiseuille flow and plane Couette flow. J. Fluid Mech. **14,** 336 (1960). [Notes for II]

Wendt, F.: Turbulente Strömung zwischen zwei rotierenden koaxialen Zylindern. Ing. Arch. **4,** 577 (1933). [38]

Wygnanski, I. J., Champagne, F. H.: On transition in a pipe. Part 1. The origin of puffs and slugs and the flow in a turbulent slug. J. Fluid Mech. **59,** 281—335 (1973). [35]

Wygnanski, I. J., Sokolov, M., Friedman, D.: On transition in a pipe. Part 2. The equilibrium puff. J. Fluid Mech. **69,** 283—305 (1975). [35]

Yakushin, V. I.: Instability of fluid motion of a liquid in a thin spherical layer. Eng. Trans.: Fluid Dyn. **4,** 83 (1969). [53]

Yakushin, V. I.: Instability of fluid motion of a liquid in a thin spherical layer. Izv. Akad. Nauk SSSR, Mekh. Zh. i Gaza, **4,** 1 (1969). (Eng. Trans. Fluid Dyn. **4,** 1 (1972).) [53]

Yakushin, V. I.: Instability of the motion of a liquid between two rotating spherical surfaces. Izv. Akad. Nauk. SSSR, Mekh. Zh. i Gaza, **5,** 4 (1970). (Eng. Trans. Fluid Dyn. **5,** 4 (1973).) [53]

Yih, C. S.: Dynamics of Nonhomogeneous Fluids. New York: Macmillan 1965. [Notes for II]

Yih, C. S.: Note on eigenvalue bounds for the Orr-Sommerfeld equation. J. Fluid Mech. **38,** 273 (1969). [44]

Yih, C. S.: Spectral theory of Taylor vortices. Part I: Structure of unstable modes. Arch. Rational Mech. Anal. **46,** 218 (1972A), [Notes for II]

Yih, C. S.: Spectral theory of Taylor vortices. Part II: Proof of non-oscillation. Arch. Rational Mech. Anal. **47,** 288 (1972B). [Notes for II, 39]

Yih, C. S.: Wave velocity in parallel flows of a viscous fluid. J. Fluid Mech. **58,** 703 (1973). [34]

Yudovich, V. I.: Stability of steady flows of viscous incompressible fluids. Soviet Physics Dokl. **10,** 293 (1965). [Notes for II]

Yudovich, V. I.: Secondary flows and fluid instability between rotating cylinders. J. Appl. Math. Mech. (PMM) **30,** 822 (1966A). [39]

Yudovich, V. I.: On the origin of convection. J. Appl. Math. Mech. (PMM) **30,** 1193 (1966B). [Notes for II]

Yudovich, V. I.: On the stability of forced oscillations of a liquid. Soviet Math. Dokl. **11,** 1473 (1970A). [7, 8]

Yudovich, V. I.: On the stability of self-oscillations of a liquid. Soviet Math. Dokl. **11,** 1543 (1970). [7, 8]

Yudovich, V. I.: The onset of auto-oscillations in a viscous fluid. J. Appl. Math. Mech. **35,** 587 (1971). [Notes for II]

Zahn, J., Toomre, J., Spiegel, E., Gough, D.: Nonlinear cellular motions in Poiseuille channel flow. J. Fluid Mech. **64,** 319 (1974). [14, Notes for II, IV]

Zierep, J., Sawatzki, O.: Three dimensional instabilities and vortices between two rotating spheres, 8th Symposium on Naval Hydrodynamics, 1970. (See also Sawatzki, Zierep, 1970). [53]

Subject Index